Disaster Risk Reduction

Methods, Approaches and Practices

Series Editor

Disaster risk reduction is a process that leads to the safety of communities and nations. After the 2005 World Conference on Disaster Reduction, held in Kobe, Japan, the Hyogo Framework for Action (HFA) was adopted as a framework for risk reduction. The academic research and higher education in disaster risk reduction has made, and continues to make, a gradual shift from pure basic research to applied, implementation-oriented research. More emphasis is being given to multi-stakeholder collaboration and multi-disciplinary research. Emerging university networks in Asia, Europe, Africa, and the Americas have urged process-oriented research in the disaster risk reduction field. With this in mind, this new series will promote the output of action research on disaster risk reduction, which will be useful for a wide range of stakeholders including academicians, professionals, practitioners, and students and researchers in related fields. The series will focus on emerging needs in the risk reduction field, starting from climate change adaptation, urban ecosystem, coastal risk reduction, education for sustainable development, community-based practices, risk communication, and human security, among other areas. Through academic review, this series will encourage young researchers and practitioners to analyze field practices and link them to theory and policies with logic, data, and evidence. In this way, the series will emphasize evidence-based risk reduction methods, approaches, and practices.

All submitted proposals are carefully reviewed by the series editor.

Publication Ethics and Publication Malpractice Statement: https://www.springernature.com/gp/policies/book-publishing-policies

Ethical, Conflict-of-Interest, and Editorial Policy Statement: https://www.springer.com/us/authors-editors/journal-author/journal-author-helpdesk/before-you-start/before-you-start/1330#c14214

Pradeep Kumar Ray · Rajib Shaw ·
Yasuhiro Soshino · Anirban Dutta ·
Teuku Aulia Geumpana
Editors

Technology Innovation for Sustainable Development of Healthcare and Disaster Management

Springer

Editors
Pradeep Kumar Ray
University of Michigan—Shanghai Jiao Tong University Joint Institute
Shanghai, China

UNSW Business
Sydney, Australia

Yasuhiro Soshino
Disaster Management Research Institute
Japanese Red Cross College of Nursing
Tokyo, Japan

Teuku Aulia Geumpana
School of Information and Physical Sciences
The University of Newcastle
Newcastle, NSW, Australia

Rajib Shaw
Graduate School of Media and Governance
Keio University
Fujisawa, Kanagawa, Japan

Anirban Dutta
Institute of Metabolism and Systems Research
University of Birmingham
Edgbaston, UK

ISSN 2196-4106 ISSN 2196-4114 (electronic)
Disaster Risk Reduction
ISBN 978-981-97-2048-4 ISBN 978-981-97-2049-1 (eBook)
https://doi.org/10.1007/978-981-97-2049-1

This Springer imprint is published by the registered company Springer Nature Singapore Pte Ltd.
The registered company address is: 152 Beach Road, #21-01/04 Gateway East, Singapore 189721, Singapore

Preface

The world is facing different challenges to the management of disasters caused by natural and other phenomena. On the other hand, technology is progressing very fast in different areas. Some of these technologies are also being used by criminals (e.g., financial scams) to cause disasters in the lives of people. On the other hand, technology innovations (e.g., drones) are indispensable in managing disasters within limited resources (cost and time), especially in the developing countries.

This book provides an overview of how technology innovation has been used in different types of disasters, such as natural disasters and those caused in healthcare (e.g., wrong surgery, severe mental illnesses, etc.). The book has three sections and 16 chapters altogether, and each section has 5 chapters. The book starts with an overview and description of technology innovation ecosystem and an outline of the chapters. The first part *Role of Technologies in Healthcare and Disaster Management* illustrates the role of technologies in disaster management through a number of case studies involving various technologies, such as digital technologies (including smartphones, social media, and drones), hydrogen technologies and evolving technological innovations. The second part *Complexities Involved* highlights some of the major challenges in the planning, development, and deployment of emerging technologies in view of the multidisciplinary nature of disaster management (involving agriculture, healthcare, law, business, social sciences, technology management, etc.). The complexity issues have been illustrated using examples of services like agriculture and public health services. The third and last part *Information Technologies Used* discusses some emerging Information Technologies used in disaster management, such as mobile phones, eye tracking, and blockchain. In particular, this section discusses emerging ways of making disaster management resilient, using deployment case studies.

Covering significant aspects of technology innovations, disaster types, this book is intended for students, researchers, academia, policymakers, and development practitioners in the fields of public health, technology management, social business, and social innovation, which have a strong relevance to Asia Pacific and beyond. This book discusses disasters (beyond natural ones) that bring miseries in the lives of common people. We will be happy if the readers find this book useful and relevant.

Shanghai, China — Pradeep Kumar Ray
Fujisawa. Japan — Rajib Shaw
Tokyo, Japan — Yasuhiro Soshino
Edgbaston, UK — Anirban Dutta
Newcastle, Australia — Teuku Aulia Geumpana

About This Book

The world has seen many disasters caused by natural and other phenomena, such as terrorism, wars, cyberattacks, financial scams, and healthcare failures. While some of these disasters are caused by unethical, malicious, and harmful use of technologies, technology innovations (e.g., drones, mobile phones, robots, and software like social media, blockchain etc.) are being used to assist in the disaster mitigation, detection and recovery. On the other hand, governments all over the world (particularly the low- and medium-income countries) are facing severe resource constraints in the development, deployment, and maintenance of disaster management solutions, given the other pressures of alleviating major issues, such as food shortage, inflation, etc. Hence, there is a strong need for technology innovations for sustainable solutions for disasters caused by both natural and other causes, especially healthcare.

This book presents recent advances, trends, challenges, and potentials of technology innovation for sustainable solutions to disaster mitigation, detection and recovery with examples of natural disasters, such as tsunamis, earthquakes, and healthcare disasters, such as severe autism, wrong surgery and Alzheimer's. Covering different aspects of technology innovations, this book is a valuable resource material for students, researchers, academia, policymakers, and development practitioners.

Contents

Editors and Contributors

About the Editors

Prof. Pradeep Kumar Ray is the founder of the Centre For Entrepreneurship (CFE) at the University of Michigan-Shanghai Jiao Tong University Joint Institute since 2017 (see http://cfe.ji.sjtu.edu.cn/index.html). He is a fellow of IET. He established the research theme of Technology Entrepreneurship for Sustainable Development that led to the recent book "Technology Entrepreneurship and Sustainable Development" by Springer Nature (eds. Pradeep Ray and Rajib Shaw), Singapore in 2022. He has also been an honorary professor in the UNSW Medicine and Health in Australia since 2016. Prior to 2016, he established and led for 10 years from 2006 the Asia-Pacific ubiquitous Healthcare Research Centre (APuHC) that was designated as the WHO Collaborating Centre (WHOCC) on eHealth at UNSW in 2013. Pradeep was a professor and the director of this WHOCC from 2013 to 2016 and led many projects on eHealth including those involving disaster management with collaborations across several countries (Bangladesh, India, Indonesia, China, Japan, EU, etc.). He founded IEEE Healthcom (eHealth Networking, Applications, and Services Conference) in 1999. In his current role in China, he has been leading multi-country projects, such as mHealth for Belt and Road Region (2018–2020) leading to a book including a Section on mHealth for disasters with four chapters led by Dr. Soshino and Dr. Geumpana "Mobile Technologies for Delivering Healthcare in Remote, Rural or Developing Regions", published by IET Press, London in 2020. His full list of publications is available at http://cfe.ji.sjtu.edu.cn/views/pradeep_bio.html.

Rajib Shaw is a professor in Graduate School of Media and Governance of Keio University, Japan. He is also a distinguished professor in the Institute of Disaster Management and Reconstruction (IDMR) of Sichuan University, China. He did his studies at Yokohama National University and Osaka City University in Japan and University of Allahabad and Burdwan University of India. He is co-founder of a Delhi-based social entrepreneur startup, Resilience Innovation Knowledge Academy (RIKA), and chair of the board of two Japanese non-government agencies: SEEDS

Asia and CWS Japan. He is the Co-chair of the United Nations Asia Pacific Science Technology Advisory Group (AP-STAG), and CLA for IPCC's 6th Assessment Report. Professor Shaw has 61 books and more than 400 research papers in the field of environment, disaster management and climate change. Professor Shaw is the recipient of the prestigious "Pravasi Bharatiya Samman Award (PBSA)" in 2021 for his contribution in education sector. Professor Shaw is also the recipient of United Nations Sasakawa Award for Disaster Risk Reduction in 2022 for his lifetime contributions in the field of disaster risk reduction. Professor Shaw is the recipient of "Norio Okada Implementation Science Award 2022" from the International Society for the Integrated Disaster Risk Management (IDRIM). More about his work can be found in https://www.rajibshaw.org.

Yasuhiro Soshino is the head of Humanitarian Technology Unit at Disaster Management Research Institute in the Japanese Red Cross College of Nursing (https://jrcdmri.jp/) and currently leading the world's first hydrogen ambulance project with Toyota Motor Corporation (https://global.toyota/en/newsroom/corporate/35008661.html). Soshino has been leading various projects on humanitarian technologies for disaster management. Soshino is focusing on supporting the self-help efforts of the disaster-affected people by using the mobile technologies. Soshino's five inventions for motivating the people toward their evacuation have been patented in Japan, US, China, and other EU countries. Soshino also invented the evacuation support system using the MaaS technology which was also patented in 2019. Soshino is the director of the International Medical Relief Department at the Japanese Red Cross Kumamoto Hospital, which is one of the hub hospitals for the Global Disaster Response by the Japanese Red Cross Society. Since 2001 Gujarat Earthquake in India, Soshino was deployed to more than 10 global disaster response missions as a team leader and a coordinator. Soshino is a member of the Field Assessment and Coordination Team (FACT) of the International Federation of Red Cross and Red Crescent Societies (IFRC).

Dr. Anirban Dutta is a biomedical engineer with an entrepreneurial background and a global career. He earned a Bachelor of Engineering from Jadavpur University, India, followed by an M.S. at the University of Florida and a Ph.D. at Case Western Reserve University, USA. In addition, he obtained an M.S. at Charité—Universitätsmedizin Berlin Stroke Research Centre, Germany. Dr. Dutta's career spans various academic and research roles globally, including positions at Tata Consultancy Services, Janelia Farm Research Campus (Howard Hughes Medical Institute), and the Rehabilitation Institute of Chicago. He continued his academic journey with roles such as Humboldt research fellow at the University Medicine Göttingen, starting research scientist at INRIA, France, and assistant/associate professor at the University at Buffalo, USA. He also held academic positions at the University of Lincoln and is currently at the University of Birmingham, UK. His group is continuing the NeuroPhys4NeuroRehab project (https://team.inria.fr/nphys4nrehab/) work to achieve a transformation in research outlook to interdisciplinary research by pursuing solution-oriented projects in neurorehabilitation that requires a broad cross-section of expertise.

Dr. Teuku Aulia Geumpana is a Computing and IT lecturer at the School of Information and Physical Sciences, the University of Newcastle (UON) Australia. Teuku has a strong interest in promoting the utilization of Information Technology in disaster management response via research and education. Teuku has published several book chapters, journals, and conference proceedings about implementing Information Technology in Disaster Management Response. During his Ph.D. at the University of New South Wales (UNSW), Teuku was an affiliated graduate researcher at the Asia Pacific ubiquitous Healthcare Research Centre (APuHC) and Data61-CSIRO. Earlier, Teuku took part in the emergency response as a technologist and IT specialist during Indonesia's 2004 Indian Ocean Tsunami and Earthquake emergency response. Under the IRC disaster response initiative program, Teuku was appointed as the head of the Health Information System manager to set up a health surveillance database in four districts of Aceh Province. In 2009, Teuku continued serving in disaster response as the MIS Task Team Leader at GHD Pty Ltd. to set up a database monitoring system for disaster rehabilitation and reconstruction projects.

Contributors

Arni Ariani Indonesian Biomedical Engineering Society, Bandung, Indonesia; Indonesian eHealth and Telemedicine Society, Bandung, Indonesia

Abhijit Das Royal Preston Hospital, Sharoe Green Lane, Preston, UK

Huiyu Duan Shanghai Jiao Tong University, Shanghai, China

Anirban Dutta University of Birmingham, Birmingham, UK

Teuku Aulia Geumpana School of Information and Physical Sciences, University of Newcastle, Newcastle, Australia

Bhumika Gupta Institute-Telecom Ecole de Management, Évry-Courcouronnes, France

Sankar Halder MUKTI, Raidighi, Kolkata, West Bengal, India

Allya Paramita Koesoema Indonesian Biomedical Engineering Society, Bandung, Indonesia;
Indonesian eHealth and Telemedicine Society, Bandung, Indonesia;
School of Electrical Engineering and Informatics, Institut Teknologi Bandung, Bandung, Indonesia

Ankita Kothiyal MUKTI, Kolkata, West Bengal, India

Akane Kurisu Disaster Management Research Institute, Japanese Red Cross College of Nursing, Shibuya, Tokyo, Japan

Chandrani Maitra Centre for Social Change and Development, Manchester, United Kingdom

Esperanza Miyake University of Strathclyde Glasgow, Glasgow, UK

Akira Miyata Japanese Red Cross Kumamoto Hospital, Kumamoto, Japan

Somnath Mukherji Association for India's Development, College Park, USA

Alivelu Mukkamala Department of Digitalization, Copenhagen Business School, Frederiksberg, Denmark

Raghava Rao Mukkamala Department of Digitalization, Copenhagen Business School, Frederiksberg, Denmark;
School of Business, The NorthCap University, Gurugram, India

Masahiro Nemoto Japanese Red Cross Hokkaido College of Nursing, Kitami, Japan;
Disaster Management Research Institute, Japanese Red Cross College of Nursing, Shibuya, Japan

Chongdan Pan Centre For Entrepreneurship (CFE), University of Michigan-Shanghai Jiao Tong University Joint Institute, Shanghai, China

Uttam Pudasaini Nepal Flying Labs, Kathmandu, Nepal;
NAXA, Kathmandu, Nepal

Colin Pye Preston, UK

Pradeep Kumar Ray University of Michigan—Shanghai Jiao Tong University Joint Institute, Shanghai, China;
UNSW Business, Sydney, Australia

Xiaoyu Ren Shanghai Jiao Tong University, Shanghai, China

Jenny Rowley Manchester Metropolitan University, Manchester, UK

Padmanava Sen Asha for Education, Covina, USA;
Barkhausen Institut, Dresden, Germany

Rajib Shaw Graduate School of Media and Governance, Keio University, Minato, Japan

Soegijardjo Soegijoko Indonesian Biomedical Engineering Society, Bandung, Indonesia;
Indonesian eHealth and Telemedicine Society, Bandung, Indonesia;
School of Electrical Engineering and Informatics, Institut Teknologi Bandung, Bandung, Indonesia

Yasuhiro Soshino Disaster Management Research Institute, Japanese Red Cross College of Nursing, Shibuya, Tokyo, Japan;
Japanese Red Cross Kumamoto Hospital, Kumamoto, Japan

Machiko Tomita School of Public Health and Health Professions, University at Buffalo, Buffalo, NY, USA

Guangtao Zhai Shanghai Jiao Tong University, Shanghai, China

Fei Zhao School of Public Health and Health Professions, University at Buffalo, Buffalo, NY, USA

Overview of Technology Innovation for Sustainable Healthcare and Disaster Management

Pradeep Kumar Ray, Rajib Shaw, Yasuhiro Soshino, Anirban Dutta, and Teuku Aulia Geumpana

1 Introduction

Disasters (e.g., cyclones, tsunami, earthquakes, volcanoes, etc.) have been disrupting the lives of people all over the world, particularly in Asia. Many technologies, such as surveillance systems, have been developed to manage disasters. The devastation caused by the COVID-19 highlights the multidisciplinary nature (including agriculture, healthcare, economics, environment, engineering, and technology) of disaster management. Most disasters have a healthcare dimension (e.g., outbreak of diseases in refugee camps) and some disasters (e.g., pandemics) are about healthcare. The pandemic also led to the development and uptake of innovative technologies, such as vaccine development, new biotech innovations, telemedicine, IoT, and mHealth all

Present Address:
P. K. Ray (✉)
University of Michigan—Shanghai Jiao Tong University Joint Institute, Shanghai, China
e-mail: p.ray@unsw.edu.au

UNSW Business, Sydney, Australia

R. Shaw
Graduate School of Media and Governance, Keio University, Minato, Japan

Y. Soshino
Disaster Management Research Institute, Japanese Red Cross College of Nursing, Shibuya, Japan
e-mail: y-soshino@jrcdmri.redcross.ac.jp

Japanese Red Cross Kumamoto Hospital, Kumamoto, Japan

A. Dutta
University of Birmingham, Birmingham, UK
e-mail: a.dutta.1@bham.ac.uk

T. A. Geumpana
School of Information and Physical Sciences, University of Newcastle, Newcastle, Australia
e-mail: Teuku.Geumpana@newcastle.edu.au

P. K. Ray et al. (eds.), *Technology Innovation for Sustainable Development of Healthcare and Disaster Management*, Disaster Risk Reduction,
https://doi.org/10.1007/978-981-97-2049-1_1

over the world. The UN (United Nations) Sustainable Development Goals (SDGs) provide an integrated approach to the development including that for a disaster management framework. For example, it needs to be addressed in a holistic perspective involving technology, healthcare, environment, economy, regulations, etc. Hence there are challenges in the development, validation, and deployment of technological innovations. This book presents a discussion of various dimensions of technological innovations in the multidisciplinary context of healthcare and disaster management with sustainable technology innovation case studies, using examples of complexities involved and an in-depth analysis of information technology perspective.

The global risk landscape changes over time. Over the last 4 years, new risks arise like infectious disease, digital power concentration and digital divide, cyber security, energy crisis, and cost of living. Therefore, uncertainties become a key challenge in the global risk landscape. In spite of all these new risks, environmental risks predominate in the risk landscape in terms of likelihood and impacts. These are disasters, extreme weather, climate inaction, biodiversity losses, etc. Thus, to cope with the new challenges, it is very important to focus on innovation, which needs to be linked to new technologies as well as existing technologies.

This book discusses the role of technology innovation in disaster management, especially in the context of healthcare with example case studies and their challenges in the implementation. Hence the book has been organized into three sections. Part 1 on the role of technology in healthcare and disaster management (how different types of technology innovations are being used). Part 2 discusses the complexities involved in the sustainable deployment of these technologies and the need for the cooperation among diverse organizations and across disciplines (public health, agriculture, information technology, management, social sciences, etc.). Part 3 on information technology delves deeper into the information technologies (e.g., Cloudlets, blockchain, eye tracking, and health technologies), the most common form of technology innovations.

This introduction chapter starts in Sect. 2 with the structure of the book and the summary of the three sections. This is followed by a summary of all chapters in Part 1 of the book in Sect. 3. The next Sect. 4 provides a summary of all chapters in Part 2 of the book. The following Sect. 5 provides a summary of all chapters in Part 3 of the book. This Introduction chapter concludes with a summary of the book in Sect. 6.

2 Structure

This book complements our previous book "Technology Entrepreneurship and Sustainable Development" (Ray and Shaw 2022) and these books are a part of the Springer book series on Disaster Risk Reduction published by Springer Nature under the leadership of our co-editor Rajib Shaw. This book brings together the related experience of the five editors as follows:

- Pradeep Kumar Ray has been leading mHealth (mobile technologies for healthcare) applications through international bodies, WHO, ITU, and IEEE over the last two decades and led an international initiative on mHealth for disaster management after the Indian Ocean Tsunami in Banda Aceh in 2005.
- Rajib Shaw has been leading many national (in Japan) and international projects on disaster management leading to many highly cited publications and the Springer book series on disaster risk reduction.
- Yasuhiro Soshino has been leading a number of global disaster response missions, involving innovative projects for the International Red Cross and Red Crescent Movement.
- Anirban Dutta is leading innovative projects on the application of emerging health technologies in the management/control of potential disasters (e.g., debilitating mental illnesses) from diabetes across USA, UK, and many developing countries.
- Teuku Aulia Geumpana was directly involved in the Indian Ocean disaster recovery in Banda Aceh on behalf of a global NGO (GIZ).

The book is organized into three Sections, each Section consisting of five chapters.

1. *Role of Technologies in Healthcare and Disaster Management*: This Section illustrates the role of technologies in disaster management through a number of case studies involving various technologies, such as digital technologies (including smartphones, social media, and drones), hydrogen technologies, and evolving technological innovations.
2. *Complexities Involved*: This Section highlights some of the major challenges in the planning, development, and deployment of emerging technologies in view of the multidisciplinary nature of disaster management (involving agriculture, healthcare, law, business, social sciences, technology management, etc.). The complexity issues have been illustrated using examples of services like agriculture and public health services.
3. *Information Technologies Used*: This Section discusses some Information Technologies used in disaster management, such as mobile phones, eye tracking, and blockchain. In particular, this Section discusses emerging ways of making disaster management resilient, using evolving information technologies.

The chapters in these three parts are summarized in the following three sections.

3 Role of Technologies in Healthcare and Disaster Management

Chapter "The Role of Social Media in Disaster Management: Opportunities, Challenges, and Future Directions" *discusses the role of social media in disaster management.* Social media is becoming an important tool for creating and disseminating news stories, especially relating to disasters where the infrastructure and communication is severely disrupted. Hence the social has become an alternative to formal communication in chaotic situations of disasters. Unfortunately, the number of man-made and natural disasters has increased in recent decades. Social media, combined with crowdsourcing, is proven to help quickly capture information on natural and human-made disasters directly from the disaster site. This chapter starts with a literature review of how social media is being used in the context of disasters. Then the authors describe the opportunities and challenges of using social media for disaster management and also shed light on future directions.

Chapter "Technology Innovations for Disaster Management in Developing Countries—Case Study for the Indian Ocean Tsunami at Banda Aceh, Indonesia and Recent Developments" *discusses a range of technology innovations for various stages of disaster management and a case study of the Indian Ocean Tsunami in Banda Aceh.* This chapter discusses the role of technology innovations for disaster management at the major stages of a disaster, namely, preparedness, response, and recovery. The chapter discusses the use of a range of technologies (e.g., smart phones, drones, 3D printing, biometrics, geospatial information systems, and specialized software including artificial intelligence) that have the potential to transform disaster management at various stages. This may include better preparedness to disasters, such as increasing community capacity through educational technologies, better immediate response, such as using aerial drones to survey disaster areas and deliver much needed supplies, as well as better recovery and risk reduction. This is followed by a case study of the Indian Ocean Tsunami in Banda Aceh, Indonesia.

Chapter "Design and Development of the Medical Fuel Cell Vehicle—The Technology Links Mitigation and Adaptation in Climate Change" *discusses the development of a Fuel Cell vehicle in the Japanese Red Cross Kumamoto Hospital in Japan.* Climate change is now recognized as a major source of disaster and there is now a worldwide effort to manage climate change by mitigating environmental pollution occurring through the carbon emissions of vehicles. Hence there is now a move toward battery and hydrogen powered electric vehicles. In contrast to other electric vehicles, Fuel Cell Electric Vehicles (FCEV) produce electricity using a fuel cell powered by hydrogen. They are environmentally friendly and produce no harmful emissions unlike petrol or diesel vehicles. They only emit water vapor and warm air. They may also act as portable external power supply during disaster response. Therefore, this chapter discusses the application of the Medical Fuel Cell Electric Vehicle (M-FCEV) as a solution for the climate change mitigation in medical services and disaster response, using a case study of the application of MFCEVs in Red Cross locations of Kumamoto and Okinawa, Japan.

Chapter "Technologies for Disaster Management: Case Studies in the Past Tsunami in Japan" *discusses the development and use of mobile technology-based location data and a case study of the* Great East Japan Earthquake and Tsunami (GEJET). The immediate search and rescue services are crucial for disaster affected people. In this respect, application of the location data from the mobile phones of the affected people is a possible solution. This chapter discusses the application of mobile phone location data (people flow data) as public assistance toward self-help efforts of affected people, using a case study of the tsunami in Japan. In addition, they help with the widespread uses of the personal floatation devices for the people living in the tsunami affected areas. Application of drones and image sharing systems also helps to identify the damage assessment and search and rescue services in emergencies in the tsunami hazard areas.

Chapter "Emerging Technologies to Avoid Preventable Death in Emergencies: A Case Study on Hypothermia Prevention in a Refugee Centre in Winter" *discusses some innovations in regular objects like sleeping bags and cardboard beds to prevent hypothermia in a refugee center during a disaster in winter.* In emergencies, especially in severe winter, the prevention of hypothermia cases is important to reduce the loss of lives and to reduce the workloads of medical facilities in the disaster affected areas. Therefore, this study aims to examine practical equipment, tools, and services to prevent hypothermia in refugee centers by reducing risk factors of heat loss and supporting efficient heating of the inmates. Refugee centers use simple technological innovations, such as sleeping bags, cardboard beds, simple sauna systems, partitions, special toilets, and kitchen. For example, even simple technology, such as a cardboard bed, helped to keep the body temperature by making some space between the body and the cold floor. This chapter describes the use of such innovations in real disasters, such as the Great East Japan Earthquake and Tsunami in Japan.

4 Complexities Involved

Chapter "Techniques and Process Development for Quick Response and Sustainable Recovery" *provides an overview of the process of quick response and sustainable recovery before and after a natural disaster (especially, floods and cyclones), with the help of technology.* This chapter provides a holistic discussion of the different aspects of cyclone-resilient livelihood, namely agriculture, food security, education, health, housing, and river embankment protection in a natural disaster-prone area. This chapter focuses on the sustainability of these solutions for the local population largely dependent on agriculture for their livelihood in the cyclone-prone rural coastal areas of eastern India. Hence flood-tolerant embankment methodologies and salt-tolerant sustainable agricultural methods are described. In addition, efficient communication strategies for preparedness and technology-aided educational support/rehabilitation are discussed. There is a description of how the role of technology has changed in the last 15 years for an efficient disaster response and recovery in Sunderbans, recognized as a UNESCO World Heritage site across India and Bangladesh.

Chapter "Towards a Housing Design That Can Withstand Natural Disasters: A Study in Sundarbans" *discusses the design of disaster-resilient housing in Sunderbans in India, frequently ravaged by cyclones and floods.* The Sundarbans region (UNESCO World Heritage Site) requires cyclone resilient housing, given the frequent and severe occurrences of cyclones and flooding. These frequent disasters cause extensive damage to the livelihood and infrastructure, particularly to the housing. The prevailing traditional housing structures in Sundarbans have significant vulnerabilities. To address this, the chapter introduces the cyclone/flood resilient housing design and construction led by MUKTI, a socio-economic development organization in the Sundarbans area in India. These disaster-resilient houses aim to achieve three major objectives of sustainability, i.e., affordability, adoptability, and disaster resilience. The structural innovation involves five major elements; pyramidal foundation strip beam, stilt system of splayed RCC columns, ferrocement staircase, shallow brick dome floor slabs and bamboo roofs.

Chapter "Innovative Sanitation System for Disaster Victims and Social Sanitation Challenges" *discusses an innovative design of a safe and portable sanitation system for disaster victims.* In emergencies, the management of toilets is one of the most important points which influences the protection of human dignity, the prevention of diseases, the security of females and children, and evacuation behaviors. Therefore, this chapter presents the development and deployments of an innovative sanitation system as part of a smart shelter concept of the Japanese Red Cross Kumamoto hospital. This concept proposes the daily uses of movable infrastructure such as batteries, water-recycling shower, sanitation kits, and movable flush toilets in public facilities which are used as refugee centers in emergencies. The smart shelter functions without relying on existing infrastructure such as the water supply, drainage, and sewage treatment system. This chapter focuses on the design and deployment of movable flush toilets, that are not only important for disaster refugee centers, they are also very useful in remote rural areas with less developed sanitation infrastructure.

Chapter "Benefits and Barriers of Using WhatsApp in Eye Health Communication in Deprived Settings in India" *discusses the role of the popular social media tool WhatsApp for eye health.* Globally, 2.2 billion people have a Vision Impairment (VI) that is a disaster for the family and society in some developing countries. For at least 1 billion, VI could have been prevented or yet to be addressed. Over recent decades, social media has emerged as an effective healthcare communication tool, but this has not been studied systematically in countries with high VI burden. In India, social media penetration is estimated to be 61.66% with 662.73 million WhatsApp users (2023 estimates). This qualitative case study used WhatsApp (as an interventional tool) to create an information resource link on basic eye care between a tertiary city-based eye healthcare provider and the deprived community, resident in the fringe of a large metropolitan city (Kolkata, Eastern India).

Chapter "Towards a Multi-dimensional Deployment Framework for Drones in Disaster Management" *discusses the complexities involved in the deployment of drones in disasters in a country like Nepal.* Drones have significant potential across all phases of disaster management as discussed in some of the previous chapters. This study narrates the case of a developing and disaster-prone country like Nepal where drones were prominently introduced for disaster response activities during the Gorkha earthquake in 2015, a mega-disaster incident. However, Nepal still struggles with scaling the use of drone technology in disaster response operations. As disaster response is a time-critical and most sensitive phase of disaster management, this research aims to systematically identify the underlying challenges that are hampering the use of drones during a disaster situation. This study draws on qualitative research and data from semi-structured interviews with 13 individuals including drone pilots, project managers, regulators, and representatives from the private sector. A thematic frame "ROTIDI" consisting of six thematic categories (resources, operational, technological, institutional, data privacy, safety, instrumental) was identified through a literature review and an iterative coding process after analyzing the response from the participants of the interview. Detailed challenges under each theme were identified and presented along with the lessons learned to deal with individual challenges based on practitioners' experience of working in real disaster scenarios.

5 Information Technologies Used

Chapter "A Generic Model to Capture Essential Information on Limited Capacity Mobile Devices during a Disaster" *discusses the design of a resilient information technology (mobile based) during a disaster*. During a disaster, most of the Information and Communication Technology (ICT) infrastructure is destroyed and hence disaster recovery requires innovative ICT solutions to assist the disaster recovery efforts. This chapter discusses such a solution based on mobile and cloud services. Priority Over Cloudlet Terminal (POCkET) framework is a mobile cloud computing architecture that aims to improve performance by prioritizing critical tasks during disaster response on cloudlets. To support this architecture, a Generic Information Model (GIMo) has been developed. GIMo is a flexible, extensible, and vendor-neutral information model that enables the integration of heterogeneous components in the POCkET framework. GIMo provides a standard way of representing and exchanging information between different components of the POCkET architecture, such as mobile devices, cloudlets, and cloud servers. This flexibility enables the POCkET framework to evolve and adapt to changing user demands and network conditions. This chapter discusses the process of developing GIMo as one of the important elements of POCkET Framework.

Chapter "Intelligent Eye Tracking for the Early Diagnosis of Autism: A Mental Health Disaster with Families" *discusses the role of eye tracking technology for the early diagnosis of autism, a social disaster for affected families.* Autism is a complex, lifelong developmental condition that typically appears during early childhood and can impact a person's social skills, communication, relationships, and self-regulation. In most cases, symptoms of autism become apparent by the age of 5 causing an emotional disaster for the family. The diagnosis of a child with autism is a form of disaster for the family and early diagnosis followed by interventions before the age of 7 may help. This chapter discusses the evolving intelligent eye-tracking technology to enable early diagnosis of autism among children.

Chapter "Blockchain Technology for Preventing Disasters in Surgery: A Case Study on Plastic Surgery" *discusses the role of the new technology of blockchains in the prevention of disasters caused by wrong surgery.* The literature is replete with examples of health disasters caused by unacceptable outcomes of elective surgeries, (e.g., plastic surgery) resulting in problems, such as no benefit from the surgery, wrong surgery operation, mistakes by hospitals/surgeons, and overlooking past health issues/medications. For example, patients are injured or die from malpractice caused by unqualified surgeons or medicine, that may be unknown to the patients and families. In many cases, these negative outcomes are covered up due to the lack of transparency in the process of surgery. This chapter discusses the role of blockchain technology to bring in more transparency in a secure manner to enable the conduct of the planning and evaluation of surgery. We illustrate the implementation of a practical blockchain system for an increasing popular form of elective surgery, plastic surgery. The blockchain system stores surgery data as traceable and immutable records. Besides, the system can serve as a crowd-sourcing information platform.

Chapter "Healthcare 4.0 (and 5.0): Digitising Health and Care" *explores the transformative impact of people flow modeling in the context of Healthcare 4.0 (and 5.0) that would help prevent some health disasters.* It suggests the links between the implementation of suitable technology and that of clinical need and for example, FND (Functional Neurological Disorder). Emerging areas such as AI (Artificial Intelligence), IoT (Internet of Things), and Data Analytics will be essential. By investigating people flow modeling in the healthcare sector, the chapter aims to elucidate its role in creating smart, responsive, and patient-centric environments in home and community settings. While the example is that of FND the same principles can be applied to other clinical areas. The chapter proposes an example "Flow" method and partnership way of working that encapsulates health sector needs, academic research and rigor and industry experience in order to deliver an accelerated and sustainable change in the way that patient-centric healthcare is delivered, taking account of patient (and person-centric) flow. Finally, while the deployment of technology presents new opportunities for the delivery of care and new ways of working it does not come without the need to acknowledge, and mitigate, the risk of depending upon specific technologies. In this respect lessons can, and must, be learned from other sectors.

Chapter "Portable Neuroimaging Based Digital Twin Model for Individualized Interventions in Type 2 Diabetes" *discusses the role of emerging digital twin model using functional Near-Infrared Spectroscopy (fNIRS) to prevent/delay the onset of Alzheimer's and Related Dementias among Type 2 Diabetes patients.* The Global Burden of Disease Study highlights diabetes' substantial impact, contributing to health disasters, such as amputation of limbs, blindness, and serious mental illnesses. Diabetes, with its elevated glucose levels, can lead to human health catastrophe with aging population in Alzheimer's disease and related dementias (AD/ADRD). Leveraging advancements in big data neuroimaging, cloud computing, and the Internet of Things, the chapter proposes a digital twin model for monitoring vascular brain health using functional near-infrared spectroscopy (fNIRS) while exergaming (technology-driven physical activities) in home settings. Here, fNIRS-based modal analysis serves as a foundation for developing digital twin models, offering a precise framework for monitoring vascular cognitive health. This technology innovation can empower clinicians and caregivers to implement timely interventions, potentially preventing or delaying neurovascular dysfunction in T2DM and progression to AD/ADRD.

6 Summary

This book has described some technological innovations in the management of both natural and other types of disasters, especially in healthcare as introduced in the first section of the book. Hence the chapters discuss a range of innovative technological solutions including information technologies, such as mobile phones, social media, drones, health technologies, and other technologies for resilient housing, shelter, and sanitation. Since the information technology solutions are the most common, Part 3 of the book is fully devoted to information technologies used. Since the deployment of these technologies present many challenges, Part 2 of the book also discusses the sustainability of these technological solutions in the context of natural disasters in different countries, such as tsunami in Indonesia and Japan and earthquakes in Nepal. While healthcare is an important element of most natural disasters, health disasters bring many miseries to families, such as severe mental illnesses (e.g., Alzheimer's), diagnosis of autism of a young child, and a surgery gone wrong. Technology innovations are now helping all these different types of disaster scenarios and this book provides some examples. However, this book does not include any specific chapter on some other types of important disasters, such as financial scams, cyberterrorism, and pandemics.

Reference

Ray PK, Shaw R (eds) (2022) Technology entrepreneurship and sustainable development. Springer, Singapore, ISBN 978-981-19-2053-0. https://www.springerprofessional.de/en/technology-entrepreneurship-and-sustainable-development/23174380. Accessed June 2022

Role of Technologies in Healthcare and Disaster Management

The Role of Social Media in Disaster Management: Opportunities, Challenges, and Future Directions

Alivelu Mukkamala, Bhumika Gupta, and Raghava Rao Mukkamala

1 Introduction

Humans are susceptible to different types of disasters: on the one hand, natural disasters like earthquakes, floods, and epidemics; on the other hand, man-made disasters such as acts of terrorism, war, and conflicts. In either case, we have to deal with the devastating consequences of a disaster, and these events disrupt a community's everyday functioning and have an impact on people's lives, the economy, and the environment. Compared to epidemic diseases (such as COVID-19) or economic crises, natural and man-made disasters occur suddenly and require quick response and fast relief activities. Therefore, disasters are typically managed through four disaster management phases: mitigation, preparedness, response, and recovery (Cozzolino 2012; Lee et al. 2017) to reduce the influence of catastrophic activities. During natural and man-made disasters, protecting lives, property, and the environment during and after the disaster is a critical component of disaster management.

Social media has become an integral part of our personal and professional lives. Platforms like Twitter, Facebook, and Instagram have revolutionized how individuals, communities, and organizations communicate and share information over the past 20 years. As per Kaplan and Haenlein (2010) and Lin (2018), social media platforms have enabled bi-directional communication where people become both consumers

A. Mukkamala · R. R. Mukkamala (✉)
Department of Digitalization, Copenhagen Business School, Frederiksberg, Denmark
e-mail: raghava.mukkamala@cbs.dk; rrm.digi@cbs.dk

A. Mukkamala
e-mail: alivelu.mukkamala@gmail.com

B. Gupta
Institute-Telecom Ecole de Management, 9 rue Charles Fourier, Évry-Courcouronnes, France
e-mail: bhumika.gupta@imt-bs.eu

R. R. Mukkamala
School of Business, The NorthCap University, Gurugram, India

P. K. Ray et al. (eds.), *Technology Innovation for Sustainable Development of Healthcare and Disaster Management*, Disaster Risk Reduction,
https://doi.org/10.1007/978-981-97-2049-1_2

and producers of information. Due to this, social media has become an important tool in disaster management, allowing for real-time information sharing, increased situational awareness, and mobilizing volunteers through public engagement (Houston et al. 2015).

1.1 Usage of Social Media in Disasters

In recent times, there has been a significant increase in the frequency and severity of natural disasters worldwide, spanning across various continents (CRED 2021; Stiegler et al. 2011). Apart from natural calamities, man-made disasters have also become a significant threat to humanity due to changing geographical conditions and geopolitical instability, making communities more vulnerable and further endangering their stability and safety globally. During such crisis and disaster situations, the availability of real-time information becomes crucial for emergency management agencies and humanitarian organizations. This immediate and real-time data is highly instrumental in making effective decisions and coordinating prompt response activities. Swift action by these organizations is vital to impactful relief efforts, enabling rescue teams to save lives, assess infrastructural damages, and estimate economic losses and casualties. The sooner the organizations react, the more lives they can save. However, the challenge lies in efficiently gathering and processing information regarding the needs of the affected people, which has been proven to be very challenging due to the problems involved in tracing and tracking the people's needs (Comfort et al. 2005). Nonetheless, the advent of the Internet, advancements in mobile communication, and, notably, the proliferation of social media have significantly reduced the time it takes to obtain information from disaster zones if we compare the present times to the 1990s (Meier 2016).

Disasters often lead to the breakdown of traditional communication channels, as vital communication infrastructures are prone to damage during disaster situations. Consequently, people in crisis situations increasingly turn to alternative methods for sharing and receiving information. Social media platforms have transformed the way disaster-related information is distributed and received during disasters. Social media platforms such as Twitter, Facebook, and others allow for real-time updates, enabling the rapid dissemination of information to a wider audience (Simon et al. 2015). It's not just individuals who are utilizing these platforms; institutions like government entities, media outlets, and non-governmental organizations are also actively participating in gathering and sharing information during crises. Instances like the Alberta floods, the Boston Marathon bombing, the Oklahoma tornado, and the West Texas explosion saw an increased reliance on social media platforms such as Twitter by affected individuals and others seeking information about these crisis situations (Olteanu et al. 2014, 2015).

On social media platforms such as Twitter and Facebook, the abundance and availability of user-generated content offers a plethora of valuable information, including rescue requests, situational updates, and coordination of relief efforts. Moreover, social media is not only a tool for information dissemination but also a platform for community support and engagement during disasters, as further elaborated in the following sections. It enables affected individuals to report their status, seek help, and connect with others, and also serves as a critical tool for communication, coordination, and community building. Its real-time nature facilitates rapid information dissemination, while its wide reach and accessibility make it an effective platform for mobilizing resources and support during and after disasters.

1.2 Usage of Social Media by Affected Communities

Moreover, affected communities use social media to share information, mobilize, and self-organize to coordinate activities among themselves (Vieweg et al. 2010; Starbird et al. 2014). Hence, disaster management agencies need to listen to the voices of affected communities to organize their relief activities because of the benefit of real-time information from the affected zones (Krstajic et al. 2012). However, social media data's volume and velocity make it challenging to monitor and extract valuable information. Therefore, it is essential to develop methodologies and techniques to automate the process to determine the rescue requests and the urgencies of the affected people's needs, including food, clothing, and shelter shortages. In this regard, we refer to the example of two case studies (Mukkamala and Beck 2016, 2018), where we applied different methods to analyze disaster social media data to extract crucial information that will be an asset to disaster management agencies. The two case studies are Hurricane Sandy, which happened in 2012 in the USA (Mukkamala and Beck 2016), and the Chennai floods, which occurred in India in 2015 (Mukkamala and Beck 2018). These case studies belong to two different geographical locations with different cultural norms, but observing the commonalities between the two different contexts is interesting. Moreover, the first disaster occurred in a developed country with advanced infrastructure and support/relief systems. The other case study is from a developing country with minimal infrastructure and relief systems. Our motivation for using these case studies is also to explore how social media is being used by people in crisis situations, in different contexts, to share and receive information.

Understanding the role of social media in disaster management is crucial for practitioners, policymakers, and researchers alike, as it offers insights into the opportunities and challenges associated with leveraging these platforms to improve disaster response and recovery efforts (Soden and Palen 2018; Palen and Liu 2007). This chapter aims to provide a comprehensive analysis of the role of social media in disaster management, examining the opportunities it presents, its challenges, and potential future directions for research and practice.

The rest of the chapter is organized as follows. First, we present a case study on the usage of Twitter during Hurricane Sandy to describe the type and nature of messages posted by the affected users in Sect. 2. In Sect. 3, we outline some of the opportunities on using the social media during disaster situations. Along with opportunities, there are quite a few challenges in using social media in disaster situations, and these are elaborated in Sect. 4. Finally, we outline the future directions in Sect. 5.

2 Use Case: Tweet Messages During Hurricane Sandy

During Hurricane Sandy in 2012, social media was used to communicate during troubled times. The hurricane started in the Caribbean waters on October 22, 2012, and eventually turned into a superstorm that affected 60 million people across 24 states in the US. On October 29, Hurricane Sandy landed, causing power outages, flooded subway systems, disrupted communication, and shortages of gasoline, commodities, and food in New York City (Zubiaga and Ji 2014). In this case study, we explored how Twitter is used for situation awareness during disasters and also looked into the different nature of the messages (Mukkamala and Beck 2016) that exhibited the situation awareness.

2.1 Data Collection

We used a Tweet dataset containing Hurricane Sandy tweet IDs (Zubiaga and Ji 2014) that can be publicly accessed (Zubiaga 2015). The original dataset contains nearly 15 million tweets posted on Twitter between October 25 and November 4, 2012, when hurricane Sandy hit the East Coast of the US. To comply with Twitter's terms of service, the researchers only shared the tweet IDs and not the actual tweets from the dataset of Hurricane Sandy. Using those tweet IDs, we retrieved complete data about tweets between May and June 2015 using the Twitter APIs (Twitter 2015). The originally shared dataset comprised nearly 15 million tweets, but when we downloaded the tweets, we could only download 11 million tweets. As we downloaded the tweets a couple of years later, some of the tweets/retweets might have been deleted, or it could also be the reason that users might have deleted their Twitter accounts. Finally, we ended up with a dataset of 11 million tweets, as further described below.

2.2 Data Pre-processing

To begin with, we utilized Tableau, a data visualization software, to understand the distribution of tweets during a specific period. Despite the large amount of data, our primary objective was to isolate tweets that Twitter users shared in the area affected along the US East Coast. The availability of geo-location information, such

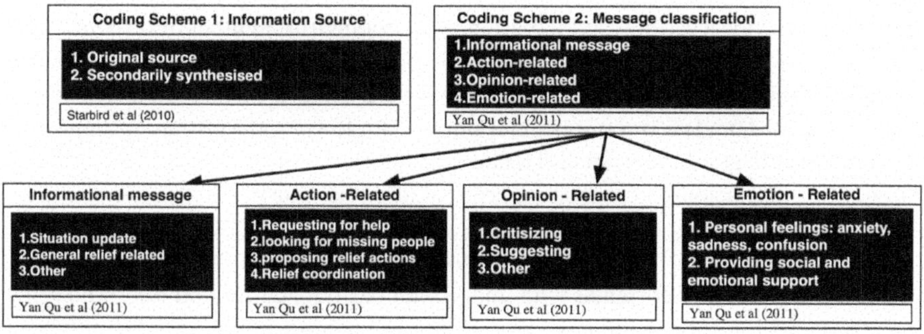

Fig. 1 Coding schemes adopted to classify the Twitter messages in Hurricane Sandy

as latitude and longitude coordinates in tweets' metadata, is primarily influenced by Twitter users' privacy settings.

The longitude and latitude coordinates can precisely indicate the location of a Twitter user when sharing information (Graham et al. 2014). To extract geo-location information from the tweets, we used Cosmos software (Burnap et al. 2015) and retrieved 115,800 tweets that contained longitudinal and latitudinal coordinates, which is roughly 1% of the initial 11 million tweets. Then, we used CartoDB[1] to narrow down the tweets geographically to the path Hurricane Sandy took along the coastline. We limited the tweets to those written in English, which further declined to 100,700 tweets. We then focused on the tweets from the hurricane-affected areas of the US East Coast, resulting in 68,800 tweets originating from the hurricane-affected area between October 25 and November 5, 2012.

2.3 *Manual Content Analysis of Tweets*

We evaluated various coding schemes to conduct manual content analysis and incorporated two coding classifications from previous studies (Qu et al. 2011; Starbird et al. 2010). These were used to develop our coding framework (Mukkamala and Beck 2016), shown in Fig. 1. Our decision to use these coding classifications was based on the fact that both were designed to analyze disaster-related Twitter data, focusing on earthquakes and floods. The coding scheme presented in a study by Qu et al. (2011) was built upon a previous scheme developed by Qu et al. (2009), Vieweg et al. (2010) for analyzing Twitter data during disasters. The purpose of these coding classifications was to assess the relevance of social media content in disaster management research.

The first coding scheme (Starbird et al. 2010) used in this analysis was developed in a bottom-up approach during the analysis of Twitter data related to floods. They classified tweets based on the information source, which could be original (reflecting

[1] https://carto.com/.

users' personal observations and experiences), secondarily synthesized (from other tweets and news sources), re-sourced (passing on other online sources), and retweets (forwarding the tweets). In our coding scheme, we merged the secondarily synthesized, re-sourced, and retweets categories into a single secondary source category to aggregate tweets at a higher level. Hence, our first level of coding scheme consists of two information source categories: original and secondary.

The coding scheme described in this study (Qu et al. 2011) was developed through a combination of two previous coding schemes (Qu et al. 2009; Vieweg et al. 2010) and further enriched with analysis of earthquake-related microblogging data. The resulting scheme consists of four major categories based on the nature of messages: informational, action-related, opinion-related, and emotion-related. We combined both coding schemes to classify Hurricane Sandy tweets into an integrated classification (Mukkamala and Beck 2016). However, we did not use the more granular coding approach of Qu et al. (2011) because our main interest was the real-time nature of the data.

Our adopted coding scheme (Mukkamala and Beck 2016) allows us to analyze tweets in two dimensions: information source and nature of the message. This helps us classify tweets based on whether they originated from the original source or a secondary source and also provides an understanding of the nature of the messages shared by individuals. Finally, the classification is based on four categories: informational, action-related, opinion-related, and emotion-related.

Our main objective in the case study was to analyze the kind and quality of information that can help emergency management agencies achieve situational awareness during a disaster. Specifically, original and informative messages that include personal experiences and observations about a given situation can significantly contribute to the situational awareness of the agencies. Furthermore, messages that express opinions and suggest actions can provide appropriate guidance to emergency agencies. Although secondary sources may not contain personal status updates, they can contribute to situation awareness by indirectly providing useful information about the situation. Specifically, emotion-related messages can reveal that individuals stand by the affected community during disasters to provide emotional support. Finally, around 677 geo-located tweets were extracted and further analyzed based on the coding scheme. The distribution of tweets along the two coding schemes during Hurricane Sandy was presented in Fig. 2.

The figure in Fig. 2 shows that information messages were the most common message type during the disaster period, both in original and secondary source categories. However, most informational messages came from the original source category only. Regarding other message categories, emotion-related tweets began to increase on October 28 and peaked on October 30. During the time the hurricane was happening, it was evident that people showed their sympathy and offered emotional and social support. Emotions are usually personal, so we could only observe them in the original source, as shown in Fig. 2. Similarly, discussions about opinions and actions are often personal; thus, they were found only in the original source category.

Our research has shown that most information shared during emergencies is situational updates, whether from original or secondary sources. However, more original

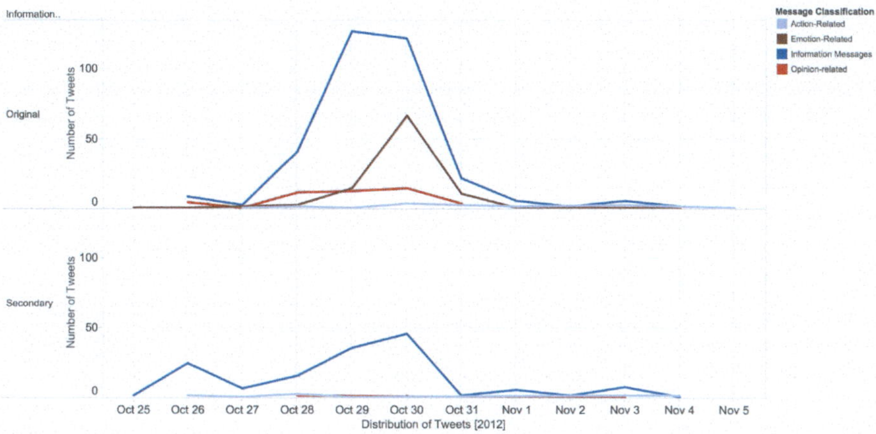

Fig. 2 Distribution of Twitter messages according to information source and message classification

tweets indicate that individuals affected by the situation share real-time information to provide updates or request immediate help. This type of information, generated from the bottom up, is crucial for emergency agencies.

3 Opportunities of Social Media in Disaster Management

This section outlines the opportunities for using social media to help in disaster management. Based on our literature review, we identified and outlined the opportunities for using social media in disaster management along four main categories:

1. Communicating Real-Time Information
2. Situational Awareness Data
3. Public Engagement and Volunteer Mobilization
4. Support and Solidarity Among Affected Communities.

3.1 Using Social Media for Dissemination of Real-Time Information

In recent decades, social media has become an essential tool for real-time information dissemination in disaster management, enabling rapid communication and coordination among affected populations, first responders, and government agencies (Houston et al. 2015). Social media presents several significant opportunities for real-time information dissemination in disaster management. Government agencies, emergency services, and aid organizations can leverage these platforms to share updates, warnings, and safety instructions, potentially reaching many people in a short period of time (Houston et al. 2015; Lindsay 2011) Social media platforms such as Twitter, Facebook, and Instagram have proven to be quite helpful and demonstrate their value in various disasters by facilitating the sharing of critical updates, as well as providing a means for affected individuals to request assistance and report local conditions. In addition, the widespread reach and accessibility of social media platforms allow for quick and efficient information dissemination, which can be crucial in times of crisis when traditional communication channels may be disrupted (Palen and Liu 2007; Lindsay 2011).

One of the significant advantages of using social media in disaster management is its capacity for real-time data collection and analysis. Administrators from emergency management agencies can monitor social media activity to gain insights into the severity and extent of a disaster, identify emerging threats, and track the progress of relief efforts (Castillo 2016). This type of information extracted from social media can then inform decision-making processes, allocate resources based on needs, and develop targeted response strategies. For instance, during Hurricane Sandy in 2012, emergency responders utilized social media data to identify areas with the greatest need for assistance and coordinate the distribution of essential supplies (Hughes et al. 2014; Mukkamala and Beck 2016).

3.1.1 Social Media as Two-Way Communication

Social media allows for two-way communication, providing affected individuals a platform to share their experiences, request help, and provide situational updates (Houston et al. 2015). This kind of user-generated content can serve as a valuable source of real-time, on-the-ground information, aiding emergency services in identifying areas of need and directing their resources effectively. During the 2010 Haiti earthquake, crisis mapping platforms like Ushahidi used crowdsourced social media data to create detailed maps of the disaster-affected areas, guiding search and rescue efforts (Valecha et al. 2013). Additionally, social media data can be analyzed to track the progression of a disaster and its aftermath. By utilizing advanced data analytics and machine learning algorithms, emergency managers can extract valuable insights from the vast amounts of data generated during a crisis (Castillo 2016). For instance, during Hurricane Harvey, researchers used machine learning to analyze Twitter data and identify emerging trends and patterns, which could inform decision-making and resource allocation (Yu et al. 2019; Reynard and Shirgaokar 2019).

Integrating social media with emerging technologies like artificial intelligence (AI), geospatial technologies, and the Internet of Things (IoT) presents a great opportunity. This integration can improve the capacity of social media platforms for real-time information dissemination and analysis. For example, AI can filter and verify social media content in real time, ensuring emergency managers can access accurate and relevant information (Nguyen et al. 2017). Similarly, geospatial technologies can provide a spatial context to social media data, enabling the creation of real-time maps that visualize the geographic spread of a disaster (Roche et al. 2013).

Lastly, social media platforms provide a space for collaboration and coordination among different stakeholders involved in disaster management. Government agencies, NGOs, private sector partners, and the general public can use these platforms to share information, coordinate efforts, and build partnerships (Meier 2015). This kind of collaborative approach can significantly enhance the efficiency and effectiveness of disaster response operations, as seen in response to the 2015 Nepal earthquake, where a consortium of volunteer and technical communities collaborated with various organizations to analyze social media data and create crisis maps (Park and Johnston 2017). Similar initiatives can also be seen in response to the Chennai floods in December 2015, where people used Twitter and other social media platforms to self-organize themselves by sharing disaster-related information, such as the collection and distribution of reliefs and also to coordinate and volunteer to take active participation in real-time response relief activities (Mukkamala and Beck 2018).

3.2 Social Media for Situational Awareness Data

Social media platforms offer numerous opportunities for data collection on situational awareness in disaster management. They provide a dynamic, real-time source of information that can inform emergency response teams about unfolding disaster situations. The first opportunity lies in the speed and volume of information shared on these platforms. In times of crisis, individuals often turn to social media to share their experiences, seek help, and stay informed about the situation (Soden and Palen 2018). By monitoring social media activity, emergency managers can gain a real-time understanding of the situation on the ground, which can be vital for effective disaster response (Yin et al. 2012; Houston et al. 2015). Moreover, social media data can be analyzed to identify and track emerging trends and patterns in situational awareness during disasters. Advanced data analytics and machine learning algorithms can be used to process the vast amounts of data generated during a disaster, helping emergency managers identify areas of need, understand the progression of the disaster, and anticipate potential challenges (Castillo 2016; Nguyen et al. 2017; Yu et al. 2019).

Social media platforms also offer the opportunity for geospatial analysis during disaster situations. Many social media posts are geotagged, providing a spatial context that can be incredibly valuable in disaster situations (Zook et al. 2010). Geospatial technologies can be used to visualize social media data on maps, allowing emergency administrators to see where a disaster is most severe or where help is most urgently needed. This was demonstrated during the 2010 Haiti earthquake, where crisis mapping platforms used crowdsourced social media data to create detailed maps of the disaster-affected areas (Zook et al. 2010). Furthermore, the IoT can provide additional real-time data that, when combined with social media data, can enhance situational awareness. For example, data from IoT devices such as weather sensors or traffic cameras can complement the information obtained from social media (Gao et al. 2011; Ray et al. 2017).

3.3 Social Media for Public Engagement and Volunteer Mobilization

In addition to playing a crucial role in disseminating real-time information and situational awareness, social media platforms also play a vital role in public engagement and mobilization of volunteers during disasters. These platforms serve as interactive communication channels, enabling agencies to disseminate critical information and engage with the public more directly and meaningfully (Kusumasari and Prabowo 2020; Houston et al. 2015). Social media can provide updates, answer queries, and solicit feedback, fostering a sense of involvement and cooperation among the public. For instance, during the Fort McMurray wildfire in Canada in 2016, social media platforms were extensively used to share information, answer public inquiries, and maintain a constant line of communication with affected communities (Boulianne et al. 2018).

Additionally, social media is a powerful tool for mobilizing volunteers in times of crisis. Platforms like Twitter and Facebook can be used to call for volunteers, coordinate volunteer efforts, and share volunteer experiences, which can encourage more people to get involved (Sharp and Carter 2020). Social media has been used by digital volunteering initiatives like the Digital Humanitarian Network (Meier 2015) to bring together virtual volunteers from different parts of the globe. These volunteers aid disaster response efforts by analyzing social media data, mapping crisis information, and performing other digital tasks. Similarly, social media facilitates the formation of spontaneous, self-organizing volunteer groups during disasters. These emergent volunteer groups can play a significant role in disaster response, carrying out search and rescue operations, distributing aid, and providing other forms of assistance. Social media platforms provide a space for these groups to coordinate their activities, share information, and connect with official response agencies (Simon et al. 2015). This was exemplified during Hurricane Sandy, where spontaneous volunteer groups used social media to organize grassroots response efforts, often filling gaps in official response operations (Hughes and Tapia 2015).

Given social media's significant role in public engagement and volunteer mobilization, various efforts have been made to formalize its use in disaster management. Governments and humanitarian organizations are increasingly incorporating social media strategies into their disaster response plans, recognizing the importance of these platforms for public engagement and volunteer coordination (Nahkur et al. 2022). The Red Cross, for example, has developed a Digital Volunteer program where trained volunteers monitor social media platforms during disasters, providing real-time updates, answering public queries, and connecting people with resources (Reuter 2022).

3.4 *Using Social Media for Support and Solidarity*

Social media have become a critical tool for facilitating support and solidarity during disaster management. These digital platforms are increasingly used to create a sense of community, provide emotional support, and mobilize resources in times of crisis. According to Houston et al. (2015), social media serve as critical platforms for exchanging disaster-related information and emotional expression, fostering collective problem-solving and resilience-building among the affected communities. In addition, it is noted that these platforms helped to create and reinforce a sense of shared experience and collective identity, which was essential for fostering mutual aid and resilience in the aftermath of a disaster.

Moreover, social media have proven to be powerful tools for coordinating volunteer efforts and mobilizing donations. For example, a study by Reuter (2022) found that emergency services were using social media not only to disseminate information but also to gather volunteers and coordinate relief efforts. Similarly, research by Merchant et al. (2011) highlighted how social media were used to mobilize aid and resources during the Haiti earthquake in 2010, demonstrating the potential of these platforms to facilitate solidarity and support in times of crisis. Hence, social media are not just platforms for communication but are becoming integral elements in fostering community resilience and solidarity during disasters.

4 Challenges of Using Social Media in Disaster Management

Based on our literature review, the following are the challenges of using social media in disaster management which are categorized as follows:

1. Challenge of Misinformation and Disinformation
2. Privacy and Ethical Concerns
3. Social Media Fatigue.

4.1 *Challenge of Misinformation and Disinformation*

Despite the benefits of real-time information dissemination, social media platforms can also be sources of misinformation and disinformation during disasters (Oh et al. 2013). Furthermore, misinformation, the unintentional spreading of false information, and disinformation, the deliberate spreading of incorrect information, can hinder disaster response efforts and deepen panic among affected populations (Starbird et al. 2014). Rumors, hoaxes, and fraudulent appeals for donations can spread rapidly on social media, making it challenging for users to discern fact from fiction (Veil et al.

2011). Researchers and practitioners have proposed strategies for combating misinformation to address this challenge, such as developing algorithms for detecting and flagging false content, promoting digital literacy, and encouraging users to verify the information before sharing (Zeng et al. 2010).

4.2 Privacy and Ethical Concerns

The use of social media in disaster management also raises privacy and ethical concerns, particularly regarding the collection, storage, and sharing of personal data (Lovari and Bowen 2020). While collecting data from social media can enhance situational awareness and inform response efforts, it may also infringe on individual privacy rights and increase the risk of surveillance, identity theft, or other malicious uses of personal information (Watson and Rodrigues 2018). To address these concerns, disaster management professionals must balance the need for timely and accurate information with the ethical responsibility to protect individual privacy. This can be achieved by implementing data protection policies, anonymizing personal data, and ensuring that only relevant and necessary information is collected and shared during disaster response efforts.

4.3 Social Media Fatigue

Finally, constant exposure to disaster-related content on social media may contribute to fatigue and desensitization, potentially undermining public engagement and response (Park and Johnston 2017; Thompson et al. 2017; Sutton et al. 2015). As users become inundated with disaster-related images, videos, and updates, they may experience feelings of helplessness, anxiety, or apathy, which could hinder their willingness or ability to take appropriate action. To mitigate the effects of social media fatigue, disaster management organizations should consider strategies for minimizing information overload, such as curating and prioritizing content, providing clear calls to action, and promoting self-care and mental health resources (Pechta et al. 2010).

Finally, addressing the challenges posed by social media use in disaster management requires continuous ongoing research, education, and policy development. By focusing on solutions to combat misinformation, bridging the digital divide, addressing privacy and ethical concerns, and mitigating social media fatigue and desensitization, disaster management professionals can harness the full potential of social media while minimizing its drawbacks.

5 Future Directions

Here are some potential ways that social media could be used for disaster management in the future.

As social media usage in disasters continues to evolve and generate massive amounts of data during disasters, the integration of artificial intelligence and machine learning can significantly enhance the effectiveness of disaster management efforts (Imran et al. 2015; Niyazi and Behnamian 2023). Machine learning algorithms can automatically identify, classify, and filter relevant disaster-related information from vast social media data, enabling decision-makers to access critical information more efficiently. Additionally, these technologies can help detect and mitigate misinformation and disinformation by automatically identifying and flagging potentially false or misleading content. Recent generative AI models such as ChatGPT (Bommasani et al. 2021) and other models can be used to detect fake and misinformation during disasters. However, there is also a challenge that these generative models, like ChatGPT, can also be used to generate fake content and misinformation at large scales by users with malicious intent. However, the recent advances in generative AI (Bommasani et al. 2021) show promising results and could be used to identify such fake content. As these ML technologies continue to evolve and advance, their application in disaster management via social media is expected to grow, offering new opportunities for improved response and recovery efforts.

Integrating geospatial mapping and visualization tools with social media data can provide valuable insights for disaster management professionals, enabling them to better understand the spatial distribution of impacts, needs, and resources during a disaster (Goodchild and Glennon 2010). By combining location-based social media data with geographic information systems and remote sensing technologies, disaster management teams can create real-time, dynamic maps that can be used for situational awareness, resource allocation, and decision-making (Huang and Xiao 2015). Future developments in geospatial mapping and visualization technologies, such as augmented reality and 3D modeling, may further enhance the utility of social media data in disaster management efforts.

As the number of social media platforms and users continues to grow, there is an increasing need for cross-platform collaboration and interoperability in disaster management efforts (Reuter 2022). Developing mechanisms for integrating data from multiple platforms, such as Twitter, Facebook, Instagram, and others, can provide a more comprehensive and accurate picture of the disaster situation. Additionally, fostering collaboration between social media platforms and disaster management organizations can help ensure the effective dissemination of accurate and timely information, as well as enhance public engagement and response efforts (Wukich et al. 2019; Houston et al. 2015).

Lastly, social media provides a way for communities to get involved in managing disasters. It enables them to share information, work together, and take collective action locally (Soden and Palen 2018). Future research should concentrate on creating ways to use social media to assist with community-led disaster management. This can include encouraging the establishment of online networks at a local level, using social media for disaster preparedness and response at the neighborhood level, and integrating social media data created by the community into official disaster management decision-making procedures (Hughes and Tapia 2015).

References

Bommasani R, Hudson DA, Adeli E, Altman R, Arora S, von Arx S, Bernstein MS, Bohg J, Bosselut A, Brunskill E, et al. (2021) On the opportunities and risks of foundation models. arXiv preprint arXiv:2108.07258

Boulianne S, Minaker J, Haney TJ (2018) Does compassion go viral? social media, caring, and the fort mcmurray wildfire. Inf, Commun Soc 21(5):697–711

Burnap P, Rana O, Williams M, Housley W, Edwards A, Morgan J, Sloan L, Conejero J (2015) Cosmos: towards an integrated and scalable service for analysing social media on demand. Int J Parallel Emergent Distrib Syst 30(2):80–100

Castillo C (2016) Big crisis data: social media in disasters and time-critical situations. Cambridge University Press, Cambridge

Comfort LK, Ko K, Zagorecki A (2005) Coordination in rapidly evolving disaster response systems: the role of information. In: Agent-based simulation: from modeling methodologies to real-world applications: post-proceedings of the third international workshop on agent-based approaches in economic and social complex systems 2004. Springer, pp 208–219

Cozzolino A (2012) Humanitarian logistics and supply chain management. In: Humanitarian logistics. Springer, Berlin

CRED (2021) Disaster year in review 2020: global trends and perspectives. Technical report, Centre for Research on the Epidemiology of Disasters, Université Catholique de Louvain Brussels Belgium

Gao H, Barbier G, Goolsby R (2011) Harnessing the crowdsourcing power of social media for disaster relief. IEEE Intell Syst 26(3):10–14

Goodchild MF, Glennon JA (2010) Crowdsourcing geographic information for disaster response: a research frontier. Int J Digit Earth 3(3):231–241

Graham M, Hale SA, Gaffney D (2014) Where in the world are you? Geolocation and language identification in twitter. Prof Geogr 66(4):568–578

Houston JB, Hawthorne J, Perreault MF, Park EH, Goldstein Hode M, Halliwell MR, Turner McGowen SE, Davis R, Vaid S, McElderry JA et al (2015) Social media and disasters: a functional framework for social media use in disaster planning, response, and research. Disasters 39(1):1–22

Huang Q, Xiao Y (2015) Geographic situational awareness: mining tweets for disaster preparedness, emergency response, impact, and recovery. ISPRS Int J Geo Inf 4(3):1549–1568

Hughes AL, Tapia AH (2015) Social media in crisis: when professional responders meet digital volunteers. J Homel Secur Emerg Manage 12(3):679–706

Hughes AL, St Denis LA, Palen L, Anderson KM (2014) Online public communications by police and fire services during the 2012 hurricane sandy. In: Proceedings of the SIGCHI conference on human factors in computing systems, pp 1505–1514

Imran M, Castillo C, Diaz F, Vieweg S (2015) Processing social media messages in mass emergency: a survey. ACM Comput Surv (CSUR) 47(4):1–38

Kaplan AM, Haenlein M (2010) Users of the world, unite! the challenges and opportunities of social media. Bus Horiz 53(1):59–68

Krstajic M, Rohrdantz C, Blumenschein M, Weiler A (2012) Getting there first: real-time detection of real-world incidents on twitter. In: Workshop on interactive visual text analytics: task-driven analysis of social media content with Visweek'12

Kusumasari B, Prabowo NPA (2020) Scraping social media data for disaster communication: how the pattern of twitter users affects disasters in asia and the pacific. Nat Hazards 103(3):3415–3435

Lee NS, Hirschmeier S, Müller S, Luz LJ (2017) Enablers in crisis information management: a literature review. In: The Hawaii international conference on system sciences (HICSS). http://hdl.handle.net/10125/41182

Lin TC (2018) Incorporating social activism. BUL Rev HeinOnline 98:1535

Lindsay BR (2011) Social media and disasters: current uses, future options, and policy considerations

Lovari A, Bowen SA (2020) Social media in disaster communication: a case study of strategies, barriers, and ethical implications. J Public Aff 20(1):e1967

Meier P (2015) Digital humanitarians: how big data is changing the face of humanitarian response. CRC Press

Meier P (2016) The value of timely information during disasters (measured in hours). https://irevolutions.org/2016/04/06/the-value-of-timely-information-during-disasters-measured-in-hours/

Merchant RM, Elmer S, Lurie N (2011) Integrating social media into emergency-preparedness efforts. N Engl J Med 365(4):289–291

Mukkamala AM, Beck R (2016) Enhancing disaster management through social media analytics to develop situation awareness: What can be learned from twitter messages about hurricane sandy? In: PACIS 2016 proceedings. AIS Electronic Library (AISeL)

Mukkamala AM, Beck R (2018) The role of social media for collective behaviour development in response to natural disasters. In: 26th European conference on information systems. AIS Electronic Library (AISeL)

Nahkur O, Orru K, Hansson S, Jukarainen P, Myllylä M, Krüger M, Max M, Savadori L, Nævestad TO, Meyer SF et al (2022) The engagement of informal volunteers in disaster management in europe. Int J Disast Risk Reduct 83:103413

Nguyen DT, Alam F, Ofli F, Imran M (2017) Automatic image filtering on social networks using deep learning and perceptual hashing during crises. arXiv preprint arXiv:1704.02602

Niyazi M, Behnamian J (2023) Application of emerging digital technologies in disaster relief operations: a systematic review. Arch Comput Methods Engin 30(3):1579–1599

Oh O, Agrawal M, Rao HR (2013) Community intelligence and social media services: a rumor theoretic analysis of tweets during social crises. MIS Quart 407–426

Olteanu A, Castillo C, Diaz F, Vieweg S (2014) Crisislex: a lexicon for collecting and filtering microblogged communications in crises. Proceedings of the international AAAI conference on web and social media 8–1:376–385

Olteanu A, Vieweg S, Castillo C (2015) What to expect when the unexpected happens: Social media communications across crises. In: Proceedings of the 18th ACM conference on computer supported cooperative work and social computing, pp 994–1009

Palen L, Liu SB (2007) Citizen communications in crisis: anticipating a future of ict-supported public participation. In: Proceedings of the SIGCHI conference on Human factors in computing systems, pp 727–736

Park CH, Johnston EW (2017) A framework for analyzing digital volunteer contributions in emergent crisis response efforts. New Media Soc 19(8):1308–1327

Pechta LE, Brandenburg DC, Seeger MW (2010) Understanding the dynamics of emergency communication: propositions for a four-channel model. J Homeland Secur Emerg Manag 7(1)

Qu Y, Wu PF, Wang X (2009) Online community response to major disaster: a study of tianya forum in the 2008 sichuan earthquake. In: 2009 42nd Hawaii international conference on system sciences. IEEE, pp 1–11

Qu Y, Huang C, Zhang P, Zhang J (2011) Microblogging after a major disaster in China: a case study of the 2010 yushu earthquake. In: Proceedings of the ACM 2011 conference on Computer supported cooperative work, pp 25–34

Ray PP, Mukherjee M, Shu L (2017) Internet of things for disaster management: state-of-the-art and prospects. IEEE Access 5:18818–18835

Reuter C (2022) A European perspective on crisis informatics: citizens' and authorities' attitudes towards social media for public safety and security. Springer Nature

Reynard D, Shirgaokar M (2019) Harnessing the power of machine learning: can twitter data be useful in guiding resource allocation decisions during a natural disaster? Transp Res Part D: Transp Environ 77:449–463

Roche S, Propeck-Zimmermann E, Mericskay B (2013) Geoweb and crisis management: issues and perspectives of volunteered geographic information. GeoJournal 78:21–40

Sharp EN, Carter H (2020) Examination of how social media can inform the management of volunteers during a flood disaster. J Flood Risk Manag 13(4):e12665

Simon T, Goldberg A, Adini B (2015) Socializing in emergencies-a review of the use of social media in emergency situations. Int J Inf Manag 35(5):609–619

Soden R, Palen L (2018) Informating crisis: expanding critical perspectives in crisis informatics. Proc ACM Human-Comput Interact 2(CSCW):1–22

Starbird K, Palen L, Hughes AL, Vieweg S (2010) Chatter on the red: what hazards threat reveals about the social life of microblogged information. In: Proceedings of the 2010 ACM conference on Computer supported cooperative work, pp 241–250

Starbird K, Maddock J, Orand M, Achterman P, Mason RM (2014) Rumors, false flags, and digital vigilantes: misinformation on twitter after the 2013 boston marathon bombing. IConference 2014 proceedings

Stiegler R, Tilley S, Parveen T (2011) Finding family and friends in the aftermath of a disaster using federated queries on social networks and websites. In: 2011 13th IEEE International Symposium on Web Systems Evolution (WSE). IEEE, pp 21–26

Sutton J, Gibson CB, Phillips NE, Spiro ES, League C, Johnson B, Fitzhugh SM, Butts CT (2015) A cross-hazard analysis of terse message retransmission on twitter. Proc Natl Acad Sci 112(48):14793–14798

Thompson RR, Garfin DR, Holman EA, Silver RC (2017) Distress, worry, and functioning following a global health crisis: a national study of americans' responses to ebola. Clin Psychol Sci 5(3):513–521

Twitter (2015) Twitter rest apis. https://dev.twitter.com/rest/public

Valecha R, Oh O, Rao R (2013) An exploration of collaboration over time in collective crisis response during the haiti 2010 earthquake. In: International conference on information systems (ICIS 2013): reshaping society through information systems design, vol 1, pp 378–387

Veil SR, Buehner T, Palenchar MJ (2011) A work-in-process literature review: incorporating social media in risk and crisis communication. J Contingenc Crisis Manag 19(2):110–122

Vieweg S, Hughes A, Starbird K, Palen L (2010) Supporting situational awareness during emergencies using microblogged information. SIGCHI 2010

Watson H, Rodrigues R (2018) Bringing privacy into the fold: Considerations for the use of social media in crisis management. J Contingenc Crisis Manag 26(1):89–98

Wukich C, Hu Q, Siciliano MD (2019) Cross-sector emergency information networks on social media: online bridging and bonding communication patterns. Am Rev Public Administr 49(7):825–839

Yin J, Lampert A, Cameron M, Robinson B, Power R (2012) Using social media to enhance emergency situation awareness. IEEE Intell Syst 27(06):52–59

Yu M, Huang Q, Qin H, Scheele C, Yang C (2019) Deep learning for real-time social media text classification for situation awareness–using hurricanes sandy, harvey, and irma as case studies. Int J Digit Earth

Zeng D, Chen H, Lusch R, Li SH (2010) Social media analytics and intelligence. IEEE Intell Syst 25(6):13–16

Zook M, Graham M, Shelton T, Gorman S (2010) Volunteered geographic information and crowdsourcing disaster relief: a case study of the haitian earthquake. World Med Health Policy 2(2):7–33

Zubiaga A (2015) Dataset: Hurricane sandy tweets. http://www.zubiaga.org/datasets/hurricane-sandy-tweets/

Zubiaga A, Ji H (2014) Tweet, but verify: epistemic study of information verification on twitter. Soc Netw Anal Min 4:1–12

Technology Innovations for Disaster Management in Developing Countries—Case Study for the Indian Ocean Tsunami at Banda Aceh, Indonesia and Recent Developments

Arni Ariani, Allya Paramita Koesoema, and Soegijardjo Soegijoko

1 Introduction

In the coming decades, the number of people who live in disaster-prone areas will continue to rise (Oktari et al. 2020). Disasters can be classified as natural, man-made, and hybrid disasters (Shaluf 2007). Natural disasters can result from natural phenomena beneath the Earth's surface (e.g., earthquakes, volcanic eruptions, and tsunamis), above the earth's surface (e.g., landslides and avalanches), biological phenomena (e.g., epidemics), or meteorological and hydrological phenomena (e.g., storms, tornados, heatwaves, and droughts). Meanwhile, man-made disasters include both socio-technical and conflict/warfare-based disasters. Socio-technical disasters include technological disasters (e.g., factory failures); transport failures; "public place" failures; and production failures (Shaluf 2007). Meanwhile, hybrid disasters result from the combination of natural forces and human error. An example includes placing residential areas at the foot of an active volcano or tsunami-prone areas, which may increase the impact of natural disasters (e.g., tsunamis or volcanic eruptions).

A. Ariani · A. P. Koesoema (✉) · S. Soegijoko
Indonesian Biomedical Engineering Society, Bandung, Indonesia
e-mail: allya.koesoema@itb.ac.id

Indonesian eHealth and Telemedicine Society, Bandung, Indonesia

S. Soegijoko
e-mail: soegi@ieee.org

A. P. Koesoema · S. Soegijoko
School of Electrical Engineering and Informatics, Institut Teknologi Bandung, Bandung, Indonesia

P. K. Ray et al. (eds.), *Technology Innovation for Sustainable Development of Healthcare and Disaster Management*, Disaster Risk Reduction,
https://doi.org/10.1007/978-981-97-2049-1_3

Disaster impact will continue to rise due to climate change, population growth, and other related factors. Further, when not managed well, disasters can lead to prolonged impact or subsequent disasters. For example, refugee camps that temporarily hold survivors of natural disasters may be stricken by epidemics related to the insufficient hygiene of the site (Pascapurnama 2018). A clear case of climate change driven disasters is increasing coastal hazard populations living in coastal area, who are increasingly vulnerable due to erosion, flooding, storm surges and tsunamis (Oktari, et al. 2020; Bunce et al. 2010; Ferro-Azcona et al. 2019; Kumar et al. 2018; Lewis 2013; Mehvar et al., 2018). The impacts are worse for coastal areas that do not have adaptation, planning, and implementation strategies (Bevacqua et al. 2018; Nunn et al. 2016).

To counter the rising incidence and hazard of disasters, there has been a concerted effort of many stakeholders around the world for better disaster management. Disaster risk management can be defined as "the application of disaster risk reduction policies and strategies to prevent new disaster risk, reduce existing disaster risk and manage residual risk, contributing to the strengthening of resilience and reduction of disaster losses" (United Nations 2016, p. 15). The literature on disaster management has many frameworks that discuss relevant models for managing disasters (e.g., Nojavan et al. 2018). One of the most commonly used frameworks is the disaster management cycle, which include factors such as mitigation, prevention, preparedness, response, recovery and reconstruction, or variations of these factors (Alexander 2019; Capgemini Consulting, the Netherlands 2019; Coppola 2011; Cuny 1998).

One of the key contributing factors in disaster management is technological innovations. Appropriate technologies have and will continue to enable much more effective use of resources in various disaster management activities, and even offer novel activities that were not previously available. In this chapter, we will discuss the role of technological innovations for disaster management throughout the phases of the disaster management cycle, focusing on the phases of preparedness, response, and recovery, reconstruction, and risk reduction (Capgemini Consulting, the Netherlands 2019). To close, we will illustrate the use of technology for disaster management using the case of the Indian Ocean Tsunami in Banda Aceh, Indonesia. Using the Banda Aceh case, we illustrate how there was initially low preparedness for the Indian Ocean Tsunami in Banda Aceh, Indonesia. The major disaster has triggered a cycle of disaster management actions utilizing various technologies that will help increase preparedness in the case of such disasters coming again.

2 Technological Innovations for Disaster Risk Reduction

Improvements in humanitarian action is one of the main goals of technological innovations (Obrecht 2016). These improvements can be achieved through the generation and collection of new ideas for developing better tools, the renewed approaches and ways of thinking for delivering better humanitarian assistance (Obrecht 2016). By

Fig. 1 Overview of humanitarian technological innovations covered in this book chapter

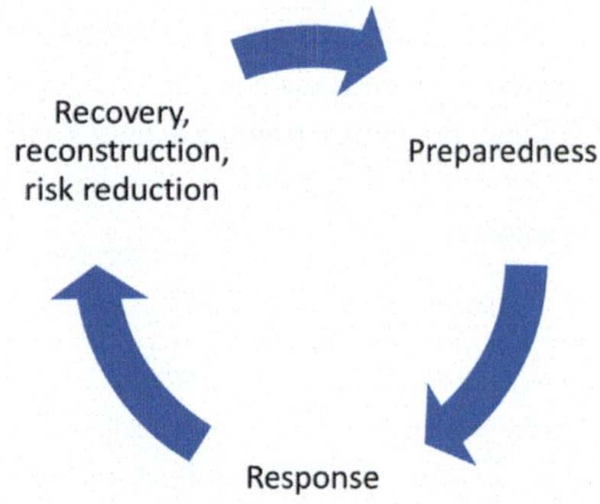

exploring many options, innovators would be able to address many challenges and issues in effective ways (Obrecht 2016).

Uncertainty and exploration in the innovation process means many failures would come and the end results may not meet the expectations (Obrecht 2016). While improvements will be made in humanitarian assistance, the stumbling blocks (e.g., reevaluation of original plans, unsuccessful pilots, and prototype trials) exist, and many have argued that organizations and donors must embrace failures as an integral part of the innovation process, otherwise fear of failures will strangle innovations in the humanitarian sector (Obrecht 2016).

In this chapter, the discussion on technology innovation for disaster risk reduction in the humanitarian sector will be divided into three subsections, namely, (1). preparedness, (2). Response, and (3). recovery, reconstruction, and risk reduction. These three subsections are based on approaches and strategies by the United Nations for disaster risk management (Capgemini Consulting, the Netherlands 2019):

- Preparedness, including activities that required assessment planning, analysis, strategy review.
- Response, including activities that targeted the implementation, monitoring, and delivery of humanitarian assistance.
- Recovery, reconstruction and disaster risk reduction, including activities for building resilience and opportunities for people that are seriously affected by disasters.

Figure 1 showed the breakdown of technologies and applications in the humanitarian sector for each subsection.

2.1 Response

This subsection focused on examples of innovations related to the response to disasters, which include assessment and monitoring systems, cash transfer program, cargo delivery and surveillance drone, 3D printing and digital IDs and biometrics. Some of the innovations aim to monitor and facilitate aid to reach the most vulnerable populations in disaster-affected areas and avoid the possibility of mismanagement. Meanwhile, other innovations aim to reduce transportation cost (i.e., cargo delivery by drone) or production cost (i.e., 3D printing), especially for essential medical supplies.

Assessment and Monitoring

The provision of relief and recovery assistance during disasters includes the availability and access to basic sanitation, food, health facilities and services, hygiene, and safe drinking-water (Izumi 2020). Sharing the right information at the right time using technologies for real-time communication can enable timely decision making and coordination by government and international relief agencies (health clusters), thus reducing lives lost and health risks. While timeliness is important, health information security and access authorization must also be taken into account, since the gathering information and its distribution must reach the authorized health actors (Izumi 2020). Assessment of health and environment status must be performed by professional personnels that not only have the required skills but also in-depth knowledge of areas where disaster has struck (Izumi 2020). Active collaboration between community nurses, health volunteers and stakeholders in the field would lead to an effective monitoring process, enabling improvements in high-quality health care delivery and the sustainability of healthy communities during and after disasters (Izumi 2020).

One of the monitoring and evaluation examples is the Epi-Nurse Nepal (Kanbara 2016). In this program, nurses were given intensive training for conducting rapid health assessments and operating open source mapping and referral systems. The surveillance systems were also established to collect information on the ground after a disaster happens and help nurses to make the right health-related interventions. Days after the earthquake struck Nepal on April 25, 2015, immediate response services were provided for people in the affected areas such as: Sindhupalchowk (Thulobhotang, Chautara, Melamchi), Jiri, Dhading. At that time, fully trained nurses were sent to deliver essential and basic needs (like clothes, food, medicine, rice, soup and water) to the affected areas. The availability of this program has given a substantial impact on people's lives and livelihoods and has become a vital part of Nepal's health system.

Cash Transfer Program

Cash Transfer Programs (CTP) are "…programs that provide non-contributory cash grants to selected beneficiaries to satisfy minimum consumption needs" (Garcia & Moore 2012, p. 18). CTP ensures people will be able to afford basic living costs and to recoup their losses, hence minimizing a chance of falling into poverty (Garcia & Moore 2012). The humanitarian sector needs to ask for permission from authorities before launching CTP (Pega et al. 2022; Arnold, Conway, & Greenslade 2011; Garcia & Moore 2012).

Recently, the humanitarian sector increasingly considers the cash transfer program (CTP) as an alternative humanitarian intervention (Barder et al. 2015). In 2004, CTP only accounted for 1% of total humanitarian aid, but the figure has been climbing steadily in recent years to nearly 15% by the end of 2017 (510. Global 2017). These facts are aligned with the World Bank prediction that the CTP will become a popular alternative for humanitarian aid after its success in helping the 2004 Indian Ocean Tsunami victims (World Bank 2012).

One form of CTP is an electronic voucher program for the affected communities. The affected communities could exchange their vouchers with things they believe important (like food, health, pharmaceuticals, sanitation, shelter, and water). Therefore, this program has pivotal roles not only in enhancing the socioeconomic status of the disaster victims but also providing better access to essential facilities (Groupe Speciale Mobile Association 2014; El-Huni 2015). The humanitarian sector also granted additional electronic vouchers to people who decided to participate to clear municipal waste in and around disaster areas, thus improving the financial freedom of the voucher's recipient and enhancing environmental health (Groupe Speciale Mobile Association 2014; El-Huni 2015).

One of the e-voucher program examples is the world food program for displaced people in Kenya, Philippines, and Lebanon (WFP 2013, 2015). The program has reported benefits including the creation of jobs that lead to economic development, financial literacy improvement, and food security enhancement (particularly for voucher recipients). However, an increased dependence on vouchers brings potential harm to the beneficiaries' financial status if the transfer process is disrupted.

Despite the highlighted benefits of CTP, there are also challenges and issues that need to be reconsidered when launching the CTP for vulnerable communities (Stevens 2018).These include technology access (Stevens 2018), lack of interoperability between vendors (Barnett & Weiss 2011), identity confirmation of recipients, especially for countries with centralized management systems (The World Bank Group 2022; Drescher 2017), risk of local economic inflation, shortages, and price distortions (La Brooy 2009). Therefore, to deliver CTP efficiently, technical infrastructures, technology accessibility and network connectivity must be made available to both humanitarian vendors and cash receivers (Capgemini 2017). They should also consider data privacy and protection (Stevens 2018) and adaptability to local context in their practices (Arnold, Conway, & Greenslade 2011). Moreover, there is a need to establish well-functioning markets to ensure the market availability of goods and services and the absorption of cash (Capgemini Consulting, the Netherlands, 2019).

Cargo Delivery and Surveillance Drones

Another technological innovation in the humanitarian sector is the aerial robot (popularly known as unmanned aerial vehicle (UAVs) or "drone") (Schroeder et al. 2016). Surveillance drones were used to take aerial photos of difficult to reach disaster areas, both immediately after the event to assess immediate damage and needs, as well as over time to monitor recovery and reconstruction efforts. Drones can also be utilized for cargo delivery, strategically providing required supplies to hard to reach areas. At the time of the Ebola crisis in West Africa, effective drone delivery systems have been used to avoid the spread of the diseases and to deliver essential health supplies for treating childhood illnesses (Schroeder et al. 2016).

Some other examples of the use of aerial robots in humanitarian projects include a. The International Organization for Migration (IOM) took aerial photos of affected areas in Haiti since 2012 for assessing the severity of the disaster (Schroeder et al. 2016); b. In 2013, Medair and Catholic Relief Services (CRS) collected aerial images to support the ongoing recovery and reconstruction activities after Typhoon Haiyan struck in the Philippines; c. In 2014, Médecins Sans Frontières (MSF) and the World Health Organization (WHO) flew small payloads of medicines and vaccines to Papua New Guinea and Bhutan via aerial robots; d. In 2015, drone mapping had affected the formation of shared situation awareness in search and rescue operations after a series of earthquakes in Nepal; e. In 2016, the UN Children's Fund (UNICEF) trialed the use of drones to air-transport medical payloads in Malawi; and f. In 2018, Vanuatu started the utilization of commercial drones for delivering temperature-sensitive vaccines to hard-to-reach villages (UNICEF 2018). Finally, the availability of unmanned delivery vehicles and robots have provided solutions when the coronavirus outbreak began, particularly to contain the spread of viruses (UNCTAD 2021). One of the examples was the use of two white rhino vehicles in Wuhan in December 2019 for delivering medical supplies and meals and reducing medical staff workloads (UNCTAD 2021).

3D Printing

With the intention of enhancing the effectiveness of disaster response, the humanitarian sector has introduced the use of 3D printing for reducing the cost of each product, simplifying logistics and supply chains while saving more lives (Jones 2015; Mis 2015). Manufacturing items in the field can help reduce logistics costs (Mis 2015), which has previously taken the lion's share of relief agencies. A computer-aided design and drafting (CADD) program can be used to create the prototype; once the design is ready for deployment, the printer gradually builds up a three-dimensional object layer by layer (Capgemini Consulting, the Netherlands, 2019).

For example, 3D printing empowered people that have undergone amputation due to war-related injuries by giving them an opportunity to design and create their own prostheses (replacement body parts) (Mesmar et al. 2016). One of the examples was Zaatari residents camp in Jordan (Daniel 2015), in which the refugees have been given extensive training on prosthesis production with 3D printing technologies. Starting in 2014, Real Hope for Haiti, Project Medishare, and Haiti Communitere collaborated for the Field Ready program in Haiti, which aims to demonstrate the potential of 3D printing of humanitarian supplies. Items printed include a butterfly needle holder, a mockup of a gas cylinder regulator, a prosthetic hand, a prototype screwdriver, three parts of printer, two prototype bottles and three prototype pipe clamps (James and James 2016). These items have helped twelve aid professionals and around sixty patients (James and James 2016).

Digital IDs and Biometrics

Digital identity can be defined as "a set of electronically captured and stored attributes and credentials that can uniquely identify a person" (Casswell 2019: 64). There are two main roles of digital identity: as foundational identity (an identity that has legal roots, like a birth certificate or passport) and as functional identity (an identity that is required for accessing services, like an automated teller machine (ATM) card or a library card). Since biometric data are stored in a central database and can be accessed by an authorized user, an individual can use biometrics to verify its identity instead of carrying a physical ID (Ucciferri et al. 2017). Most of the time, however, the verification of identity required both physical ID and the biometrics data of the person (Holloway et al. 2021).

Biometrics data includes biological or physiological characteristics such as facial structures, fingerprints, retinal patterns, or voice recognition (Holloway et al. 2021). They can be used to greatly enhance information security and significantly reduce the risk of a fraudulent identity, particularly with the rise of cash and voucher assistance (CVA) in humanitarian aid (Holloway et al. 2021). However, it also poses serious security concerns when biometrics data leakage occurs due to its uniqueness and immutability (Holloway et al. 2021). Furthermore, the immutability of biometric information may lead to other challenges relating to fundamental human rights, such as individual choice, informed consent, privacy and confidentiality (Holloway et al. 2021). This is further exacerbated with structural issues such as inequality of bargaining power under traditional donor-recipient relationship (Holloway et al. 2021). Therefore, the humanitarian sector must provide a highly advanced secure system in place and minimize these risks (Holloway et al. 2021).

2.2 Recovery, Reconstruction, and Risk Reduction

This subsection focuses on smart cities and educational technologies as examples of technologies that facilitate recovery, reconstruction and risk reduction in disaster management. Smart city technologies play a role in building back better, resulting in more resilient cities that can monitor and mitigate the risk of future disasters. Meanwhile, educational technologies are essential for capacity building in the population to make sure that they have a high awareness and preparedness of any potential hazards and disasters.

Smart Cities

The increasing incidence of disasters increased the need to develop resilient cities that can adapt to the challenges posed by extreme natural disasters, particularly for disaster information sharing, management, and coordination across governmental organizations and other relevant agencies.

Building smart cities is one path to city resilience. A smart city can be defined as a city in which data collection is done fusing various types of sensors and devices (Capgemini Consulting, the Netherlands 2019). Smart City technologies allows real-time monitoring of urban growth and its impacts on the environment, enabling better decision making (Capgemini Consulting, the Netherlands 2019). The Internet of Things (IoT) has been used by city officials to monitor buildings, traffic parameters, weather situations and vibrations due to traffic loads on bridge structures (Quek, 2016). Despite the smart city technology being considered "promising", the full potential of smart cities is not yet known (Capgemini Consulting, the Netherlands 2019). Therefore, highly qualified software engineers and developers are needed to unleash the smart city, developing frontier technologies that can support inclusive innovations and safeguarding data privacy (Capgemini Consulting, the Netherlands 2019).

In Japan, the government's vision of establishing society 5.0 required the merging of cyberspace (virtual space) and physical space (real world) (Fukuyama 2018, Prime Minister of Japan et al., 2019). This principle applies also in their city disaster resilience. As landslides, earthquakes and other natural disasters happen yearly in Japan, NIED and Hitachi Ltd. has collaborated to perform research and development of the Sharing Information Platform for Disaster Management (SIP4D) (Usuda et al. 2017). The SIPD4 gathers information from various organizations and systems and processes it for decision-making in disaster response. Another key example of the use of smart city technologies for disaster management is the Jakarta flood control system. Jakarta, the capital of Indonesia, has always been beset by floods due to its coastal area and infrastructure. Prior to the use of smart city systems, monitoring activities such as river flow and levels were based on manual human observations, resulting in limited and delayed monitoring and coordination. The current Jakarta flood control system is a collaboration between Jakarta Smart City and the Jakarta Water Resources Service (DSDA). It applies IoT and AI for flood risk management in Jakarta, with sensors installed in 178 pump houses and sluice gates in Jakarta—data from sensors will be used to produce visualization of the condition, potential, and prediction of flooding in Jakarta. In the longer run, the data can also be used for data-based policy development for flood management planning, preparedness and risk reduction in Jakarta (Smartcity Jakarta 2022).

Educational Technologies

In terms of disaster recovery, questions naturally arise on how to prepare for future disasters and become resilient communities. It is crucial to protect the emotional and mental health of the affected communities before and after a disaster and empower them by providing educational programs related to disaster risk reduction (Da Silva 2011). Implementation of educational programs related to disasters in public schools can be quite challenging due to lack of guidance, limited capacity, and limited cooperation between schools, families and societies to deliver the occupational safety and health capacity building program (Hee 2005).

Although every country has their own publicly available early-warning system for disasters, awareness is often low (Basher 2013; How et al. 2015). Thus, there is a need to circulate that information to the public through alternative channels, like educational initiatives (How et al. 2015). Furthermore, these community-based disaster risk reduction programs must be available not only for people in educational institutions but also the general communities (How et al., 2015). Availability of low-cost computers and digital classrooms has been an important tool to deliver educational materials (particularly health education) to disaster-affected populations in low-resource settings (Collins 2015). One of the educational technology examples was an information platform that enables users to access contents related to protection from violence, exploitation and abuse and also general computing skills; for example Ideas Box (Gilmore 2015). Creative expression through digital platforms not only empowers young children to articulate their feelings and thoughts but also enhances their digital literacy skills; these factors will ultimately lead to better mental health, wellbeing, and emotional resilience in children for example Voices Beyond Walls (Sawhney 2009). Users of these programs have given feedback on how to improve things by introducing simple activities in the beginning and increase the complexity gradually as the children complete their tasks (Sawhney 2009).

2.3 Preparedness

In this subsection, we focus on data preparedness, citizen reporting with mobile devices, crisis mapping with geospatial information, and the use of Internet of Things. All these technologies may play a crucial role in disaster preparedness and help facilitate quick, efficient response should disaster strike.

Data Preparedness

The digital revolution has had a huge impact on how humanitarian aid is provided; it has the capability to generate, transmit and process large amounts and different types of observation data from disaster areas at a very high speed (Raymond and Harrity 2016). Even so, there are potential risks of technologies; initially, these risks centered on who has the authorization to generate, access and transmit those important information (Raymond and Harrity 2016). Recent development has made the matter more complex, which includes when and where the authorized personnel can send and receive the generated information (Raymond and Harrity 2016).

Al Achkar and Raymond (2016) have argued that there is a gap between decision making and emerging big data. In the event of disaster, the humanitarian sectors must ensure the accuracy of the collected data and the validity of the analytics findings so that the decision makers would be able to quickly make the right decisions and carry out the right actions (Al Achkar and Raymond 2016). In other words, any potential risks related to data gathering and analyzing must be eliminated or minimized to avoid big data disasters (Al Achkar and Raymond 2016).

Citizens Reporting with Mobile Devices

Social media (such as blogs) and messaging sites (such as Facebook, Instagram, Twitter, and WhatsApp) have been used to distribute information when disasters occur (Kuchai et al. 2020). These include the participation of the surrounding community during post-disaster rehabilitation. Community participation aids in understanding the situations from different points of view—which lead to different approaches to build disaster resilient societies (Hosein and Nyst 2013; Alexander 2014). For instance, the distribution of humanitarian assistance can reach the disaster-affected communities based on the information broadcasted by the disaster survivors through their social media. Moreover, sharing information via social media has encouraged individual organizations to work as a team rather than as opponents in the time of disaster (Yates and Paquette 2011).

Mobile networks are less likely to go down compared to landlines when disaster strikes, allowing communication between people in disaster-affected areas and outside world (like emergency crews and first responders) through social media and their mobile phones (Sung 2011). Twitter has successfully shared emergency news to the public when disasters occur, such as in the case of floods in Queensland, Australia in 2011 and an earthquake and tsunami in 2011 (Sung 2011; Bruns and Liang 2012). Typhoon Morakot struck Taiwan in 2009 and a popular Taiwanese social network (PTT) has successfully broadcasted information regarding the typhoon ahead of the government and mass media (Sung 2011). Presently, disaster message board service gives people in Japan a way to notify their families and friends that they are safe after a natural disaster (Sung 2011).

Sufficient spectrum plays a vital role in resilient networks so that the networks can still operate in disasters, when prolonged power failure happens or unexpected high call volumes occur (Smith 2013; West and Valentini 2013). Further, when a major disaster or public safety crisis occurs, lack of interoperability can make it hard for them to collaborate and communicate effectively with one another. In these circumstances, they are more likely to turn to commercial networks, as it has more reliable and ubiquitous service (Slabodkin 2012). In some cases, international organizations International Telecommunications Union (ITU) have provided humanitarian assistance re-establishing basic infrastructure communication through the deployment of forty satellite terminals and a base station when a destructive earthquake occurred in Haiti (ITU 2010).

Despite positive impacts offered by social media for humanitarian assistance, there are also negative effects that need to be considered, such as the spread of online hate speech and attempts to delegitimize humanitarian organizations (Kabra et al. 2015). Furthermore, inappropriate social media activity may lead to violation of ethical rules (Hammon and Hippner 2012).

Crisis Mapping with Geospatial Information

Volunteers have gathered earth observation imagery, geographic information systems (GIS) and social media information for mapping crisis areas (Hunt et al. 2016). As a result, static geographic representations have shifted to dynamic maps that represented near real-time information related to epidemiological data and humanitarian emergencies, such as the current situations in the disaster-affected areas, the spread of infectious diseases, and the trend of population displacement (Fornace et al. 2014; Roche et al. 2011). This technology can also be used to search and identify the missing family members, who accidentally separated during an emergency (Sato et al. 2016).

Expectations concerning the use of social sensing for analysis, detection, and observation of natural disasters have risen in recent years, but reliability still needs to be improved (Sato et al. 2016). A study has been conducted to assess a geo-information and content analysis of three million tweets posted by Japanese users before and after the earthquakes strike in 2011 found that while they manage to detect tweets from disaster areas, the lower number of geo-tagged tweets and insufficient information content (e.g., damage) from the tweets makes location-based analysis unreliable.

Ethical considerations have arisen with the use of geopositioning to locate things and track people (Greenough et al. 2009), and this may be even more dangerous if the generated data is used to discriminate and stigmatize certain individuals, particularly when data are traced back to individuals (Donini & Maxwell 2013). Even when no conflict exists, the gathered data still has the potential to harm people (Lüge, 2014). For instance, the use of contact tracing maps can lead to social stigma for people living in suspected infectious disease transmission hotspots (Hunt et al. 2016). Moreover, the maps can also include operational information, like the identification of a village where the community is hostile towards the humanitarian workers; hence the residents may be exposed to "collective punishment" by the authorities (Hunt et al. 2016).

Presently, there is an emerging threat to human health, which is related to inadequate water, sanitation, and hygiene (WASH) (UNCTAD, 2021). Water quality must be monitored continuously at fixed location to characterize water and assess the water quality trends over time, identify the existing or potential problems of water quality, collect information that may be helpful for designing pollution prevention or intervention strategies, verify whether pollution control program has successfully achieved its goal or not, and finally to provide emergency response like floods and spills (UNCTAD 2021). Satellite-based remote-sensing technology has offered a solution to this matter, enabling the authority to monitor water quality in wider areas like lakes, reservoirs, rivers and dams (UNCTAD 2021). Hence, provides the capability of detecting things like chlorophyll levels, lake eutrophication, light penetration, phytoplankton bloom, turbidity, and so on (UNCTAD 2021).

Internet of Things

The transformation of telemedicine has happened over the past decade in various humanitarian crises (Wootton et al. 2012; Latifi 2014). It started with simple cellphone consultations and developed to include general practitioners wearing glasses with built-in cameras during triage (Cicero et al. 2015). The availability of IoT has made it possible for clinicians to request advice from distant doctors during crises (Hunt et al. 2016).

As a self-contained unit, mobile clinics with a team of medical workers offer great solutions for providing the required health treatments to isolated and vulnerable groups (Izumi 2020). Other benefits of mobile clinics are the implementation programs for reducing disaster risk, the promotion of reproductive health and family-planning, immunization and the provision of other health services (Izumi 2020). Mobile clinics have successfully eliminated barriers in healthcare access, establish trust to reduce health disparities, lower the risk of serious health problems, and reduce the cost for healthcare treatments (Izumi 2020).

With the intention of decreasing development costs and increasing the overall system productivity, Commercial Off-The-Shelf (COTS) solutions are becoming increasingly popular in the healthcare industry (Al-Tarawneh and Althunibat 2019). Thus, standard guidelines are needed for the evaluation and the selection of COTS products, particularly for the software one (Al-Tarawneh and Althunibat 2019). Meanwhile, technology convergence has been accelerated by the proliferation of COTS-enabled hardware platforms in medical industries (Cohen 2004). One of the examples is the use of commercial wireless biosensors for personal health monitoring (Mahfouz et al. 2013). The drawbacks of using wired implants when compared to the implantable wireless medical devices (IWMDs) are the risks of infection due to catheterization (Trautner et al., 2004) and motion artifacts induced by physical movements of the patients (Johnson et al. 2017). Hence, more patients chose to have wireless medical implants than tethered implants (Khan et al. 2022).

Artificial Intelligence

Artificial intelligence (AI) has given the computer the ability to identify patterns, anomalies, or insights in datasets, which are then become reference for the future datasets to lead to better quality decisions (Chaturvedi & Schaeffer 2020). The use of Artificial Intelligence (AI) for medical technologies and services in clinical practice has become one of the strategic movements across the key players in global health in the following countries like China, India, Japan, Norway, the European Union, and the United States (Chaturvedi & Schaeffer 2020).

Along with traditional statistics, the availability of modern data science (including AI and machine learning) allows us to carry out pattern recognition and prediction based on large training datasets (Chaturvedi & Schaeffer 2020). Some examples in the general medical field include the capability to accurately predict trends of hospital readmission (Bae et al. 2016) and blood glucose monitoring (Lim et al. 2015). In disasters, the Japanese 2011 tsunami has spurred development in tsunami prediction and impact of tsunami disaster' impacts and the prioritization of activities related to disaster response and relief (Koshimura 2020). In Japan, there are three different types of tsunamis forecasting methods, namely, tsunami scenario, data assimilation, and real-time forward simulation approaches. The first method (tsunami scenario bank approach) has been applied for field trials in few coastal areas, while the third method (real-time forward simulation approach) has been integrated to the emergency response and support system in Japan.

Overall, feedback loops connecting health workers, researchers, and patients via digital technologies will promote better engagement and increase health outcomes (Chaturvedi & Schaeffer 2020). Further, the availability of AI technology has enabled healthcare improvements not only in high-income settings but also in more remote and regional areas (UNCTAD 2021).

Despite the potential of using AI for personalized medicine, there are some challenges that need to be overcomed, namely, safety and equity, acceptance and trust, evaluation, implementation and value demonstration (Chaturvedi & Schaeffer 2020). Notably, researchers must be aware of the potential risks of bias or discrimination in AI systems. Therefore, they need to carefully choose the right technologies and apply the right datasets to the system (UNCTAD 2021). Safety has become one of the biggest challenges for AI in the healthcare sector (UNCTAD 2021). For instance, AI algorithms have been used by IBM Watson for Oncology to analyze patients' medical records and to predict the best cancer treatments for their cancer patients.

However, this study has been severely criticized for its decision to use only a few "synthetic" cancer cases instead of real data when training the AI models.

Lastly, quality and accuracy of health data is of utmost importance (UNCTAD 2021). For example, to minimize the spread of misleading information, websites from trusted public-health organizations and technology companies like the WHO have been allocated on the top of the search results page (Budd 2020). Data transparency must also be extended outside of the government itself to the public, including epidemiological data and risk factors, making the data by various researchers and practitioners (Budd 2020). Also, more peer-reviewed, clinical evaluations that demonstrate the effectiveness of digital health solutions for future project improvements are needed (Budd 2020).

Data Visualization

The availability of data has strengthened our situation awareness which can lead to good decision making (Garcia-Orosa and Pérez-Seijo 2020). However, the more effective way to reach people's mind is by providing real-time visual insights (like via infographics, photos, videos, and also virtual reality (VR) (Garcia-Orosa and Pérez-Seijo 2020).

When users wear virtual reality headsets, they will be immersed in the new virtual reality world. Virtual reality enables people to encounter the experiences of disadvantaged and suffering people and allows them to feel other people's pain—this can lead to a sense of personal obligation for the welfare of others (Nash 2018, 1). People are more empathetic towards others and are sensitive to the needs of distant others (Soler-Adillon and Sora 2018; Hardee 2016). A study reveals the influence of virtual reality content on people's attitudes, for instance when the audience experiences the aftermath of a natural disaster in an intense 360-degree virtual reality environment (Fraustino et al. 2018, 331). Another research suggested that people showed more intense empathetic responses when viewing the events through the VR technology than seeing static photos or text messages and a higher likelihood they will participate in political or social actions once the research is done (Archer and Finger 2018). Findings also showed a higher level of donation intention, interactivity, social presence, and vividness among audiences after viewing an immersive 360-degree video of a Syrian refugee camp in a study that involved VR glasses and tablets (Yoo and Drumwright 2018).

Innovative Financing

An example of innovative financing in the humanitarian sector is Forecast-based Financing (FbF). This approach has addressed gaps in the humanitarian system by releasing a pool of funds based on in-depth assessment of current and future climate and disaster risks (Future Humanitarian Financing, n.d.). The Bangladesh FbF project issued an early action protocol ahead of the flood season in July/August 2017 and distributed unconditional cash grants to affected communities (Hossain 2017). This program enabled the community to buy food and fodder, to evacuate their livestock, to fortify their shelters, and to stock up on medicine before the flooding happens (Hossain 2017). Subsequently, a survey was conducted to compare the outcome differences between post-disaster recovery and an early warning intervention program (Hossain 2017). Moreover, this approach has the biggest potential of closing the gap between DRR activities and post disaster interventions (Hossain 2017). Hence, provides positive mindsets among government officials towards the promotion of anticipatory actions for disaster management (Hossain 2017).

2.4 Challenges in the Application of Innovative Technologies in Humanitarian Actions

Innovative technologies have given fresh hope for humanitarian actions in recent years. The availability of artificial intelligence has enhanced analytical capabilities by resolving even the most complex humanitarian datasets (Arendt-Cassetta 2021). Mobile-based messaging apps and social media allows for much more immediate feedback loops from the disasters' victims (West and Valentini 2013). The use of remote sensing and unmanned aerial vehicles (UAVs) has given the capabilities to speed up the vulnerability assessment, mapping and monitoring process (Schroeder et al., 2016). Digital payment solutions give disasters' victims more flexibility and control over their money (Hossain 2017). Biometrics is an ideal gateway to verify people's authenticity and could help reconnect separated families (Kaurin 2019). In summary, all these innovative technologies provide faster, effective and efficient rapid humanitarian response to those in need (Arendt-Cassetta 2021).

However, there are also barriers, challenges, and disadvantages of applying innovative technologies in humanitarian actions. Inadequate data management may lead to data breaches and may impede the principled delivery of humanitarian aid (Holloway et al. 2021). Inequality in access to the Internet and low digital literacy can worsen gender inequality and women's disadvantage (Sebastián et al. 2022). Missing datasets unfortunately could lead to digital discrimination per se (UNCTAD 2021). In other words, there are many possibilities of technologies' failures that could break down users' trust, and ultimately cause distrust (Arendt-Cassetta 2021). The technology's potential can only be reached if all the supporting key points (such as underlying data sources, decision-making process, political buy-in and user distribution) are in the right places (Arendt-Cassetta 2021).

3 Challenges of Developing and Implementing Innovative Technologies for Disaster Management in Developing Countries

Developing countries face special challenges for developing and implementing innovative technologies for disaster management. The low level of infrastructure and preparedness, combined with lower level of resources, unprepared, bureaucratic organizations, and pervasive trends such as rapid urbanization and environmental degradation may make disasters more catastrophic and hinder efficient deployment of technologies (Provitolo 2012; Henderson 2004). However, while there are challenges, the opportunities are also vast. In this section, we will focus on the challenges and opportunities centered around ICT infrastructure and use, which are almost always key to the deployment of technology in disaster situations.

3.1 ICT Infrastructure

While in general the level of access and distribution of ICT infrastructure is lower in developing countries recently many countries have made efforts to upgrade and distribute their ICT infrastructure better (Kallal et al. 2021), and even leapfrog into newer versions of standards and devices, which may be more difficult for advanced countries with legacy infrastructure (Goldemberg 2011). In addition to physical infrastructure, the development of data standards and systems that utilize the physical infrastructure is just as important. Fragmented and incompatible data/information systems between government institutions or health institutions hinder timely and efficient disaster response. Meanwhile, an integrated national online database for health or other information would be exceedingly useful in disaster preparedness, mitigation and response. As an example, Indonesia's Digital Health Transformation Strategy includes developing an integrated database for health in Indonesia for resilience.

3.2 ICT & Mobile Users: Rising Numbers, Uneven Distribution, and Low Literacy

The number of mobile and internet users in developing countries has been rising significantly and steadily throughout the past decade. However, access in rural areas are often still scarce, for instance, approximately 15% of the total villagers in Indonesia lives without the Internet (MCIT 2020). Further, the level of digital literacy is on average still low (Asongu et al. 2023), which may hinder application of ICT for disaster management.

A key example is the use of mobile applications (apps) and social media, which can serve as double edged swords for disaster management. Increasing numbers of ICT users and the strengthening of communications infrastructure opens up the potential of mobile apps and social media as tools in disaster management. Mobile apps can be effective for disaster risk awareness and education, early warning systems, as well as a fast response indicator. Similarly, social media may be highly effective for public announcement for early warning, as well as for early detection or situational surveillance in the case of disasters using data analytics and geospatial analysis. In

Indonesia, the Bureau for Disaster Management has a mobile app called INARISK, increase awareness by explaining the local disaster risk in the user's current area. Pedulilindungi, Indonesia's national tracking system for COVID-19, has been highly instrumental in controlling the COVID-19 pandemic throughout 2021 and 2022.

While it has many potential benefits, mobile apps and social media also pose several important risks. Social media can be a significant source of misinformation. Through the COVID-19 pandemic, numerous hoaxes that exacerbate the pandemic impact spread through various social media unchecked, undermining official and research-based information that can help mitigate the disaster. Further, there are potential ethics and privacy issues in the use of mobile apps and social media. Surveillance-based mobile apps such as Pedulilindungi must be held accountable to guard the privacy of its users, both to keep the data safe from potential hackers or commercial users, as well as the user's right to privacy and freedoms related to government access to their data.

4 Indonesia's Disaster Management System

From the Indonesian Disaster Information Data (DIBI-BNPB) record, Indonesia is regularly hit by extreme climate events and disasters (BNPB 2022). When considering the absolute total of disaster deaths in the last two decades, Indonesia is ranked second worldwide with 41.6 total deaths per million population (278 events) (Center for Research on the Epidemiology of Disasters—CRED, 2021). Importantly, several studies have noted that poor and vulnerable populations are most heavily affected by extreme weather and natural disasters, therefore these people need adaptive solutions to revitalize their life (Banwell et al. 2018; Benevolenza & DeRigne 2018; Keim 2021).

In Indonesia, BNPB and BPBD have the mandate to coordinate all DRR programs and actions (Brown et al., n.d.). These include substantial resources' planning, preparation, coordination, and mobilization (Brown et al., n.d.). BNPB and BPBD are primarily responsible in coordinating response preparedness, which include contingency planning, simulations, disaster management plan testings, risk assessments, early warning system, evacuation signage and evacuation site installations, resource mobilization strategies and stockpiling and updating emergency supplies (Brown et al. n.d.). Throughout the recovery process, BNPB and BPBD facilitate the needs of affected communities (such as basic services, shelter, and protection) and improving their health and mental status (Brown et al., n.d.). This could be done through seed funding provision which helps to reactivate different rural–urban business ventures, homes and critical infrastructures (like schools, hospitals, and clinics) reconstruction, lastly basic service continuation (like education and health) (Brown et al. n.d.).

To effectively create disaster resilience, risk mitigation, and recovery programs, we need to learn from past disaster events (Brown et al., n.d.). The aims of these programs are the reduction of deaths and economic losses caused by disasters through the provision of education for at-risk communities, the enhancement of survival opportunities (via warning systems and disaster resilient infrastructure), the protection of livelihoods (economic gain, protection, and a sense of safety) (Brown et al., n.d.).

Two key issues for disaster impact reduction in Indonesia and other developing countries include insufficient Public Private Partnership (PPP) (Bisri 2017) and communication (Van Rossum and Krukkert 2010). Private actors in these countries tend to passively participate in disaster management (DM) (Auzzir et al. 2014). In contrast, private actors in most developed countries have a tendency to be actively involved in DM (Sawada and Zen 2014). Therefore, BNPB has focused its attention on the establishment of PPP coordination (Chinthongprasert and Parisot 2020), involving private actors in providing governmental infrastructure and services for disaster victims, which in return could increase economic efficiency (Chinthongprasert and Parisot 2020).

4.1 The 2004 Tsunami in Aceh

The Aceh province (with the official name "Nangroe Aceh Darussalam") is located in the northern part of Sumatra, one of the major islands in Indonesia. The province's capital city is Banda Aceh. The majority of Aceh's population choose to live near the sea because of better transportation infrastructure (Dasselaar 2007). Meanwhile, forests can be found in the inner parts of Aceh, similar to the majority parts of Sumatra Island (Dasselaar 2007).

When the 2004 tsunami struck, Banda Aceh was severely affected. The heights of the waves reached roughly 15 m to the east coast and approximately 20 m to the west coast (Lavigne et al. 2009). The inundation of tsunami waves reached a maximum extent of 4 km and hit the mainland; caused the destruction of half of the city (Ghobarah et al. 2006). The 2004 tsunami showed how vulnerable Banda Aceh was to tsunami threats. Other studies also confirmed the northern end of Sumatra Island, namely, Banda Aceh and Aceh Besar District, has been hit by 11 tsunamis from the period of 7,400 to 2,900 years BP (Rubin et al. 2017). The estimated shortest tsunami travel time from the subduction zone where part of the Indo-Australian Plate (IAP) to the coastal areas is around 35 min (Syamsidik et al. 2015). As shown in Fig. 2, there are nine districts and 90 villages in Banda Aceh, with four subdistricts placed directly on the shoreline, i.e., Kutaraja, Kuta Alam, Meuraxa, and Syiah Kuala (Syamsidik et al. 2017; Oktari et al. 2020). When the 2004 earthquake-induced tsunami occurred, these four districts were heavily damaged along with other five subdistricts, i.e., Baiturrahman, Banda Raya, Jaya Baru, Lueng Bata, and Ulee Kareng (Syamsidik et al. 2017).

The 2004 destructive tsunami had massive consequences in economic, environmental and social aspects and Aceh was in dire need of humanitarian support (Dasselaar 2007). Within months, collaborative efforts have taken place to ensure that vulnerable people were supported and that reconstruction of critical infrastructures

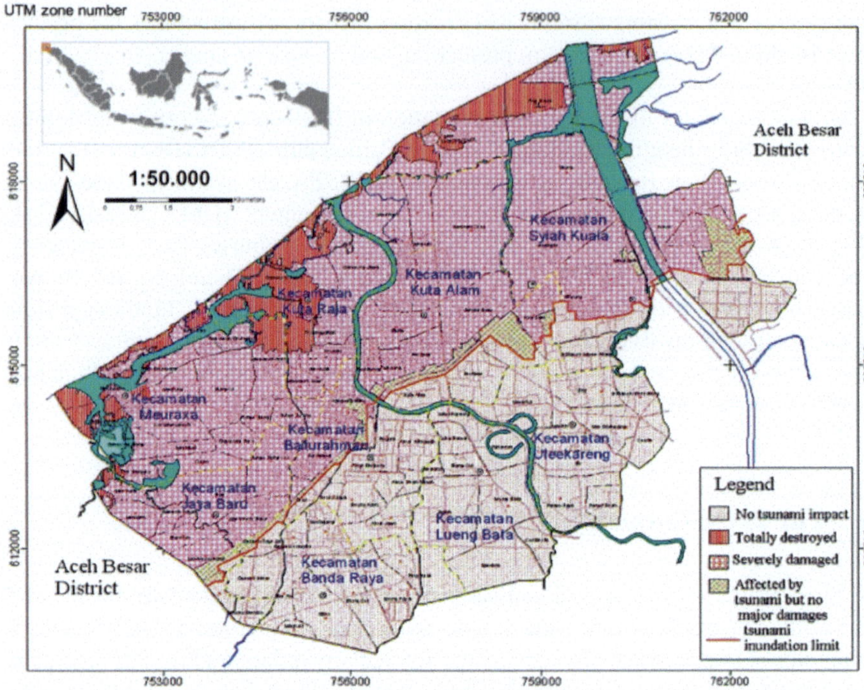

Fig. 2 The illustration of areas in Banda Aceh that were affected by the 2004 earthquake-induced tsunami (geographic coordinate system is UTM). From City Resilience towards Coastal Hazards: An Integrated Bottom-Up and Top-Down Assessment (page 4), by Oktari et al. 2020, Water. Copyright 2020 by Rina Suryani Oktari and Colleagues. Articles are licensed under an open access Creative Commons CC BY 4.0 license (https://www.mdpi.com/openaccess)

and clean up after a tsunami were completed on time (Dasselaar 2007). This collaborative work involved generous support from a number of donors and organizations (430 local NGOs and 124 international NGOs), various military and government agencies and various other parties (Dasselaar 2007).

4.2 *Studies in Aceh*

Disaster studies in Aceh just started when the tsunami struck in 2004. Before the tsunami, the province didn't have very strong preparation in place in terms of technology and evacuation plan. The tsunami disaster kickstarted a strong disaster response, and later a stronger groundwork on recovery, reconstruction, risk reduction and preparedness in Aceh and Indonesia in general.

The Tsunami and Disaster Mitigation Research Center (TDMRC) of Universitas Syiah Kuala in Aceh is the primary national center for disaster research in

Indonesia, and it was established in 2006 in response to the Aceh Indian Ocean Tsunami (TDMRC n.d.). Under several projects, TDMRC has been involved in disaster risk reduction through science and technology approaches. These include institutional capacity building, Development of Disaster Regulation Frameworks, Community's Preparedness, and disaster risk mapping are part of deliverable items. Various other research and development activities, both national and international, have also continued to develop, ensuring a stronger level of preparedness in Aceh. In this subsection, we illustrate the use of technologies and other measures in the disaster management cycle in Aceh.

Response

The e-voucher program example is the "cash for work" program for 2004 tsunami victims in Indonesia (Action Against Hunger 2009). Households or groups performed carpentry or welding jobs to rehabilitate drainage systems and earned monetary vouchers that can be exchanged for goods and services (Action Against Hunger 2009).

Recovery, Reconstruction, and Risk Reduction

In 2009, a collaboration was formed between LIPI, TDMRC and other stakeholders in Banda Aceh to deliver formal curriculums related to disaster prevention and preparedness, for students, their families, and neighborhoods around SDN 2 (Irawati and Maryam 2012). It is hoped that by the end of this collaboration, SDN 2 will become a model for disaster education, especially for other schools facing similar problems (Irawati and Maryam 2012). Some of the curricula are the following: disaster learning media, early warning systems, emergency evacuation maps, standard operating procedures (SOP) preparation, school and its community improvement, simulation-based learning in disaster Management, etc. (Irawati and Maryam 2012).

The tsunami that struck Aceh in 2004 washed away all essential infrastructure, including schools. Tsunami survivors grapple with psychological trauma. However, instead of moving to other locations, the post-tsunami reconstruction of public elementary and Islamic schools was still mostly allocated in the 2004 Indian Ocean Tsunami areas (56% of total schools in Banda Aceh). As of 2014, there were a total of 21,301 students living within the shoreline (Sakurai et al. 2017), meaning that a considerable proportion of people in Banda Aceh are exposed to tsunami hazards (Sakurai et al. 2017). Throughout post-disaster recovery assistance, external donors have invested enormous energy and resources to deliver programs including the construction of two-story school buildings with reinforced concrete systems, creation of maps and evacuation routes, implementation of evacuation drills and finally the preparation and the provision of disaster education. But since the local education agency was not fully involved in the programs, the projects simply ended once financial support ended (Sakurai et al. 2017). Local education agencies, schools, headmasters, and teachers need to understand the importance of disaster education (Sakurai et al. 2017). These are their responsibilities to ensure the continuation of the project rather than completely depending on the external donors (Sakurai et al.

2017). The curriculum for disaster education must be updated and the disaster risk reduction practices must be implemented not only in the schools but also in the community as a part of preparing for future disasters (Sakurai et al. 2017).

Preparedness

Natural disasters frequently occur in Indonesia and cause widespread infrastructure damage, therefore Indonesia requires an ad hoc communication system like a low altitude platform (LAP) system to establish direct communication link with front-line or emergency responders (Qiantori 2011).

There are three subsystems in the LAP system: namely ground station, sky station and access clients (Qiantori 2011). The ground station is equipped with the telemetry system for the platform and radio monitoring technology (Qiantori 2011). The ground station can be located in medical posts, field hospitals, or installed in a mobile/ad hoc system operated by the search and rescue team. The sky station consisted of two hydrogen balloons and various communication equipment and the access point requires a 2.4 GHz connection (Qiantori 2011). In the field trial, the clients were able to communicate within the radius of 5 KM from the balloons and share voice, data, and video information (Qiantori 2011).

Telemedicine also has a strong potential to support disaster preparedness in Indonesia. A low-cost mobile telemedicine system for disaster reliefs in Indonesia has been developed (Sutjiredjeki et al. 2009). A key aspect of of this system is that it provides integrated health services that can run both in normal situations and in disasters—it also provide multiple communication channel options, which can be especially useful in disasters (Sutjiredjeki et al. 2009).

According to a 2005 WHO report, most of healthcare facilities like hospitals and health centers were destroyed in the tsunamis and this event has also severely affected the physical and mental health of healthcare staff (WHO SEARO 2005). To help the recovery process, WHO distributed approximately 20 metric tons of medical goods per day (WHO SEARO 2005). Meanwhile, Banda Aceh and Meulaboh have received TBC medications. Moreover, vaccine and drug cold chain systems and services have also been established onsite (WHO SEARO 2005). For analysis purposes, every medical case has been stored into a database by the epidemiologist from various healthcare facilities, namely, hospitals, health centers, and mobile clinics (WHO SEARO 2005).

Learning from that report, our research team believed there is an urgency to develop and implement an integrated electronic prescription system and e-health system for tuberculosis (TB) management for healthcare facilities in Indonesia. For our e-Prescription system, there are few features that we built specifically with local content or themes in mind, including adverse drug event alerts, automatic medical report generation, electronic medical record, and paperless prescription (Puspitasari and Soegijoko 2009). The system has been clinically trialed for a period of nine months in an Indonesian Community Health Center (Soegijoko et al. 2011a) and further implemented to support our mother–child programs (Soegijoko et al. 2011b).

For the second system, the TB management e-health system, our research team has focused on five different areas, namely, measurement, recording, evaluation, reminder, and reporting functions (Koesoema et al. 2007; Rachmat et al. 2009). First, with the intention of improving the quality of TB diagnosis services and educational research, the medical report has been stored into the web-based patient database (Rachmat et al. 2009). Up to 16 automatic SMS alerts were sent to the patients as reminders for medication intake or medical visit during the 6 months trial (Rachmat et al. 2009). Furthermore, the teleconsultation feature has enabled patients to receive medical advice related to TB treatments through SMS facility (Rachmat et al. 2009). For further development, a teleconsultation feature will be completed with the choice to perform video calls with pulmonologists or other medical specialists at patients' preferred hospitals (Rachmat et al. 2009). Our research team believed that the implementation of the TB management e-health system is crucial, especially to increase theTB treatment completion and to enhance the health service quality (Rachmat et al. 2009).

4.3 Challenges in the Application of Innovative Technologies in Indonesia After the 2004 Tsunami

As stated by Granstrand & Holgersson in their 2020 publication, "An innovation ecosystem is the evolving set of actors, activities, and artifacts, and the institutions and relations, including complementary and substitute relations, that are important for the innovative performance of an actor or a population of actors" (Granstrand & Holgersson 2020). Therefore, coordinated actions between national and regional governments are needed to achieve the equality of the distribution of health resources, services, and facilities across regions in Indonesia, particularly in the event of disaster (Oktari et al. 2022).

Competition and lack of coordination between different agencies interested in Aceh's reconstruction has resulted in inefficiencies, duplication of innovation efforts and failures to deliver innovation technologies that match with the needs of the disaster-affected population (Masyrafah & McKeon 2008). Dysfunctional and inefficient financing has also been found in Aceh's reconstruction. For instance, in 2005, few of the hard-hit areas in Aceh province received limited resources and fundings to rebuild their areas (Masyrafah & McKeon 2008). Later, the issues were resolved after the improvement of access to those areas (Masyrafah & McKeon 2008). Factors like politics or infrastructure also slow innovation locally (Jayasuriya & McCawley 2010). Finally, sustainability is also a key issue, as projects were stopped after donor funding had ended, preventing them from reaching their full potential, for instance, the end of disaster education implementation in Aceh's school (Sakurai et al. 2017).

5 Discussion

While there are general principles for technologies in disasters, each disaster has its own characteristics and challenges and may require different actions to keep residents of affected communities safe (Raymond and Harrity 2016). Therefore, individual cases of technology deployment must be examined to enhance situational awareness in disaster response (Raymond and Harrity 2016). Some examples of situational awareness technologies that has recently become more accessible and less costly include crowdsourced crisis maps, mobile survey platforms, satellite imagery analysis, social media tracking (Raymond and Harrity, 2016) as well as drone technology (Zhongming et al. 2020), connected cameras and strategically placed monitoring sensors (Zhongming et al. 2020).

A second key issue in disaster technologies is insufficient comprehensive standards, ethical framework and governing principles (Raymond and Harrity 2016). For example, crowdsourcing involves a large group of untrained contributors and may end up with an overwhelming amount of useless data. Inequality access to social media may lead to the exclusion of vulnerable communities when disasters strike (Raymond and Harrity 2016).

Finally, synergies between innovations and innovative people/platforms can pave the way to provide effective disaster risk management and financing (Zhongming et al. 2020). For instance, the benefits of big data analytics will reach the masses only if all the required infrastructure facilities and resources are available for big data tasks (Zhongming et al. 2020). Therefore, trained professionals must know what their goals are and how they want to measure the data for different types of disasters (Raymond and Harrity 2016). Then, they must gather raw data specific to the disaster type and transform it into clear insights to enable actionable and strategic decision-making for humanitarian response (Raymond and Harrity 2016). As internet access and smartphone use expand globally, the trend in using social media is showing no signs of slowing down (Zhongming et al. 2020). This means a huge amount of air photos, geospatial data, maps, satellite, and radar imagery has been gathered due to the large number of volunteers involved (Zhongming et al. 2020). More importantly, those data can be used as an additional source that complements and augments existing data (Zhongming et al. 2020).

Humanitarian actors can explore different technology options as part of an approach to deliver quality services at affordable cost and to manage the risks attached to the innovation process (Raymond and Harrity 2016). Nowadays, communities are choosing the cloud over traditional ways of computing and data storage for instance to perform big data analytics in the cloud (Zhongming et al. 2020).

Lastly, although local actors and communities have been considered as an efficient, effective and feasible approach in humanitarian innovation, it still seems under-researched and suffers extensively from limited guidelines for supporting the practice (Bruder et al. 2022). Therefore, future research must be more focused on localized humanitarian innovation (Bruder et al. 2022). Particularly, the exploration on how the knowledge transfer and conversion process in the local context proceeds and generally on how capacities in disaster context further develop (Bruder et al. 2022).

6 Conclusion

Over the next decade, the impact of natural disasters is expected to rise significantly due to factors like climate change and population growth. Advancements in technology for DRR have completely transformed the management of disasters and hazards at all stages, from preparedness, response, to recovery. In this chapter we have briefly discussed the three categories and then focused on the implementation of disaster management in Indonesia and particularly Banda Aceh. From this review, we can make several conclusions related to the way forward.

First, transparency and direction from leaders to the frontline is required in every stage of design, development, and deployment of technologies (Arendt-Cassetta 2021). Secure and consistent data collection processes may lead to meaningful solutions—from simple to complex cases (Bryant 2022). More secure systems are required to safeguard the acquired data from cyber-attacks (Holloway et al. 2021). Staffs' knowledge and skills must be leveraged so that they could operate the system effectively and find the best decision-making approach for each case (Masyrafah & McKeon 2008). Second, long term orientation strengthens the case and increases sustainability in humanitarian operations (Sakurai et al. 2017). In terms of competence, narrowing gaps in digital literacy and access requires the government to develop the ICT infrastructure necessary to provide digital access to most marginalized and vulnerable citizens, particularly women (Sebastián et al. 2022). For aid, donors should gradually withdraw from delivering services to local communities to permit local actors to take control of humanitarian activities in disaster areas (Masyrafah & McKeon 2008).

Finally, while historically studies have been outcome based, lately there has been more attention to the interplay between multiple actors and humanitarian innovation (Bruder et al. 2022). This results in more research needed on the innovation process from the system perspective (Bruder et al. 2022). In this perspective, innovation in the humanitarian sector requires collaboration between all related actors, from the supply side (finance, policies, skills, sources of knowledge, etc.), middle-side (connecting agencies) to the demand-side (donors, end-users, etc.) (Bruder et al. 2022).

References

Achkar ZA, Raymond N (2016) Data preparedness: connecting data, decision making and humanitarian response. hhi.harvard.edu. Available online: https://hhi.harvard.edu/publications/data-preparedness-connecting-data-decision-making-and

Action Against Hunger (2009). Income Generating Activities: a key concept in sustainable food security. ACF International

Alexander DE (2014) Social media in disaster risk reduction and crisis management. Science and Engineering Ethics 20(3):717–733. https://doi.org/10.1007/s11948-013-9502-z

Alexander DE (2019) L'Aquila, central Italy, and the 'disaster cycle', 2009–2017. Disaster Prevention and Management 28(2):272–285

Al-Tarawneh FH, Althunibat A (2019) Pilot study of healthcare COTS software evaluation and selection. In Proceedings of the 2019 IEEE Jordan International Joint Conference on Electrical Engineering and Information Technology (JEEIT) (pp 311–314)

Archer D, Finger K (2018) Walking in another's virtual shoes: Do 360-degree video news stories generate empathy in viewers? Columbia: Tow Center for Digital Journalism. Available online: https://academiccommons.columbia.edu/doi/https://doi.org/10.7916/D8669W5C. (accessed on 30 August 2022)

Arendt-Cassetta L (2021) From digital promise to frontline practice: new and emerging technologies in humanitarian action. New York: United Nations Office for the Coordination for Humanitarian Affairs (OCHA)

Arnold C, Conway T, Greenslade M (2011) Cash transfers literature review. Department for International Development, London

Asongu SA, Odhiambo NM, Rahman M (2023) Information Technology, Inequality, and Adult Literacy in Developing Countries. J Knowl Econ 1–19. https://doi.org/10.1007/s13132-023-01307-8

Auzzir ZA, Haigh RP, Amaratunga D (2014) Public-private Partnerships (PPP) in disaster management in developing countries: a conceptual framework. Procedia Econ Financ 18:807–814. https://doi.org/10.1016/s2212-5671(14)01006-5

Bae S, Dey AK, Low CA (2016). Using passively collected sedentary behavior to predict hospital readmission. In Proceedings of the 2016 ACM International Joint Conference on Pervasive and Ubiquitous Computing (pp. 616–621)

Banwell N, Rutherford S, Mackey B, Chu C (2018) Towards improved linkage of disaster risk reduction and climate change adaptation in health: a review. Int J Environ Res Public Health 15(4):793. https://doi.org/10.3390/ijerph15040793

Barder O, Blattmann C, Cameron L, Egeland J, Elmi M, Faye M, Woodman L (2015) Doing cash differently: how cash transfers can transform humanitarian aid. Report of High Level Panel on Humanitarian Cash Transfer

Barnett MN, Weiss TG (2011) Humanitarianism contested : where angels fear to tread. Routledge

Basher R (2013) Science and technology for disaster risk reduction: A review of application and coordination needs. UNISDR Report.

Benevolenza MA, DeRigne L (2018) The impact of climate change and natural disasters on vulnerable populations: a systematic review of literature. J Hum Behav Soc Environ 29(2):266–281. https://doi.org/10.1080/10911359.2018.1527739

Betts A, Bloom L (2014) Humanitarian innovation: the state of the art. United Nations Office for Coordination of Humanitarian Affairs (OCHA)

Bevacqua A, Yu D, Zhang Y (2018) Coastal vulnerability: evolving concepts in understanding vulnerable people and places. Environ Sci Policy 82:19–29. https://doi.org/10.1016/j.envsci.2018.01.006

Bisri MBF (2017) Inter-organizational network in Indonesia during disasters: Examples and research agenda on disaster management. In IOP Conference Series: Earth and Environmental Science (56(1), p 012023). IOP Publishing. https://doi.org/10.1088/1755-1315/56/1/012023

BNPB (2022) Data Informasi Bencana Indonesia. dibi.bnpb.go.id; Badan Nasional Penanggulangan Bencana (BNPB). Available online: https://dibi.bnpb.go.id/ (accessed on 02 August 2022)

Brown J, Holden DL, Dawson S, BNPB (n.d.) Australia-Indonesia facility for disaster reduction—Phase 2, AIFDR-2 Design Document. Australian Government—Department of Foreign Affairs and Trade

Bruder M, Baar T, Ramani S, Santos CG (2022) Innovation in humanitarian assistance. United Nations University—Maastricht Economic and Social Research Institute on Innovation and Technology (UNU-MERIT)

Bruns A, Liang YE (2012) Tools and methods for capturing Twitter data during natural disasters. First Monday, 17(4). https://doi.org/10.5210/fm.v17i4.3937

Bryant, J. (2022). Digital technologies and inclusion in humanitarian response. HPG report, London: ODI, available at: https://cdn.odi.org/media/documents/Digital_inclusion_synthesis.pdf

Budd J, Miller BS, Manning EM, Lampos V, Zhuang M, Edelstein M, McKendry RA (2020) Digital technologies in the public-health response to COVID-19. Nat Med 26(8):1183–1192

Bunce M, Brown K, Rosendo S (2010) Policy misfits, climate change and cross-scale vulnerability in coastal Africa: how development projects undermine resilience. Environ Sci Policy 13(6):485–497. https://doi.org/10.1016/j.envsci.2010.06.003

Capgemini Consulting, the Netherlands. (2019). Technological innovation for humanitarian aid and assistance. Scientific Foresight Unit (STOA)

Capgemini, Report on the High Level-Expert Meeting on Technology and Humanitarian Aid, November (2017)

Casswell J (2019) The digital lives of refugees: how displaced populations use mobile phones and what gets in the way. London: GSMA. Available online: www.gsma.com/mobilefordevelopment/blog/the-digital-lives-of-refugees-how-displaced-populations-use-mobile-phones-and-what-gets-in-the-way. (accessed on 20 August 2022)

Centre for Research on the Epidemiology of Disasters—CRED (2021, May). Disaster year in review 2020: global trends and perspectives. CRED Crunch 62:1–2

Chaturvedi R, Schaeffer LD (2020) Artificial Intelligence, Big Data, and Precision Medicine. In: Harold SW, Brunelle G, Chaturvedi R, Hornung JW, Koshimura S, Osoba OA, Suga C (eds) Proceedings from a Pair of Conferences on the Impact of Artificial Intelligence on Work, Health, and Data Privacy and on Disaster Prediction, Resilience, and Recovery. RAND Corporation, pp 19–28

Chinthongprasert P, Parisot X (2020) Public-Private Indonesian Ecosystem for Disaster management: a review. In ISPIM Conference Proceedings (pp. 1–19). The International Society for Professional Innovation Management (ISPIM)

Cicero MX, Walsh B, Solad Y, Whitfill T, Paesano G, Kim K, Baum CR, Cone DC (2015) Do you see what I see? insights from using google glass for disaster telemedicine triage. Prehosp Disaster Med 30(1):4–8. https://doi.org/10.1017/s1049023x1400140x

Cohen T (2004) Medical and information technologies converge. IEEE Eng Med Biol Mag 23(3):59–65. https://doi.org/10.1109/MEMB.2004.1317983

Collins K (2015) Vodafone 'Instant Classroom is digital school in a box for refugees: Wired. CO. UK

Coppola D (2011) Introduction to international disaster management. Elsevier, London

Cuny F (1998) Principles of management: Introduction to disaster management. University of Wisconsin-Madison Press, Madison, WI

Da Silva J, Maynard V, Parker E, Twigg J, Kabir R, Chan G, Kervel A (2011) Characteristics of a safe and resilient community: community based disaster risk reduction study. International Federation of Red Cross and Red Crescent Societies, Geneva

Daniel A (2015) The remarkable way 3D printers are saving lives in refugee camp. upworthy.com; Good Worldwide Inc. Available online: http://www.upworthy.com/the-remarkable-way-3d-printers-are-saving-lives-in-refugee-camps (accessed 30 August 2022)

Dasselaar RI (2007) Diffusion of innovation in disaster areas the post-tsunami reconstruction effort in Aceh, Indonesia (pp. I–36) [Master Thesis]

Donini A, Maxwell D (2013) From face-to-face to face-to-screen: remote management, effectiveness and accountability of humanitarian action in insecure environments. Int Rev Red Cross 95(890):383–413. https://doi.org/10.1017/s1816383114000265

Drescher D (2017). Blockchain basics : a non-technical introduction in 25 steps. Apress

El-Huni EM (2015) WFP e-voucher programme in Lebanon: Field Exchange. Available online: http://www.ennonline.net/fex/48/wfpevoucher (Accessed 30 August 2022)

Ferro-Azcona H, Espinoza-Tenorio A, Calderón-Contreras R, Ramenzoni VC, de Gómez País M, las M., & Mesa-Jurado, M. A. (2019) Adaptive capacity and social-ecological resilience of coastal areas: a systematic review. Ocean Coast Manag 173:36–51. https://doi.org/10.1016/j.ocecoaman.2019.01.005

Fornace KM, Drakeley CJ, William T, Espino F, Cox J (2014) Mapping infectious disease landscapes: unmanned aerial vehicles and epidemiology. Trends Parasitol 30(11):514–519. https://doi.org/10.1016/j.pt.2014.09.001

Fraustino JD, Lee JY, Lee SY, Ahn H (2018) Effects of 360° video on attitudes toward disaster communication: Mediating and moderating roles of spatial presence and prior disaster media involvement. Public Relations Review 44(3):331–341. https://doi.org/10.1016/j.pubrev.2018.02.003

Fukuyama M (2018) Society 5.0: Aiming for a new human-centered society. Japan Spotlight, 27(5), 47–50

Future Humanitarian Financing. (n.d.). Innovative Financing. Available online: https://futurehumanitarianfinancing.wordpress.com/visioning-the-future/cross-sectoral-dialogues/dialogue-resources/briefing-document/where-are-we-looking-for-solutions/innovative-financing/. (Accessed on 30 August 2022)

Garcia M, Moore CG, Moore CM (2012) The cash dividend: the rise of cash transfer programs in sub-Saharan Africa. World Bank Publications

Garcia-Orosa B, Pérez-Seijo S (2020) The use of 360 video by international humanitarian aid organizations to spread social messages and increase engagement. VOLUNTAS: Int J Volunt Nonprofit Organ 31(6):1311–1329. https://doi.org/10.1007/s11266-020-00280-z

Ghobarah A, Saatcioglu M, Nistor I (2006) The impact of the 26 December 2004 earthquake and tsunami on structures and infrastructure. Eng Struct 28(2):312–326. https://doi.org/10.1016/j.engstruct.2005.09.028

Gilmore L (2015) Hope for humanity: How tech is making a difference. TNW | Insider. Available online : https://thenextweb.com/news/10-ways-tech-is-delivering-hope-for-humanity. (accessed on 20 August 2022)

510.Global (2017) Internal Docs: peer-to-peer cash transfers for early warning early action systems

Goldemberg J (2011) Technological leapfrogging in the developing world. Georg J Int Aff 12(1):135–141. http://www.jstor.org/stable/43133873

Granstrand O, Holgersson M (2020) Innovation ecosystems: A conceptual review and a new definition. Technovation 90–91:102098. https://doi.org/10.1016/j.technovation.2019.102098

Greenough PG, Chan JL, Meier P, Bateman L, Dutta S (2009) Applied technologies in humanitarian assistance: report of the 2009 applied technology working group. Prehosp Disaster Med 24(S2):s206–s209. https://doi.org/10.1017/s1049023x00021609

Groupe Speciale Mobile Association (2014) Disaster response: mobile money for the displaced. Groupe Speciale Mobile Association

Hammon L, Hippner H (2012) Crowdsourcing. Wirtschaftsinformatik 54(3):165–168. https://doi.org/10.1007/s11576-012-0321-7

Hardee GM (2016) Immersive journalism in VR: four theoretical domains for researching a narrative design framework. In proceedings of the 8th international conference on virtual, augmented and mixed reality (pp 679–690). Cham: Springer. https://doi.org/10.1007/978-3-319-39907-2_65

Hee TF (2005) Strategic tripartite alliance in establishing a safe school programme in Malaysia. School Safety and Security: Lesson in Danger, 118–215

Henderson LJ (2004) Emergency and disaster: Pervasive risk and public bureaucracy in developing nations. Public Organization Review 4(2):103–119

Holloway, K., Al Masri, R., & Abu Yahi, A. (2021). Digital identity, biometrics and inclusion in humanitarian responses to refugee crises. London. ODI-HPG. Available online : https://cdn.odi.org/media/documents/Digital_IP_Biometrics_case_study_web.pdf. (Accessed 30 August 2022)

Hosein G, Nyst C (2013) Aiding surveillance: an exploration of how development and humanitarian aid initiatives are enabling surveillance in developing countries. Available at SSRN 2326229.

Hossain MB (2017) Forecast-based financing: effective early actions to reduce flood impacts. Geneva: International Federation of Red Cross and Red Crescent Societies

How V, Peter PJ, Martinus KB, Dzahir NB, M., Ali, A. I. Bt., Dzahiruddin, N. D. A. Bt. A., & Aripin, M. F. B. Z. (2015) A case study of campus-based initiatives to disaster awareness program: the road to resilience. Open Access Library Journal 02(10):e1987. https://doi.org/10.4236/oalib.1101987

Hunt M, Pringle J, Christen M, Eckenwiler L, Schwartz L, Davé A (2016) Ethics of emergent information and communication technology applications in humanitarian medical assistance. Int Health 8(4):239–245. https://doi.org/10.1093/inthealth/ihw028

International Telecommunication Union. (2010). ITU joins international effort to assist Haiti: Satellite links to re-establish communications in earthquake-hit country. Available online: https://reliefweb.int/report/haiti/itu-joins-international-effort-assist-haiti-satellite-links-re-establish-communications (Accessed on August 16, 2022)

Irawati N, Maryam (2012) SEAMEO-Japan ESD Award

Izumi T, Shaw R, Ishiwatari M, Djalante R, Komino T, Sukhwani V, Adu Gyamfi B (2020) 30 Innovations Linking Disaster Risk Reduction with Sustainable Development Goals. Keio University, the University of Tokyo, UNU-IAS, CWS Japan, Japan, 97 pages

James E, James L (2016) 3D printing humanitarian supplies in the field. Humanitarian Exchange 66:43–45

Jayasuriya SK, McCawley P (2010) The Asian tsunami: aid and reconstruction after a disaster. Edward Elgar Publishing Limited, UK

Johnson BC, Gambini S, Izyumin I, Moin A, Zhou A, Alexandrov G, Muller R (2017). An implantable 700μW 64-channel neuromodulation IC for simultaneous recording and stimulation with rapid artifact recovery. In Proceedings of 2017 Symposium on VLSI Circuits (pp. C48-C49). IEEE. https://doi.org/10.23919/VLSIC.2017.8008543

Jones S (2015) When disaster strikes, it's time to fly in the 3D printers. theguardian.com; The Guardian. Available online: https://www.theguardian.com/global-development/2015/dec/30/disaster-emergency-3d-printing-humanitarian-relief-nepal-earthquake (Accessed on 5 August 2022)

Kabra G, Ramesh A, Arshinder K (2015) Identification and prioritization of coordination barriers in humanitarian supply chain management. Int J Disaster Risk Reduct 13:128–138. https://doi.org/10.1016/j.ijdrr.2015.01.011

Kallal R, Haddaji A, Ftiti Z (2021) ICT diffusion and economic growth: Evidence from the sectorial analysis of a periphery country. Technol Forecast Soc Chang 162:120403. https://doi.org/10.1016/j.techfore.2020.120403

Kanbara S (2016) Participatory surveillance on evacuation sites by local nurses and ICT in Nepal. Asia Science Technology status for disaster risk reduction. IRDR, Future Earth and ASTAAG, Beijing, 83

Kaurin D (2019) Data protection and digital agency for refugees. In World Refugee Council Research Paper No. 12—May 2019

Keim M, E. (2021) Climate-Related Disasters: The Role of Prevention for Managing Health Risk. In: Lemery J, Knowlton K, Sorensen C (eds) Global Climate Change and Human Health: From Science to Practice. John Wiley & Sons Inc., pp 25–46

Khan SR, Mugisha AJ, Tsiamis, A., & Mitra, S. (2022). Commercial Off-the-Shelf Components (COTS) in Realizing Miniature Implantable Wireless Medical Devices: A Review. Sensors, 22(10), 3635. https://doi.org/10.3390/s22103635

Koesoema, A. P., Irawan, Y. S., & Soegijoko, S. (2007). Preliminary design of a community telemedicine system for tuberculosis control. In World Congress on Medical Physics and Biomedical Engineering 2006 (pp. 366–369). Springer, Berlin, Heidelberg

Koshimura S (2020) How AI-Enabled Modeling and Simulation Can Improve Coastal Communities' Preparation, Defense, and Recovery from Disaster: Real-Time Tsunami Inundation Forecasting. In: Harold SW, Brunelle G, Chaturvedi R, Hornung JW, Koshimura S, Osoba OA, Suga C (eds) Proceedings from a Pair of Conferences on the Impact of Artificial Intelligence on Work, Health, and Data Privacy and on Disaster Prediction, Resilience, and Recovery. RAND Corporation, pp 53–59

Kuchai N, Shepherd P, Calabria-Holley J, Copping A, Matard A, Coley, D (2020) The potential for computational IT tools in disaster relief and shelter design. J Int HumItarian Action 5(1). https://doi.org/10.1186/s41018-020-00069-1

Kumar L, Eliot I, Nunn PD, Stul T, McLean R (2018) An indicative index of physical susceptibility of small islands to coastal erosion induced by climate change: an application to the Pacific islands. Geomat Nat Haz Risk 9(1):691–702. https://doi.org/10.1080/19475705.2018.1455749

La Brooy J (2009) A weighted welfare analysis of local price inflation: disaster relief and cash transfer programming in developing countries. J Hum Secur 5(2):65–82. https://doi.org/10.3316/jhs0502065

Latifi MD, FACS, R., & Tilley, PhD, EH (2014) Telemedicine for disaster management: can it transform chaos into an organized, structured care from the distance? Am J Disaster Med 9(1):25–37. https://doi.org/10.5055/ajdm.2014.0139

Lavigne F, Paris R, Grancher D, Wassmer P, Brunstein D, Vautier F, Leone F, Flohic F, De Coster B, Gunawan T, Gomez C, Setiawan A, Cahyadi R, Fachrizal (2009) Reconstruction of Tsunami inland propagation on December 26, 2004 in Banda Aceh, Indonesia, through Field Investigations. Pure Appl Geophys 166(1–2):259–281. https://doi.org/10.1007/s00024-008-0431-8

Lewis J (2013) Some realities of resilience: an updated case study of storms and flooding at Chiswell, Dorset. Disaster Prev Manag: Int J 22(4):300–311. https://doi.org/10.1108/dpm-03-2013-0053

Lim S, Kang SM, Kim KM, Moon JH, Choi SH, Hwang H, Jung HS, Park KS, Ryu JO, Jang HC (2015) Multifactorial intervention in diabetes care using real-time monitoring and tailored feedback in type 2 diabetes. Acta Diabetologica, 53(2), 189–198. https://doi.org/10.1007/s00592-015-0754-8

Lüge T, Sudhoff RS, Lessard-Fontaine A, de la Borderie S, Soupart M (2014) GIS support for the MSF Ebola response in Guinea in 2014 Case study. Médecins Sans Frontieres Operational Center, Geneva

Mahfouz MR, Kuhn MJ, To G (2013) Wireless medical devices: A review of current research and commercial systems. In Proceedings of 2013 IEEE Topical Conference on Biomedical Wireless Technologies, Networks, and Sensing Systems (pp. 16–18). IEEE. https://doi.org/10.1109/biowireless.2013.6613660

Masyrafah H, McKeon JMJA (2008) Post-tsunami aid effectiveness in Aceh. Wolfensohn Centre for Development, USA

MCIT (2020) Annual Report 2020 - Connected Indonesia more digital, more advanced. Ministry of Communications and Informatics of The Republic of Indonesia

Mehvar S, Filatova T, Dastgheib A, de Ruyter van Steveninck E, & Ranasinghe R (2018) Quantifying economic value of coastal ecosystem services: a review. J Mar Sci Eng 6(1):5. https://doi.org/10.3390/jmse6010005

Mesmar S, Talhouk R, Akik C, Olivier P, Elhajj IH, Elbassuoni S, Ghattas H (2016) The impact of digital technology on health of populations affected by humanitarian crises: recent innovations and current gaps. J Public Health Policy 37(2):167–200

Mis M (2015) How 3D printing is transforming aid. weforum.org; World Economic Forum. Available online: https://www.weforum.org/agenda/2015/06/how-3d-printing-is-transforming-aid/ (Accessed 30 August 2022)

Nash K (2018) Virtual reality witness: Exploring the ethics of mediated presence. Documentary Film 12(2):119–131. https://doi.org/10.1080/17503280.2017.1340796

Nojavan M, Salehi E, Omidvar B (2018) Conceptual change of disaster management models: A thematic analysis, Jàmbá: Journal of Disaster Risk Studies, 10(1), a451

Nunn PD, Runman J, Falanruw M, Kumar R (2016) Culturally grounded responses to coastal change on islands in the Federated States of Micronesia, northwest Pacific Ocean. Reg Environ Change 17(4):959–971. https://doi.org/10.1007/s10113-016-0950-2

Obrecht A (2016) Separating the “good” failure from the ‘bad’: three success criteria for innovation. Humltarian Exch 66:7–10

Oktari RS, Idroes R, Sofyan H, Munadi K (2020) City resilience towards coastal hazards: An integrated bottom-up and top-down assessment. Water 12(10):2823. https://doi.org/10.3390/w12102823

Oktari RS, Dwirahmadi F, Gan CCR, Darundiyah K, Nugroho PC, Wibowo A, Chu C (2022) Indonesia’s Climate-Related disasters and health adaptation policy in the Build-Up to COP26 and Beyond. Sustainability 14(2):1006. https://doi.org/10.3390/su14021006

Pascapurnama DN, Murakami A, Chagan-Yasutan H, Hattori T, Sasaki H, Egawa S (2018) Integrated health education in disaster risk reduction: lesson learned from disease outbreak following natural disasters in Indonesia. Int J Disaster Risk Reduct 29:94–102

Pega F, Pabayo R, Benny C, Lee EY, Lhachimi SK, Liu SY (2022) Unconditional cash transfers for reducing poverty and vulnerabilities: effect on use of health services and health outcomes in low-and middle-income countries. Cochrane Database Syst Rev (3):1465–1858. https://doi.org/10.1002/14651858.CD011135.pub3

Prime minister of japan and his cabinet website. Action plan of the growth strategy. 2019. Available online: https://www.kantei.go.jp/jp/singi/keizaisaisei/pdf/ap2019en.pdf (Accessed on 30 August 2022)

Provitolo D (2012) The contribution of science and technology to meeting the challenge of risk and disaster reduction in developing countries: from concrete examples to the proposal of a conceptual model of “resiliencery vulnerability”. In Technologies and innovations for development (pp 131–153). Springer, Paris

Puspitasari IM, Soegijoko S (2009) Development of an integrated ePrescription system with adverse drug events alert for community health center in Indonesia. In Med-e-Tel 2009 Electronic Proceedings of the International eHealth, Telemedicine and Health ICT Forum for Educational, Networking and Business (pp 345–346)

Qiantori A (2011) Adoption of information and communication technology In Indonesia: Focusing On Emergency Medical Communication Systems And Mobile TV Acceptance (pp. I–cix) [PhD Thesis]

Quek T (2016) The advantages and disadvantages of Internet of Things (IoT). LinkedIn 4(5):31–16

Rachmat HH, Octovia LI, Soegijoko S (2009) Development of a simple e-health system for tuberculosis management at community health center level in Indonesia. In Med-e-Tel 2009 Electronic Proceedings of the International eHealth, Telemedicine and Health ICT Forum for Educational, Networking and Business (pp 384–385).

RAN (2018) Indonesia’s rainforests: biodiversity and endangered species. Available online: https://www.ran.org/indonesia_s_rainforests_biodiversity_and_endangered_species/ (Accessed on 30 August 2022)

Raymond NA, Harrity CS (2016) Addressing the “doctrine gap”: Professionalizing the use of information communication technologies in humanitarian action. Humanitarian Exchange 66:10–13

Roche S, Propeck-Zimmermann E, Mericskay B (2011) GeoWeb and crisis management: issues and perspectives of volunteered geographic information. GeoJournal 78(1):21–40. https://doi.org/10.1007/s10708-011-9423-9

Rubin CM, Horton BP, Sieh K, Pilarczyk JE, Daly P, Ismail N, Parnell AC (2017) Highly variable recurrence of tsunamis in the 7,400 years before the 2004 Indian Ocean tsunami. Nat Commun 8(1):16019. https://doi.org/10.1038/ncomms16019

Sakurai A, Bisri MBF, Oda T, Oktari RS, Murayama Y (2017) Assessing school disaster preparedness by applying a comprehensive school safety framework: A case of elementary schools in Banda Aceh City. IOP Conference Series: Earth and Environmental Science 56:012021. https://doi.org/10.1088/1755-1315/56/1/012021

Sato S, Hanaoka K, Okumura M, Koshimura S (2016) Grasp of disaster situation and support need inside affected area with social sensing—an analysis of twitter data before and after the 2011 great east Japan earthquake disaster occurring –. J Disaster Res 11(2):198–206. https://doi.org/10.20965/jdr.2016.p0198
Sawada Y, Zen F (2014) Disaster management in ASEAN. ERIA Discussion Paper Series, 3(s1)
Sawhney N (2009) Voices beyond walls: the role of digital storytelling for empowering marginalized youth in refugee camps. In Proceedings of the 8th International Conference on Interaction Design and Children (pp. 302–305)
Schroeder A, Meier P (2016) Automation for the people: opportunities and challenges of humanitarian robotics. Humanitarian Practice Network 66(33):23–26
Smartcity Jakarta (2022) Flood control system sensor in Jakarta. Available online: https://smartcity.jakarta.go.id/en/blog/mengintip-sensor-flood-control-system/ (Accessed 115 October 2022)
WHO SEARO (2005) Tsunami and after: WHO's role. WHO Regional Office for South-East Asia. Available online: https://apps.who.int/iris/handle/10665/206049
Sebastián S, Khan S, Vinck P (2022) Technologies in humanitarian settings: engagement and inclusion of women. Harvard Humanitarian Initiative
Shaluf IM (2007) An overview on disasters. Disaster prevention and management: An Int J 16(5):687–703
Slabodkin G (2012) National wireless communications system for first responders years away. Fierce Mobile Healthcare. Available online: http://www.fiercemobilehealthcare.com/story/national-wireless-communications-system-first-responders-years-away/2012-12-10. (2020)
Smith G (2013) Oklahoma City Area Hit by Phone, Internet Outages After Tornado, The Huffington Post. Available online: http://www.huffingtonpost.com/2013/05/21/oklahoma-city-phone-internet-outages_n_3312790.html. (Accessed on August 20, 2022)
Soegijoko S, Puspitasari IM, Jani ID, Suparno S (2011a) Implementation experience of e-Prescription system in developing country: An Indonesian community health center case. In Med-e-Tel 2011 Electronic Proceedings of the International eHealth, Telemedicine and Health ICT Forum for Educational, Networking and Business (pp 606–610)
Soegijoko S, Puspitasari IM, Aridarma A, Jani ID (2011b) e-health for improving community healthcare: Encouraging clinical experience of simple e-prescription system and m-health system development for mother and childcare. In Proceedings of 2011 IEEE 13th International Conference on e-Health Networking, Applications and Services (pp 102–105)
Soler-Adillon J, Sora C (2018) Immersive journalism and virtual reality. En Pérez- Montoro, M. (Ed.), Interaction in Digital News Media. From principles to practice (pp 55–83). Palgrave Macmillan, Cham. https://doi.org/10.1007/978-3-319-96253-5
Stevens L (2018) Self-Sovereign identities for scaling up cash transfer projects: designing a blockchain based digital identity system (pp. I–147) [Master Thesis]
Sung SJ (2011) How can we use mobile apps for disaster communications in Taiwan: Problems and possible practice. In Proceedings of 8th Asia-Pacific Regional Conference of the International Telecommunications Society (ITS): "Convergence in the Digital Age" (pp. 1–15)
Sutjiredjeki E, Soegijoko S, Mengko TLR, Tjondronegoro S, Astami K, Muhammad HU (2009) Application of a mobile telemedicine system with multi communication links for disaster reliefs in indonesia. In Proceedings of World Congress on Medical Physics and Biomedical Engineering (pp. 344–347)
Syamsidik RTM, Kato S (2015) Development of accurate tsunami estimated times of arrival for tsunami-prone cities in Aceh, Indonesia. Int J Disaster Risk Reduct 14:403–410. https://doi.org/10.1016/j.ijdrr.2015.09.006
Syamsidik, Oktari RS, Munadi K, Arief S, Fajri IZ (2017) Changes in coastal land use and the reasons for selecting places to live in Banda Aceh 10 years after the 2004 Indian Ocean tsunami. Nat Hazards 88:1503–1521
TDMRC (n.d.). History of TDMRC. Available online at https://tdmrc.usk.ac.id/history/. (Accessed on 2 August 2024)
The World Bank Group (2022) ID4D & G2Px annual report 2021

Trautner BW, Darouiche RO (2004) Catheter-associated infections: pathogenesis affects prevention. Arch Intern Med 164(8):842–850. https://doi.org/10.1001/archinte.164.8.842

Ucciferri L, Milanes V, Ferreyra E, Segarra A (2017) The identity we can't change: how biometrics undermine our human rights. Buenos Aires: Asociación por los Derechos Civilies. Available online: https://adc.org.ar/en/reports/the-identity-we-cant-change-how-biometrics-undermine-our-human-rights/. (Accessed on 20 August 2022)

UNCTAD-United Nations Commission on Science and Technology for Development (2021). Issues paper on using science, technology, and innovation (STI) to close the gap on SDG 3, good health and well-being

UNICEF (2018). Child given world's first drone-delivered vaccine in Vanuatu - UNICEF. unicef.org. Available online: https://www.unicef.org/press-releases/child-given-worlds-first-drone-delivered-vaccine-vanuatu-unicef (Accessed on 05 August 2022)

United Nations (2016) Sustainable development: disaster risk reduction. preventionweb.net. Available online: https://www.preventionweb.net/files/50683_oiewgreportenglish.pdf (Accessed on 05 August 2022)

Usuda Y, Tanimoto K, Ono I, Matsui T (2017) Enhancements to resilient functions for preventing and minimizing disasters. Development of Sharing Information Platform for Disaster Management. Hitachi Review, 66(7), 108–114

Van Rossum J, Krukkert R (2010) Disaster management in Indonesia: logistical coordination and cooperation to create effective relief operations. J Tek Ind 12(1):25–32

Vollmer M (2021) Implementing innovations in disaster management to increase resilience–laws, policies, and organizational determinants (Doctoral dissertation, Universität Wuppertal, Fakultät für Maschinenbau und Sicherheitstechnik» Sicherheitstechnik» Dissertationen)

West DM, Valentini E (2013) How mobile devices are transforming disaster relief and public safety (pp. 1–18). Center for Technology Innovation at Brookings Institute

WFP (2013) World food program's electronic vouchers: An innovative transfer modality for social safety net programmes in Lebanon

WFP (2015) Cash and vouchers stories: World food programme

Wootton R, Geissbuhler A, Jethwani K, Kovarik C, Person D, Vladzymyrskyy A, Zanaboni P, Zolfo M (2012) Long-running telemedicine networks delivering humanitarian services: experience, performance and scientific output. Bull World Health Organ 90(5):341–347. https://doi.org/10.2471/blt.11.099143

World Bank (2012) Managing risk, promoting growth: developing systems for social protection in Africa. In the World Bank's Africa Social Protection Strategy 2012–2022, Human Development Africa pp 77

Yates D, Paquette S (2011) Emergency knowledge management and social media technologies: a case study of the 2010 Haitian earthquake. Int J Inf Manage 31(1):6–13. https://doi.org/10.1016/j.ijinfomgt.2010.10.001

Yoo SC, Drumwright M (2018) Nonprofit fundraising with virtual reality. Nonprofit Manag LeadSh 29(1):11–27. https://doi.org/10.1002/nml.21315

Zhongming Z, Linong L, Xiaona Y, Wangqiang Z, Wei L (2020) Leveraging technology and innovation for disaster risk management and financing. Co-publication of the Asian Development Bank and OECD

Design and Development of the Medical Fuel Cell Vehicle—The Technology Links Mitigation and Adaptation in Climate Change

Yasuhiro Soshino and Akira Miyata

1 Introduction

In the history of mankind, natural disasters have caused serious human and economic loss, and severe damage to infrastructure around the world. Even in the twenty-first century, weather-related disasters such as Hurricane Katrina (2005), Cyclone Nargis in Myanmar (2008), Typhoon Haiyan in Philippines (2013) caused catastrophic damage in the disaster affected countries. In 40 years since 1970, weather, climate and water hazards occupied 50% of all disasters which resulted in 45% of all reported death and 74% of all reported economic losses. Among them more than 91% of these deaths were caused in developing countries (WMO, 2021).

The frequent occurrences of such meteorological disasters urged the progress in climate change countermeasures by the international community. Since 1970s scientists started to point out the problems of global warming due to the dramatic expansion of human activities since the Industrial Revolution. In 1988, the Intergovernmental Panel on Climate Change (IPCC) was established that contributed to the foundation of the United Nations Framework Convention on Climate Change (UNFCCC) (Paglia and Parker 2021). In an early period, the UNFCCC process focused on energy as mitigation policy including the Kyoto Protocol. In the beginning of 2000s, a number of scholars started to emphasize the important role of adaptation in climate policy that resulted in the 2001 Marrakesh Acoords that focus on adaptation as a policy response to climate change (Schipper 2006). While the recognition of the adaptation efforts increases, various studies pointed out the importance of comprehensive efforts through mitigation and adaptation. Klein (2005) pointed out the increasing

Y. Soshino (✉)
Disaster Management Research Institute, Japanese Red Cross College of Nursing, Tokyo, Japan
e-mail: y-soshino@jrcdmri.redcross.ac.jp

Y. Soshino · A. Miyata
Japanese Red Cross Kumamoto Hospital, Kumamoto, Japan

P. K. Ray et al. (eds.), *Technology Innovation for Sustainable Development of Healthcare and Disaster Management*, Disaster Risk Reduction,
https://doi.org/10.1007/978-981-97-2049-1_4

calls for research for optimizing the mix of mitigation and adaptation while creating the synergies between the mitigation and adaptation has become a recent focus of both climate research and policy. Agroforestry has the potential to develop synergies between effort for mitigation and help the vulnerable people adapt to the negative impact of climate change (Verchot 2007). In the water sector, Mata (2007) presented some areas of complementary approaches between mitigation and adaptation in the long period, where levels of mitigation determine the levels of adaptation needs. In terms of the sea level rise issues in coastal areas, a combination of adaptation and mitigation policies are needed as this provides a robust response to human induced climate change than each independent policy (Nicholls and Lowe 2004). For the capacity building to adapt the climate change in vulnerable communities, comprehensive response that links adaptation and mitigation results in the most sustainable outcomes (Laukkonen et al., 2009). The Paris Agreement fosters adaptation by widening the normative framing around adaptation, involving stronger adaptation commitments from stake actors, being clear about the multilevel nature of adaptation governance, and building transparent mechanism on assessing adaptation progress (Lesnikowski et al., 2017).

In May 2021, the International Committee of Red Cross (ICRC) and International Federation of Red Cross and Red Crescent Societies (IFRC) adopted the Climate and Environmental Charter for Humanitarian Organizations. The charter's seven high-level commitments guide humanitarian organizations' response to the climate and environmental crisis by calling for stepping up to adapt the growing humanitarian needs of the people who are most at risk to the impacts of the climate and environmental crises while maximizing the environmental sustainability of humanitarian organizations for mitigation (ICRC and IFRC, 2021). In this way, mitigation and adaptation are commonly recognized as the inseparable pair in the climate change countermeasures.

Therefore, this study discusses the application of the Fuel Cell Electric Vehicle (hereafter "FCEV") as a new solution for the climate change mitigation and adaptation in the medical services and disaster response. FCEVs contribute to the climate change mitigation as zero-emission vehicles by using hydrogen. In addition, FCEVs also contribute to the climate change adaptation by generating and supplying electricity for humanitarian purposes to cope with weather-related disasters. The widespread uses of FCEVs, which can be compared to "Movable Power Plants," mean not only the mitigation efforts but adaptation efforts by enhancing the resilient to power outage caused by the weather-related disasters and other emergencies.

The structure of this study is as follows. Following the introduction, section two introduces the background of the M-FCEV project. Section three explains the feature of the M-FCEV and section four describes the field tests on the M-FCEV. Section five summarizes the application of fuel cell vehicles for adaptation and mitigation purposes and section six presents the discussions and section seven concludes this study.

2 Background of the Medical Fuel Cell Vehicle Project

In 2000, the Japanese Red Cross Kumamoto Hospital was designated as the first international medical relief hub hospital in the Japanese Red Cross Society. Since then, the hospital has been conducting the research and development of the field hospital equipment for the emergency response by the International Federation of Red Cross and Red Crescent Societies (IFRC). In order to provide emergency medical services in disaster affected areas, where the basin infrastructure is damaged or lost, the field hospital equipment contains various equipment to establish the hospital infrastructure such as tents, water purifiers, telecommunications, toilets, kitchen equipment, and generators. In other words, the field hospital is compared to the convergence in movable infrastructures. In the past relief operations, the field hospital equipment was transported from the Japanese Red Cross Kumamoto Hospital to respond to the 2005 Pakistan Earthquake, the 2010 Chile Earthquake, the 2013 typhoon Haiyan in the Philippines (Fig. 1).

On the other hand, this field hospital equipment was also used for disaster response in Japan (Soshino et al. 2017). In response to the 2004 Niigata Prefecture Chuetsu Earthquake, a temporary shelter was set up using the field hospital equipment (Fig. 2).

2.1 The Smart Design Shelter Concept

In the 2011 Great East Japan Earthquake, field hospital equipment was used to sustain the medical services in the disaster affected hospital. Due to the tremendous expansion of the disaster affected areas, the Japanese Red Cross Kumamoto Hospital faced the limitation in covering the wide geographic areas by using only one whole set of the field hospital equipment. Therefore, in 2013, the hospital proposed the "Smart Design Shelter Concept" (Fig. 3) to prepare for the future catastrophic disasters

Fig. 1 The red cross field hospital

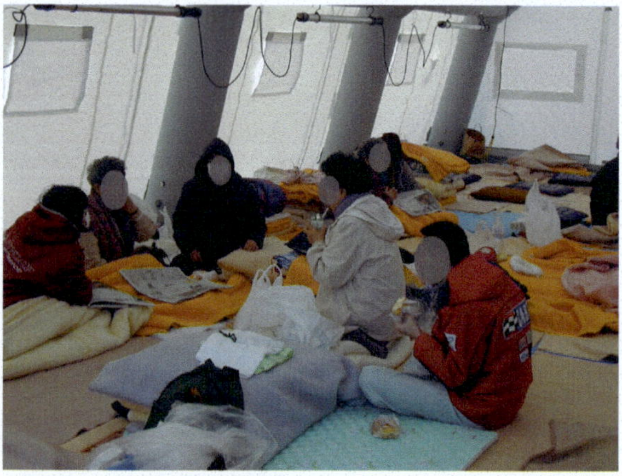

Fig. 2 A Shelter by using the relief equipment

in Japan. This concept proposes the daily uses of the movable infrastructures such as photovoltaic systems (Fig. 4), water-recycling shower system (Fig. 5), and the water-recycling toilet system in schools, stadiums, hospitals, parks and other public facilities to be used as refugee centers in emergencies. In case of disasters, these movable infrastructures can be deployed and installed at the refugee centers and hospitals in the disaster affected areas to support the recovery. This concept also aims widespread uses of movable infrastructures in other words "the environmentally friendly technologies" in peacetime toward the resilient societies.

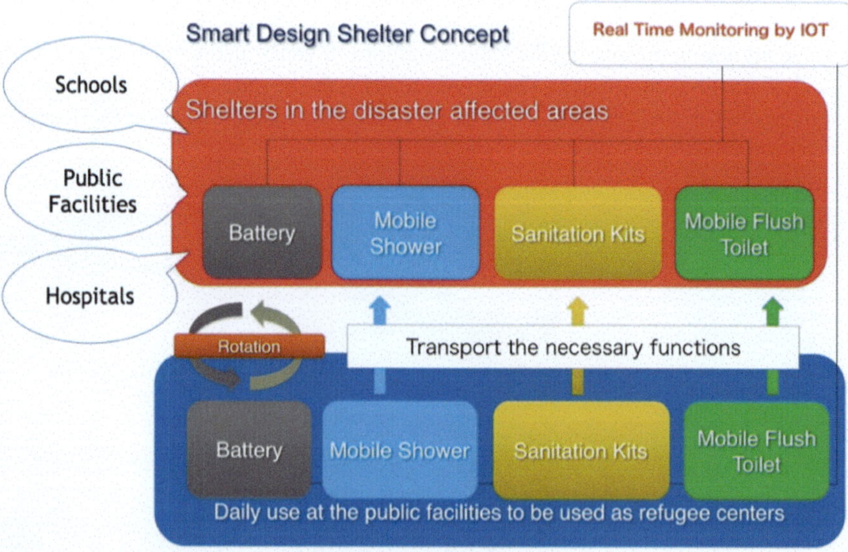

Fig. 3 The smart design shelter concept

Fig. 4 Photovoltaic system

Fig. 5 Water recycling shower

2.2 A Case Study of the Movable Infrastructure in Emergencies

Since the 2014 heavy rain in Hiroshima that killed 74 people, the linear rainfall belts have been causing the serious disasters in Japan. In response to the 2017 heavy rain in northern Kyushu, the 2018 floods in the western Japan, and the 2020 floods in southern Kumamoto, the Japanese Red Cross Kumamoto Hospital conducted relief activities by using movable flush toilets developed based on the Smart Design Shelter concept. After the 2017 Northern Kyushu heavy rain disaster, a prototype model of the movable flush toilet was installed at a volunteer center of Haki in Fukuoka, Japan. In the 2018 western Japan floods, a commercial model of the movable flush toilet is installed in a flooded hospital to support the restoration workers (Fig. 6). In 2020, the movable flush toilet was introduced to support the village of Kuma

Fig. 6 Support to the flooded hospital

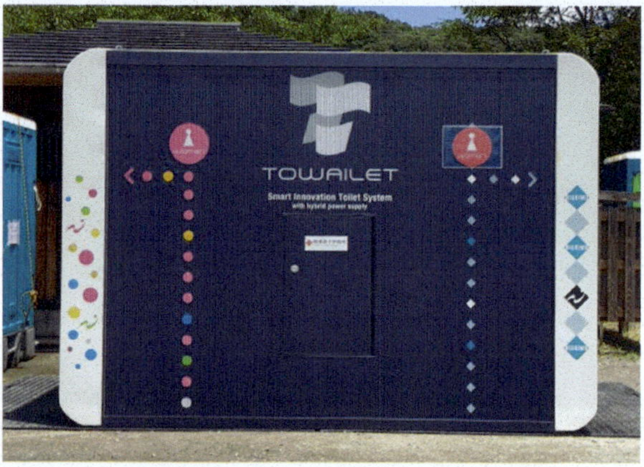

Fig. 7 Support to the relief workers in Kuma Village

(Fig. 7), where the historical rain caused the severe human loss and serious damage to the infrastructure. In this way, off-grid movable infrastructure has been utilized to respond the weather-related disasters.

2.3 Contribution to the Environment in Peacetime

In 2020, the Movable Flush Toilet was commercialized in Japan. Currently, the toilets were installed in the island of Yushima, Kumamoto Prefecture, and the island of Tsushima, Nagasaki Prefecture, Japan. The local authorities in both islands had the technical challenge in installing the flush toilets due to the lack of sewage treatment system in the islands while both authorities were trying to attract the tourists to encourage the local economy. The movable flush toilet is a suitable solution for both authorities since the toilets can be installed without relying on the water supply and sewage system. In this way, the movable flush toilet, which was initially developed with the aim of improving relief activities, has been highly evaluated for its feature of self-contained water treatment system that contributes both for environmental measures in peacetime and disaster response in emergencies. In 2020, between the Japanese Red Cross Kumamoto Hospital and Toyota Motor Corp, the idea of the widespread uses of movable infrastructures to build up the resilient society was discussed to find that this idea can be applied to the widespread uses of the fuel cell electric vehicles. This shared value between the Japanese Red Cross Kumamoto Hospital and Toyota Motor Corps has become the basis of the M-FCEV project started in 2021.

3 Medical Fuel Cell Vehicle Project

3.1 Hydrogen and Fuel Cell Vehicles

3.1.1 Hydrogen

In the global energy consumption, transport occupies about 19% while it accounts for 23% of energy-related carbon dioxide emissions expected to rise in the future. Fossil-fuel vehicles, which play a major role in transportation, are responsible for a significant amount of air pollution in the world, which is estimated to have caused six point five million premature deaths in 2012 (International Energy Agency 2009). Therefore, the automotive industry is conducting a lot of projects to develop alternative solutions such as Electric Vehicles, Hybrid Electric Vehicles, and FCEVs. FCEV use hydrogen that is the most abundant resource in the atmosphere and water. Fuel cell electric vehicles and rockets introduces hydrogen as a fuel because of its high energy potential. Hydrogen creates zero harmful emissions, which is one of the most attractive as an alternative energy to fossil fuels. (Offer et al. 2010) Even the use of natural gas-derived hydrogen, carbon free drivings of fuel cell electric vehicles still saves 30% CO_2 compared with diesel and petrol vehicles (California Fuel Cell Partnership 2013 and International Energy Agency 2016.)

3.1.2 Fuel Cell Electric Vehicles

A fuel cell is a device that generates electricity through a chemical reaction between hydrogen and oxygen in the air. Hydrogen is supplied to the electrodes of the fuel cell, and the hydrogen is decomposed into hydrogen ions and electrons by the catalytic action of the electrodes. When these electrons flow through the fuel cell circuit, electricity is generated. On the other hand, hydrogen ions move from the negative electrode to the air side, and are combined with oxygen under the influence of the catalyst on the air side to synthesize water. The chemical reaction between hydrogen and oxygen occurs in a thin flat device with two plates called a fuel cell, and multiple fuel cells are integrated in the FC Stack to produce high power (Jawad et al. 2022 and Yoshida 2015). The electricity generated by the FC stack is stored in the battery and used to drive the motor. This power is also supplied to the external power supply terminal. FCEVs and EVs are the same in terms of motor driving, but compared to EVs, FCEVs can be charged in a short time. In addition, FCEVs have a longer cruising range than EVs (Offer et al. 2010).

3.2 The Medical FCEV Project

In June 2021, Toyota Motor Corp and the Japanese Red Cross Kumamoto Hospital started joint field test of the world's first Medical FCEV (Toyota, 2021). This field test aimed to verify the effectiveness of the FCEV in the areas of medical services in peacetime and disaster response in emergencies. For this field test, the medical FCEV (hereafter "M-FCEV") "Doctor Car NEO" was developed based on a Toyota Coaster Minibus by using the Toyota Fuel Cell System for the FCEV "Mirai" as its power source (Fig. 8). Compared to another doctor car using a Toyota Coaster Minibus with the diesel engine, which the Japanese Red Cross Kumamoto Hospital has been operating for more than ten years, the M-FCEV enjoys a low-noise and low-vibration driving experience. The M-FCEV equips with the external power supply function (9 kW maximum output, approximately 90 kWh supply capacity) to supply electricity with other devices such as medical equipment and telecommunication equipment in emergency response. To supply the better quality of electricity to operate the sensitive medical equipment, the M-FCEV has installed an UPS in the power supply system.

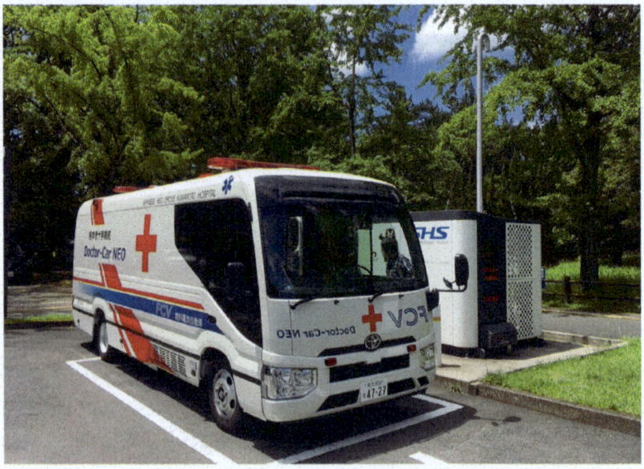

Fig. 8 The world first medical FCEV "Doctor Car NEO"

3.3 Power Supply Function

Figure 9 presents the configuration of the M-FCEV. In this system, hydrogen is supplied from the H2 Tank to the FC Stack where the oxygen in the air is sent for generating electricity through the chemical reaction between hydrogen and oxygen. The electricity is stored in the battery to drive the motor and supply the power to the external power supply functions in the M-FCEV. In the both sides of the M-FCEV, there are external outlets (Fig. 10) for the power supply.

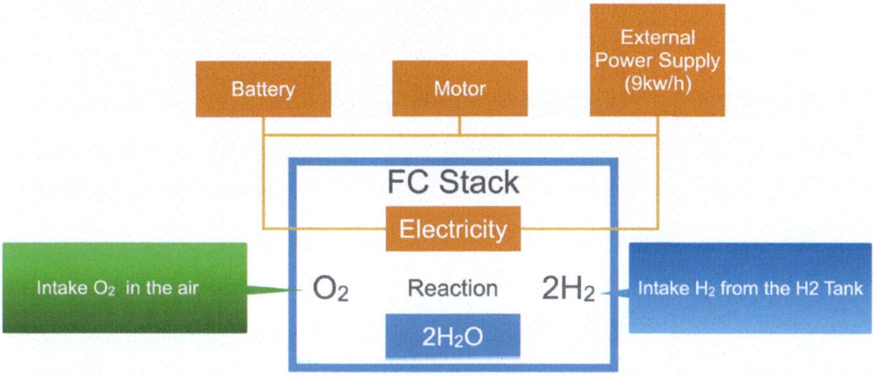

Fig. 9 The power supply function

Fig. 10 The external power supply terminal

Fig. 11 The monitor showing various information on the M-FCEV

3.4 Telecommunication

3.4.1 Image Sharing System

The M-FCEV has the image sharing system to share the images taken by a front-side camera installed behind the windscreen to capture the images outside of the vehicle and an inside camera to capture the image of the patient in a treatment space. The large monitor in the M-FCEV (Fig. 11) shows the real-time location of the vehicle, the estimate time of arrival, the images of outside of the vehicle, and the physical information of the patient collected by the medical monitor. The M-FCEV also has a Wifi-router for the internet communication. In emergencies, the M-FCEV can supply the sufficient amount of electricity to operate the satellite internet communication system such as Inmarsat and VSAT (Very Small Aperture Terminal). In addition, the M-FCEV is expected to function as an on-site drone operation room by charging the batteries of drones, sharing images taken by drone with the headquarters of the relief operation, and providing accommodation space for the emergency drone operators.

3.4.2 Vehicle Management Application

The real-time condition and location of the M-FCEV can be monitored for the smooth operation of the vehicle through the vehicle management application. This application (Fig. 12) monitors and shares the information on the real-time location, destination, amount of fuel, the current consumption of the hydrogen, the available period of the external power supply, and the nearest location of the Hydrogen Station. In addition, this application functions for matching the M-FCEV and the disaster affected people who need the power supply support by the vehicle.

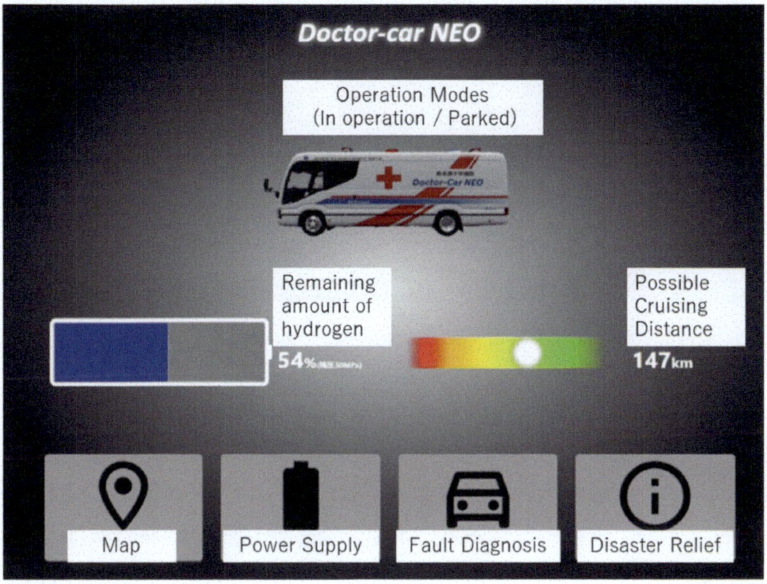

Fig. 12 The vehicle management application

3.5 *Field Test on Power Supply Function*

The field tests on the external power supply function of the M-FCEV were conducted in Kumamoto and Okinawa, Japan from October and November 2021. Firstly, the field test in Kumamoto assumed the power supply from the M-FCEV to medical equipment installed at a field hospital and the house of the home-based care patients using the medical equipment. The M-FCEV supplied electricity with a ventilator and a biological monitor in a tent (Fig. 13). This test assessed the current waveform supplied from the vehicle, the voltage drops due to the usage of other electric devices, and the status of the electric leakage. In addition, this study assessed the effectiveness of the medical UPS to provide the better quality of the electricity. The field test found that it is ideal to use a medical UPS to stabilize the voltage and current of the electricity supplied by the vehicle for the safe operation of the medical equipment. Secondly, the field test in Okinawa, the M-FCEV was operated as an emergency vehicle to establish the on-site operation room at the disaster affected areas. In this test, the M-FCEV utilized the external power supply function to operate the blower for inflating the air tent of the first aid station (Fig. 14). In addition, the M-FCEV was used as an on-site information station by projecting the image taken by an assessment drone, and the real time location of the assessment drone (Fig. 15) and the other external power supply vehicles (Fig. 16).

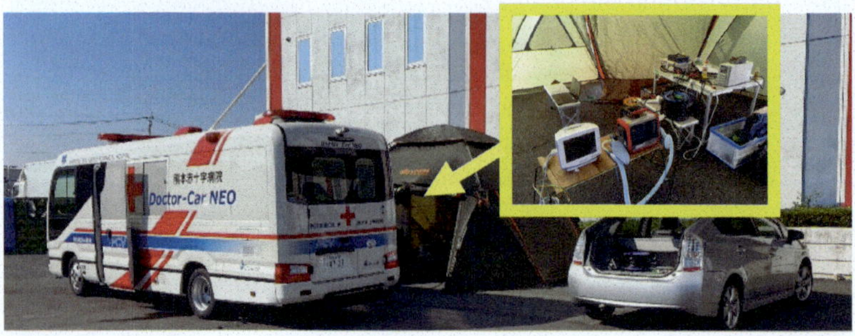

Fig. 13 The power supply to the medical equipment

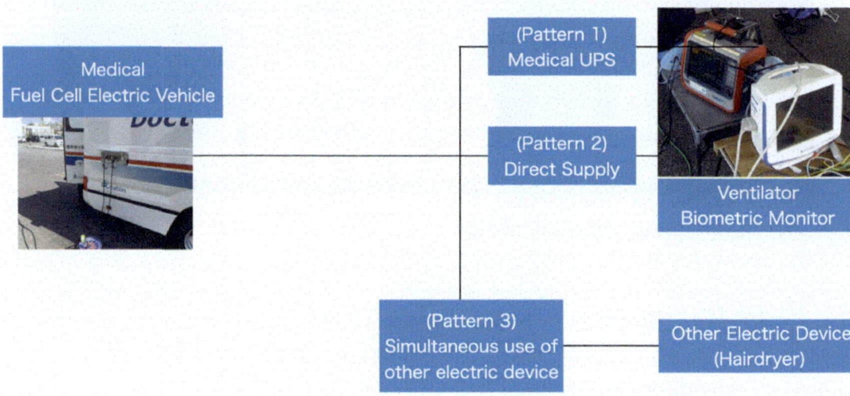

Fig. 14 The brief overview of the field test on the power supply to the medical equipment

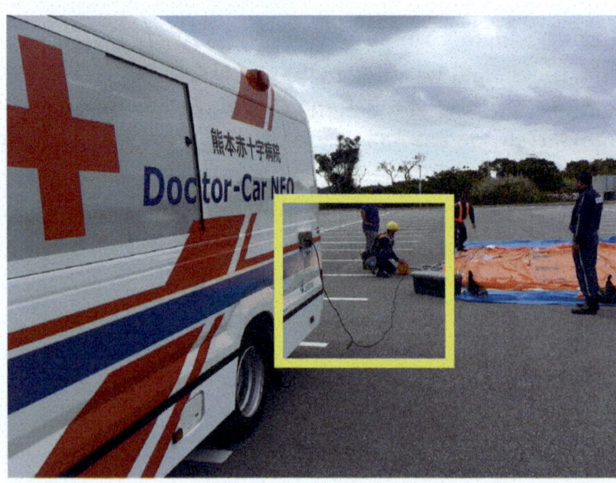

Fig. 15 Power supply to the air-blower to establish a command post tent

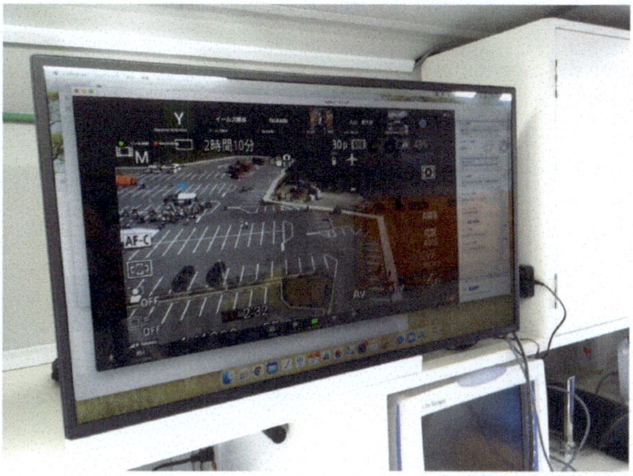

Fig. 16 The monitor shows the images taken by an assessment drone

4 FCEVs in Climate Change

This section describes the possible contribution by the applications of FCEVs to the adaptation and mitigation in the climate change. In terms of adaptation, FCEVs solve the present challenges in disaster response. On the other hand, FCEVs also contributes to mitigate the negative impact on the environment whilst FCEVs solve the current challenges of the ambulances.

4.1 FCEVs in Adaptation

In the recent meteorological disasters in Japan, the 2017 northern Kyushu heavy rain disaster, the 2018 western Japan heavy rain disaster, and the 2020 torrential rain disaster in the southern part of Kumamoto Prefecture, have caused large-scale heavy rain disasters due to the occurrence of linear rainfall belts (Japan Cabinet Office, 2018, 2019, 2021). In addition, the 2019 typhoon Faxai disaster caused serious damage to the Kanto region of Japan. In such meteorological disasters, power outages are common challenges. Power outages during disasters seriously affect not only medical facilities and nursing homes where patients using medical equipment are hospitalized, but the patients at home (Sekiguchi et al. 2022). For example, in the 2018 heavy rain disaster in western Japan, a hospital in Mabi was flooded and people were forced to be isolated for a long time. In the 2020 Kumamoto Floods, a patient receiving home oxygen therapy evacuated to the junior high school gymnasium where the frequent power outages occurred due to the limited capacity of the power supply.

In addition to the negative effect on the patients using the medical equipment, weather-related blackout becomes a cause of hypothermia, in particular, when the blackouts occur in winter. A prolonged power outage under the severe winter storms could cause the danger of hypothermia (Molarius et al. 2014). In fact, In the February

FIg. 17 Power supply to the heater in a refugee center by using the plug-in hybrid vehicle

8–19, 2021 Electricity Blackouts caused by an extreme winter storm and extended cold weather caused the blackout in Texas and the central United States. This blackout resulted in the death cases by hypothermia (Mann et al. 2021). Recently, in Japan, the application of the hybrid vehicles for supplying electricity to the refugee centers (Fig. 17).

Furthermore, In the weather-related disasters, Carbon Monoxide is one of the most dangerous hazards caused by the blackouts. Carbon Monoxide is a fatal gas without color and odor, which is made by the incomplete burning of various fuels. Equipment such as generators, vehicle can also produce CO (CPSC, 2023). Carbon monoxide may be responsible for more than one-half of the death poisoning reported in a lot of countries (Raub, 2000). The hazards in winter such as ice storms, snowstorms, and extreme cold were the leading ones that come to the situation causing CO poisoning death.2000 to 2019, in the United States, more than 8,300 CO death cases occurred and 17% of those fatalities with weather perils. These weather-related CO poisoning increase arises from lack of knowledge on generator safety and alarms in the enclosed locations (Stevens et al., 2022). In the 2011 Great East Japan Earthquake and Tsunami, it was reported that carbon monoxide poisoning was caused by operating generators indoors to survive the cold caused by blackouts (Nakajima et al. 2019) (Iseki et al. 2013), while the challenges in taking care of home oxygen therapy patients in the disaster affected areas due to the blackout (Kobayashi et al. 2012).

In this way, the weather-related disasters threaten the lives of the patients using medical equipment, kills the people due to hypothermia and carbon monoxide by disturbing the power supply. Therefore, the FCEV which can generate large amount of electricity with zero emission, is the future possible solution to prepare for the weather-related blackout. In this sense, widespread uses of FECVs means the prepositions of movable zero-emission power plants in the community, which enhance the community's resilience against the meteorological disasters.

4.2 *Fuel Cell Vehicle in Mitigation*

Applications of FCEVs in the health sector is a possible solution toward the decarbonization in the health sector. Furthermore, FCEVs could solve the challenges in the existing ambulances by the sufficient power supply capacity to operate medical devices in transporting the patients.

Firstly, in the health sector, carbon footprint is commonly categorized into three scopes. Scope 1 contains direct emissions by healthcare facilities such as the emission from the emergency vehicles for patient transport, anesthetic gases, and consumptions of fuels for air conditioning in the health facilities. Scope 1 targeted emissions can be reduced by the efforts of the health facilities. In direct emissions such as the emissions to produce the electricity purchased by the health facilities is categorized as Scope 2. Indirect emissions incurred from purchase of pharmaceuticals, medical supplies, and downstream emissions from medical waste is defined as Scope 3 (Hu et al. 2022).

In this way, the emission from medical vehicles is categorized as Scope 1. Some studies assessed the CO_2 emissions in ambulance services. Blanchard et al., (2011) estimated that in the North American emergency medical service agencies, 75% of emissions were from the fossil fuel such as diesel and gasoline (Blanchard et al. 2011). In the Australian ambulance services, approximately 20% of emissions were from direct consumption of vehicle fuels and aircraft fuels while the electricity consumption accounts for 22% emissions and Scope 3 such as supply chain and medical waste generated 58% of emissions. Introducing alternative fuels and more efficient vehicles into Australian ambulance services could reduce their direct emissions while other efforts are needed to substantially reduce their total carbon footprint (Lawrence et al., 2012). In addition, Lawrence et al. (2012) estimated the greenhouse gas emissions in the Australian ambulance services for one year and pointed out that vehicle fuels occupied 56% of the emissions from ground ambulance operations. In this way, while these studies presented the potential CO_2 reductions in the transportation of the patients by emergency vehicles using the fossil fuel.

Secondly, in addition to the contribution toward the decarbonization in the health sector, the abundant power supply capacity of FCEVs generates additional value by solving the problem on the power supply capacity of the existing ambulances. In Japan, a lot of studies reported the challenges in ensuring the electricity to operate the medical equipment in transporting the patients by emergency vehicles (Takahashi et al., 2021), (Kurosawa et al., 2014). For the land transport of the patient with an intra-aortic balloon pumping (IABP) that needs the sufficient amount of electricity (20A) the medical staff used a disaster response vehicle which is the large bus (Takeuchi et al., 2012). The electricity consumption in the standard ambulance in Japan is limited under 3.0A which is not sufficient to operate an intra-aortic balloon pumping (IABP) and a percutaneous cardiopulmonary support system (PCPS) at the same time. Even operation of IABP is not possible since the standard ambulance does not have enough capacity to provide the electricity to IABP. In addition, the batteries of the medical equipment should be removed to prioritize the use of electricity (Ito et al., 2009). In planning the long-distance transportation of the patient with venous-arterial extracorporeal membrane oxygenation (VA-ECMO) for ischemic cardiomyopathy, estimation of the total consumption of electricity was conducted to ensuring the safe operation in an ambulance (Shibuya et al., 2022). In this way, toward the decarbonization in the health sector, application of new types of vehicles that use alternative fuels replacing the vehicles with internal combustion engines using fossil fuels. Application of FCEVs for the future ambulances is an effective solution not only for decarbonization but to solve the present challenges in transportations of the patients who require medical equipment that consumes a lot of electricity.

5 Discussions

This study discussed the application of FCEV as a new solution for the climate change mitigation and adaptation and pointed out that FCEV's unique characteristics of zero emission and external power supply function are the new solutions for mitigation in the health services and adaptation to weather-related disasters. In addition to the FCEVs' contribution to both mitigation and adaptation against climate change, FCEVs also show their additional value such as power supply in the transportation of the patients with medical equipment, and preventing carbon monoxide in the weather-related blackouts, which have been the challenges in the uses of the vehicles with internal combustion engines. From the viewpoint of community development, health sector is the suitable area for the application of fuel cell technologies. In the hospital, instead of operating generators by fossil fuels, fuel cell can be used to support the hospital infrastructure. Fuel cell technologies will be used for various medical mobility such as vehicles, helicopters, and medial drones. As this study tried to supply electricity with medical equipment, FCEVs will be integrated as a component of the hospital power management system.

While the past study discussed the roles of climate policy to foster the successful integration and implementation of mitigation and adaptation (Klein 2005), or, the challenges in the less recognition of the importance of the links, synergies and inclusive implementation between mitigation and adaptation (Swart and Raes, 2007), in this study, FCEVs presented their possibility as the solutions to link between mitigation and adaptation. In the past, a few studies pointed out the effectiveness of the application of technologies to contribute to both mitigation and adaptation. Shifting off-grid energy generation by using environmentally friendly technology is not only a mitigation approach but a mitigation strategy for reducing the risks associated with widespread power loss caused by severe storm evens (Hamin et al. 2008). In terms of the application of Information and Communication Technologies (ICT), Ajwang and Nambiro (2022) pointed out that technology advances benefited by innovation in ICT have been understood as effective solutions to cope with climate change by identifying signs of changing climate, analyzing and modeling, and implementing mitigation and adaptation approaches that enhance resilience. In most of the past studies, FCEVs have been recognized as a sustainable solution for mitigation (Chapman, 2007). Or, FCEVs are recognized as a solution for adaptation by utilizing their external power supply functions for ensuring the telecommunication services in emergencies (Soshino, 2022). In this way, while the past studies on FCEVs discussed their contribution to either mitigation or adaptation, this study pointed out that FCEVs are the solutions which contribute to both mitigation and adaptation in the climate change actions. In addition to the FCEVs contributions to the climate change, they also contribute to the ongoing effort for achieving the sustainable development goals. Kelman (2017) studied the relation between the Paris Agreement for climate change and disaster risk reduction and the sustainable development goals to find that climate change mitigation is placed within the existing pollution prevention framework while climate change adaptation is included within the existing disaster risk reduction framework. If so, FCEVs contribute to pollution prevention by their unique features of zero emission while contribute to disaster risk reduction by the external power supply functions as the movable power plants.

Further studies are required to seek for other application models of technology that solves the challenges both in mitigation and adaptation. Through FCEVs are the promising solutions for adaptation and mitigation in climate change, difficult challenges remain in their widespread uses. The development of hydrogen infrastructure is a crucial barrier (Staffel et al., 2019). The hydrogen recharging infrastructure in the world reached to 814 in 2022(H2 Stations.org, 2022). As of February 2022, in Japan, 137 Hydrogens station started the services, while the cumulative number of fuel cell vehicles remains 4,600 (Japan Agency for Natural Resources and Energy, 2021). Toward the realization of hydrogen society, all the components of the hydrogen economy: production, storage, transportation and distribution should be established in balance (Brandon et al., 2017).

For this reason, application of EVs and Hybrid EVs with external power supply functions, which offer relatively lower CO emissions compare to in internal combustion engine vehicles, are also realistic and practical solutions in disaster response and emerging countries. In other words, in order to promote adaptation and mitigation on a global scale, applications of multiple solutions are needed according to the circumstances of each country and region.

6 Conclusion

This chapter discussed the application of the FCEVs as a solution for the medical service and disaster response. The development of the Medical FCEV was presented. In peacetime, application of FCEVs in medical service reduces the greenhouse gas emission in transporting the patients. In addition, the application o FCEVs for medical vehicles solves the challenges in the power supply to the medical devices in the existing ambulances. In emergencies, FCEVs can supply electricity to support the relief operations without producing poisoning gas such as carbon monoxide as "Movable Power Plants." In this way, FCEVs contribute both mitigation and adaptation in the climate change. In other words, FCEVs contribute pollution prevention and disaster risk reduction toward the achievement of the sustainable development goals. To realize this goal, all the components of the hydrogen economy should be established in balance.

References

Ajwang SO, Nambiro AW (2022) Climate change adaptation and mitigation using information and communication technology. Int J Comput Sci Res 6:1046–1063

Blanchard EI, Brown HL (2011) North American EMS emissions study group, Carbon footprinting of North American emergency medical services systems. Prehospital Emerg Care, 15(1): 23-29, https://doi.org/10.3109/10903127.2010.519818

Brandon NP, Kurban Z (2017) Clean energy and the hydrogen economy. Philos Trans R Soc A 375

Brown HL, Buettner GP, Canyon VD, Crawford MJ, Judd J (2012a) Estimating the life cycle greenhouse gas emissions of Australian ambulance services. J Clean Prod 37:135–141

Brown HL, Canyon VD, Buettner GP, Crawford MJ, Judd J (2012b) The carbon footprint of Australian ambulance operations. Emerg Med Australas 24:657–662

Cabinet Office, Government of Japan (2021) White Paper on disaster management Available at: https://www.bousai.go.jp/en/documentation/white_paper/2021.html, Accessed 19 April 2023

California Fuel Cell Partnership (2013) Air, climate, energy, water security: a guide to understanding the well-to-wheels impact of fuel cell electric vehicles. Available at: https://test.h2fcp.org/sites/default/files/W2W-2014_Final.pdf, Accessed 14 September 2023

Chapman L (2007) Transport and climate change: a review. J Transp Geogr 15(5):354–367

Toyota Motor Corp., (2021) Japanese red cross kumamoto hospital and toyota to begin utilization demonstration of the world's first fuel cell electric vehicle mobile clinic, news release. Available at: https://global.toyota/en/newsroom/corporate/35008661.html, Accessed 19 April 2023

CPSC, Carbon Monoxide Fact Sheet 2023, [Online]. Available at: https://www.cpsc.gov/safety-education/safety-guides/carbon-monoxide/carbon-monoxide-fact-sheet. Accessed 9 March 2023

Hamin EM, Gurran N (2009) Urban form and climate change: Balancing adaptation and mitigation in the US and Australia. Habitat Int 33(3):238–245

Hu H, Cohen G, Sharma B, Yin H, McConnell R (2022) Sustainability in Health Care. Annual Reviw of Envirionment and Resources 47:173–196

ICRC, IFRC (2021) the climate and environment charter for humanitarian organizations [Online]. Avaulable at: https://www.climate-charter.org/wp-content/uploads/2022/05/ClimateEnvironmentCharter-EN.pdf, Accessed 19 April 2023

International Energy Agency. (2009) Transport, energy and CO_2: moving toward sustainability. Paris, France: International Energy Agency

International Energy Agency. (2016) Energy technology perspectives 2016: towards sustainable urban energy systems. Paris, France: International Energy Agency

Iseki K, Hayashida A, Shikama Y, Goto K, Tase C (2013) An outbreak of carbon monoxide poisoning in yamagata prefecture following the great East Japan Earthquake, Asia Pac J Med Toxicol, 37–41

Ito K, Nagao T, Tsuboi H, Ide Y, Tsuruyama K, Hayami Y (2009) Assessment of the power supply capacity during emergency transportation usinga an ambulance. Heart 41(10):1094–1101

Japan Agency for Natural Resources and Energy, Ministry of Economy Trade and Industry (2021) FCV and Hydrogen Station Business Available at: https://www.meti.go.jp/shingikai/energy_environment/suiso_nenryo/pdf/024_01_00.pdf, Accessed 19 April 2023

Jawad NH, Yahya AA, Al-Shathr AR, Salih HG, Rashid KT, Al-Saadi S, AbdulRazak AA, Salih IK, Zrelli A, Alsalhy QF (2022) Fuel cell types, properties of membrane, and operating conditions: a review. Sustainability 14:14653. https://doi.org/10.3390/su142114653

Kelman I (2017) Linking disaster risk reduction, climate change, and the sustainable development goals. Disaster Prev Manag: Int J 26(3):254–258

Klein RJ, Schipper ELF, Dessai S (2005a) Integrating mitigation and adaptation into climate and development policy: three research questions. Environ Sci Policy 8(6):579–588

Klein RJT, Lisa E, Schipper F, Dessai S (2005b) Integrating mitigation and adaptation into climate and development policy: three research questions. Environ Sci Policy 8:579–588

Kobayashi S, Hanagama M, Yamanda S, Yanai M (2012) Home Oxygen therapy during natural disasters: Lessons from the great East Japan earthquake. Eur Respir J, 39(4):1047-48, https://doi.org/10.1183/09031936.00149111

Kurosawa K, Ishii K, Tanabe S, Nakashima T, Noguchi T (2014) A patient with cardiogenic cardiac arrest who returned to society following percutaneous cardiopulmonary support delivered by ambulance-type emergency response vehicle: A case report. J Jpn Assoc Acute Med 25:319–324

Laukkonen J, Blanco KP, Lenhart J, Keiner M, Cavric B, Kinuthia-Njenga C (2009) Combating climate change adaptation and mitigation measures at the local level. Habitat Int 33(3):287–292

Lesnikowski A, Ford J, Biesbroek R, Berrang-Ford L, Maillet M, Araos M, Austin ES (2017) What does the Paris Agreement mean for adaptation? Climate Policy 17:825–831

Mann NW, Biegel K, Stauff EN, Dixon B (2021) Electricity blackouts and natural gas shortages in texas, preliminary analysis of texas 2021 power outrages, energy systems division, nuclear science and engineering division, agronne national laboratory. Available at https://publications.anl.gov/anlpubs/2021/07/169454.pdf, Accessed 9 March 2023

Mata JL, Budhooram J (2007) Complementary between mitigation and adaptation: the water sector. Adapt Strat Glob Chang 12:799–807

Molarius R, Tuomaala P, Piira K, Räikkönen M, Aubrecht C, Polese M, Zuccaro G, Pilli-Sihvola K, Rannat K, (2014) A framework for comprehensive impact assessment in the case of an extreme winter scenario, considering integrative aspects of systemic vulnerability and resilience. In: Conference Paper, Second International Conference on Vulnerability and Risk Analysis and Management (ICVRAM) and the Sixth International Symposium on Uncertainty, Modeling, and Analysis (ISUMA)

Nakajima M, Aso S, Matsui H, Fushimi K, Yamaguchi Y, Yasunaga H (2019) Disaster-related carbon monoxide poisoning after the Great East Japan Earthquake, 2011:a nationwide observational study. Acute Medicine & Surgery 6:294–300

Nocholls RJ, Lowe JA (2004) Benefits of mitigation of climate change for coastal areas. Glob Environ Chang 14(3):229–244

Offer GJ, Howey D, Contestabile M, Clague R, Brandon NP (2010) Comparative analysis of battery electric, hydrogen fuel cell and hybrid vehicles in a future sustainable road transport system. Energy Policy 38:24–29

Paglia E, Parker C (2021). The Intergovernmental Panel on Climate Change: Guardian of Climate Science. In: Boin, A., Fahy, L.A., 't Hart, P. (eds) Guardians of Public Value. Palgrave Macmillan, Cham. https://doi.org/10.1007/978-3-030-51701-4_12

Raub AJ, Mathieu-Nolf M, Hampson BN, Thom RS (2000) Carbon monoxide poisoning- a public health perspective. Toxicology 145:1–14

Schipper L (2006) Conceptual history of adaptation in the UNFCCC process. Rev Eur Community Int Environ Law 15(1):82–92. https://doi.org/10.1111/j.1467-9388.2006.00501.x

Sekiguchi K, Harimoto H, Aoshima A, Izukura H, Fujie S (2022) Typhoon No.15 and home oxygen therapy in Tateyama city, Chiba, Japan Journal of Disaster Medicine, 27(3):161–164

Shibuya M, Shoya K, Hirano A, Umezu A, Sueishi T, Yoshida T, Yoshimura Y, Saito O (2022) Mobile ECMO for long distance ground transport in a pediatric cardiac patient with central VA-ECMO in Japan. Pediatr Cardiol Card Surg 38(4):254–261

Soshino Y, Kuroda A, Miyata A, (2017) Review of the domestic utilization of pre-positioned international medical relief equipment in Japan. J Nat Disaster Sci, 38(2)

Soshino Y (2022) Application of the fuel cell vehicle to support ICT in Emergency Response. In International Conference on Information Technology in Disaster Risk Reduction (pp. 89–100). Cham: Springer Nature Switzerland

Staffel I, Scamman D, Abad VA, Balcombe P, Dodds EP, Ekins P, Shah N, Ward RK (2019) The role of hydrogen and fuel cells in the global energy system. Energy Environ Sci 12:463–491
H2 Stations. org, last accessed 6 Mar 2023, https://www.h2stations.org/wp-content/uploads/Development-HRS-worldwide-2022-t.jpg, (Accessed) 18 April 2023
Stevens BR, Ashley WS (2022) Fatal Weather-Related carbon monoxide poisonings in the United States. Weather, Climate, and Society 14(2):373–386
Swart ROB, Raes F (2007) Making integration of adaptation and mitigation work: mainstreaming into sustainable development policies?. In Integrating climate change actions into local development (pp 288–303). Routledge
Takahashi M, Oi Y, Shirasawa A, Yamagata H, Michishita T, Imaki S, Takeuchi I (2021) Challenges in transportation of critically ill patients by ordinary ambulance. KANTO J Jpn Assoc Acute Med, 42(2)
Takeuchi I, Imaki R, Sato N, Inomata T, Niwano S, Izumi T, Soma K (2012) Land Transport of patients with mechanical support in disaster support vehicles. J Jpn Assoc Acute Med 23:856–860
Verchot VL, Noordwijk VM, Kandji S, Tomich T, Ong C, Albrecht A, Macensen J, Bantilan C, Anupama VK, Palm C (2007) Climate change: linking adaptation and mitigation through agroforestry. Mitig Adapt Strat Glob Change 12:901–918
World Meteorological Organization, (2021) WMO Atlas of Mortality and Economic Losses from Weather, Climate and Water Extremes, WMO-No.1267, ISBN: 978–92–63–11267–5
Yoshida T, Kojima K (2015) Toyota MIRAI fuel cell vehicle and progress toward a future hydrogen society. Electrochem Soc Inte 24(2). https://doi.org/10.1149/2.F03152if

Technologies for Disaster Management: Case Studies in the Past Tsunami in Japan

Akane Kurisu and Yasuhiro Soshino

1 Introduction

Throughout human history, natural hazards have threatened the peaceful lives of humans. Among the various natural hazards such as earthquakes, typhoons, floods, snow storms and heatwaves, tsunamis cause catastrophic damages in the disaster affected areas and influence the prosperity of a nation. In 365, an earthquake occurred near Crete Island and caused catastrophic tsunami that struck the southern and eastern coastal areas in the Mediterranean Ocean (Stiros 2001). On November 1, 1755, a strong earthquake hit Lisbon, the capital of Portugal followed by a tsunami that created additional victims and damage. The city of Lisbon also experienced major fire lasted several days destroyed a lot of buildings not damaged by the earthquake. In addition to an estimated the lower 10,000 to highest 90,000 victims, a large number of books and precious art works collected from around the world were also lost. Even though various efforts for recovery were made, Portugal's influence in Europe declined (Dynes 2003). Tragic history of tsunami disasters repeated even in the twenty-first century. On December 24, 2004, a magnitude 9.1 earthquake struck off the coast of Sumatra. This caused a huge tsunami, which caused damage not only in Indonesia, but also in India, Sri Lanka, the Kingdom of Thailand, Malaysia, the Maldives, Madagascar, Somalia, and other parts of Southeast Asia along the Indian Ocean coast, as well as in East Africa. Due to this earthquake and tsunami, more than

A. Kurisu · Y. Soshino (✉)
Disaster Management Research Institute, Japanese Red Cross College of Nursing, Shibuya 150-0012, Tokyo, Japan
e-mail: y-soshino@jrcdmri.redcross.ac.jp

A. Kurisu
e-mail: a-kurisu@jrcdmri.redcross.ac.jp

Y. Soshino
Japanese Red Cross Kumamoto Hospital, Kumamoto 861-8520, Japan

P. K. Ray et al. (eds.), *Technology Innovation for Sustainable Development of Healthcare and Disaster Management*, Disaster Risk Reduction,
https://doi.org/10.1007/978-981-97-2049-1_5

two hundred thousand people lost their lives (IFRC 2013). Furthermore, on March 11, 2011, an earthquake of magnitude 9.0 struck off the coast of Sanriku, Japan and caused catastrophic damage in eastern Pacific Ocean side of Japan called the Great East Japan Earthquake and Tsunami (hereafter the GEJET). After the GEJET, the Japan Cabinet Office established various working groups on the preparedness against future national catastrophes such as Tokyo Inland Earthquake, Nankai Trough Earthquake and Tsunami, and the arrival of super-typhoon in mega-city of Japan. In addition, there is a growing concern about the occurrence of Japan trench and Chishima trench earthquake and tsunami which may strike the large coastal areas in northern eastern Pacific Ocean side of Japan (Japan Cabinet Office 2021). To reduce the human loss in this disaster, prevention of drowning and hypothermia is imperative. In fact, the GEJET, which occurred during the latter part of winter, revealed the challenges in preventing drowning and hypothermia. Firstly, in the GEJET, more than 90% of the victims lost their lives by drowning (Seto et al. 2020). While most of the tsunami exposed people lost their lives by drownings, a tsunami exposed person who were drifted in the ocean for 43 h after the disaster was found and rescued by a vessel of the Japan Self-Defense Forces (NHK 2022). A couple was drifted on the roof of their house and successfully landed before washed away to the ocean. During the drift, a wife sent an email by using her mobile phone (Saito 2020). This means, floating material and devices are the minimum condition in saving lives in tsunami disaster and mobile phones can be used even from the surface of water during being drifted by tsunami. Secondly, in the GEJET, the majority of the hypothermia patients who were hospitalized to the emergency hospitals in Miyagi Prefecture within 72 h were caused by the exposure to the tsunami (Furukawa et al. 2014). Generally, hypothermia is defined as an unintentional drop of core body temperature below 35 degrees Celsius. Since the thermal conductivity of water is about 25 times higher than that of air, the effect of water temperature on body temperature is extremely high. Hypothermia occurs when the body temperature drops below 35 degrees Celsius when the body is immersed in the sea, and when it is below 32 degrees Celsius. The ability to regulate body temperature is lost, and the drop in body temperature further increases, creating a very dangerous situation. Therefore, in tsunami disasters, search and rescue services play a critical role in preventing hypothermia. To find the drifted survivor or to be found by the rescue workers in tsunami before the survivors fell into hypothermia, application of mobile phones technology is one of the possible solutions. Therefore, this chapter discusses the application of affordable and familiar technologies for preventing human loss in future tsunami disasters. The structure of this chapter is as follows: following introduction, section two describes the application of mobile technologies in disasters. Section three reviews the past experiences of drifts by tsunami. Section four reviews the applications of mobile devices to identify the location of victims in sea accidents and natural disasters. Section five examines the effectiveness of lifejackets in tsunami. Section six describes the discussion and section seven concludes this chapter.

2 Mobile Communication and the Challenges in the GEJET

2.1 An Emerging Solution in Humanitarian Action

Nowadays, mobile apps and artificial intelligence are new and emerging technology applications in humanitarian action (UN OCHA 2021). On the other hand, in the past, mobile phones are the tools of dream. The first voice call from a mobile phone was made on April 3, 1973 by Mr. Marty Cooper, an engineer at Motorola (BBC 2023). The users of mobile internet services reached 55 percent of the total population in the world. By the end of 2021, four point three billion people were the users of mobile internet. The rapid expansion of mobile internet users is driven by people in low- and middle-income countries (GSMA 2022). As the recognition of mobile phones changes from the tool of dream to the partners of the users, in addition to the communication tools, mobile phones have become sensors in our societies. Data collected from mobile phones begun to be utilized to solve various social challenges. Ratti et al. (2006) presented the 'Mobile Landscapes' project that aims to apply the geographical mapping of cell phone usage in the metropolitan area of Milan and discussed their future applications for urban studies. Calabrese et al. (2011) presented a new real-time urban monitoring system based on the monitoring of mobile cellular networks. This study also described the real-time monitoring of urban mobility such as buses and taxi. Deville et al. (2014) pointed out the significant effectiveness of the application of mobile phone data for population mapping. In humanitarian services, the application of the data of mobile phones was highlighted in the 2010 Haiti earthquake. Bengtsson et al. (2011) estimated the population movements by following daily locations of SIM cards in the mobile phones 42 days before the earthquake and 158 days after the earthquake. This study also covered the cholera outbreak analysis covering eight days. Their estimation of the population movements can be rapidly delivered and accurate in areas where the mobile phones are highly utilized. Wilson et al. (2016) analyzed the evolution of mobility patterns of the people affected by the 2015 Nepal earthquake based on the 12 million of mobile phone users. In this way, understanding the movement status of evacuees based on mobile phone location information is being used in the field as a new solution for disaster response.

2.2 The Challenges in the Application of Mobile Devices in the GEJET

Apart from the previous past disasters in Japan including the 1995 Great Hanshin-Awaji earthquake, when the GEJET occurred, mobile phones had been in widespread uses as a basic infrastructure to support the daily business by the government, local authorities, medical facilities, and daily lives in the communities. Mobile phones had established their important presences as the key components of social infrastructure. The GEJET may be one of the first mega-disasters which struck in a developed country where a lot of people uses internet and smartphones. Due to the enormous telecommunication demands in the disaster affected areas, the 90% of the calls in the most devastated areas were blocked by the telephone operators (Aizu 2011). In the GEJET, the mobile phones were not available due to the convergence of massive communication demands, the physical damage of hub stations by tsunami, and the power outages. The widespread uses of mobile phones before the GEJET created the reliance on the mobile phones in Japan. Hence, the little availability of mobile phones in the disaster affected areas caused the significant negative impact on the disaster affected people and relief service providers just after the occurrence of the earthquake. The tsunami affected people were not able to communicate with their families. A lot of people returned to their homes to pick up their family members. In the search and rescue operations, searchers were not able to ring the mobile phones of the missing persons. Kudo et al. (2014) assessed the availability of telecommunication system after the occurrence of the GEJET. Among the medical facilities in the disaster affected areas, mobile phones, landlines, and the internet did not correctly function during the first four days since the occurrence of the earthquake. Upon the occurrence of the earthquake on March 11, 2011, the Japanese Red Cross Kumamoto Hospital deployed medical relief convoys to the disaster affected areas. Due to the limited availability of the mobile phones, the dispatched teams communicated with the headquarters in the Kumamoto Hospital by using a satellite phone (Fig. 1).

It is obvious that the availability of the information network service is most crucial for the users who evacuated for saving their lives and relied on the remaining communication services (Shibata et al. 2014). The management of refugee centers was also seriously affected by the mobile phone outages. Due to the lack of mobile phone services, there was little communication tools between local authorities that manage the refugee centers and the refugees without the direct voices and the information bulletin boards.

While the mobile phones outages disrupted the communication in the disaster affected areas, the information of the evacuee's location sent by a mobile phone contributed to the successful rescue operation in the GEJET (Inose 2015). By using a mobile phone, a director of the Children's facility sent an email to her son living in

Fig. 1 A Red Cross Medical Team communicate with the hospital by a satellite phone in the GEJET

London informing that she evacuated the roof of the community center with children. Her son tweeted that his mother and a dozen of children evacuated to the third floors at the community center in Kesennuma City of Miyagi prefecture. The lieutenant governor of Tokyo saw his tweet and this resulted in the successful evacuation from the isolated community center by the helicopter of Tokyo Fire Department. This case is considered to demonstrate the effectiveness of mobile phones in the search and rescue phase in tsunami.

3 Usage of Lifejackets and Other Items in the Tsunami Drifts

In the GEJET, little study discussed the people who were exposed and drifted by tsunami while a lot of studies targeted the evacuation processes and the situations at the refugee centers.

The Nihonkai Chubu Earthquake Tsunami (hereafter "NCET") occurred on May 26, 1983 and Magnitude seven-point seven earthquake killed four people and caused tsunami that killed 100 people Aomori and Akita prefectures in northeast region of Japan. Among the 100 NCET tsunami victims, there were 34 workers at the premises of Noshiro Port Power Plant, seven port workers at other port, 18 fish anglers, 13 elementary school students on a school trip, and 14 inhabitants in the tsunami exposed areas. In the Noshiro Port where tsunami reached approximately tsunami reached after 25 min from the occurrence of the earthquake and little evacuation efforts were made due to the lack of the alert system. In the Noshiro Port, there were approximately 400 workers and limited workers wore lifejackets. Among the worker with their lifejackets on, only one worker was killed due to the damage of the lifejacket (Ushiyama et al. 2004) (Akita Anglers Association 1983).

Table 1 presents the tsunami exposed and drifted people in the 1983 NCET and Table 2 presents the experiences of survivors in the 2011 GEJET. In the NCET, tsunami exposed people pointed out the effectiveness of lifejackets based on their experiences.

On the other hand, in the GEJET, the survivors were drifted on the collapsed houses and floating debris. The NCET and the GEJET occurred in the different seasons which meant the air and seawater temperature were different. However, in both cases, the survivor reported that they felt cold after their exposed and drifted by tsunami. In this way, past experience points out that wearing a life jacket is important in preventing tsunami drowning. In addition, preventing hypothermia is required for saving lives of tsunami exposed people. In this sense, people living in the coastal areas where the sufficient lead time for tsunami evacuation cannot be ensured be assured should prepare for lifejackets as one of the efforts in self-help. In contrast, relief agencies are required to detect the tsunami drifted people to prevent the people from falling into hypothermia cases. To realize this, application of the location information from mobile phones of the tsunami affected people is expected.

Table 1 The records of the tsunami survivors in the NCET

Cases	Buoyancy	Feeling the cold after exposure
"The tsunami collapsed the seawall used for land reclamation at Noshiro Port, and many employees were thrown into the sea. The state of the vortex was well understood, and it was just like a person and a stone for revetment were rotating in an electric washing machine. All but one of the people who wore life jackets survived the many deaths. One of the deceased had an inflatable (gas) lifejacket with a split in the back, leaking gas. "	Lifejackets	Not mentioned
"Again, I am under the water, and this time I cannot float. Thinking that this would kill me, I let go of my hand's pole and reached out to catch the cooler from below the water surface."	Cooler box	Not mentioned
"With my life jacket in my hand, I thrashed my arms and legs and finally came out to the surface. The sea was swirling, choppy, and barely breathing."	Lifejackets	"When I got on land from the ship, I thought, "I'm really saved by this," and suddenly my body started shivering from the cold. I took off my wet clothes, put on the work clothes that I had been lent to me, and was warming myself by the stove in the hut at the boat dock."

Source NCET (Akita Fishing Association 1983)

4 Application of Mobile Technologies in Emergencies in the Sea

4.1 Accident Information Sharing System in the Sea

In Japan, as the usages of mobile phones and satellite phones become popularized, such communication devices are introduced in the accident reporting system in the ocean. Wada (2006) proposed the rescue system for man-overboard accidents in the ocean. His study developed and evaluated the rescue system consists of three different communication tools such as the satellite phone, the radio device, and the mobile phone.

Marine accidents and personal injury accidents in Japan are under the jurisdiction of the Japan Coast Guard. With the spread of mobile phones in the 2000s, the Japan Coast Guard launched a "location information notification system" in 2007 that can quickly ascertain the location of an emergency call to 118 from a mobile phone. This system receives location information notifications along with voice calls in the event of an emergency call to 118 from a mobile phone and displays them on an

Table 2 The records of the tsunami survivors in the GEJET

Cases	Buoyancy	Feeling the cold after exposure
"In the distance, I could see a person who looked like a young man or a child riding on the roof and drifting offshore, but there was nothing I could do about it. Then, when the man approached the bank below, he jumped skillfully and was relieved."	Debris of a house	Not Confirmed
"A couple in the neighborhood also drifted in with their house, still upstairs. Then, after the first wave receded, my face came out of the window, so my son called out, "No, no, there's another big wave coming, so hurry up!" We were soaking wet, but we were able to pull them both out"	Debris of a house	"We were soaking wet, but we were able to pull them both out. After that, the refrigerated truck on the road even took me to Hanawa. I was wearing thick clothes, so I shared the clothes with the wet couple."
"I was caught by a tire that was drifting, but I drank water several times. The tire was heavy and I would sink even if I grabbed it by myself, so I swam with my hands and feet. (Omitted) I was tired from swimming on my stomach for a while, so I floated like a backstroke (Omitted) I didn't feel cold in the water. Since it was winter, I wore a down jacket and escaped, but that might have been a good thing. "	A Tire	"The waves calmed down and I swam quietly towards my house. At that time, I thought that I didn't want to die in front of my house, so I finally reached the stairs leading up to my house It was so cold that I shivered violently, and I thought I was going to die if I continued like this."

Source (Miyako City 2020)

electronic nautical chart at the command center at each Regional Coast Guard Headquarters Operations Command Center nationwide. The target phones are mobile phones (models called third generation) and IP phones (excluding mobile types). In the case of reports from mobile phones, GPS positioning information is notified for GPS compatible models, and other location information calculated from the location of the base station is notified (Japan Cabinet Office 2007). Recently, as the progress of the communication technologies, Personal Location Beacon (PLB) and Automatic Identification System—Man Overboard (AIS MOB) have been developed and utilized to improve the rescue operation following the overboard of the crews on board. In addition, a system that notifies crew members and anglers that they have fallen into the water using specified power-saving radio or mobile phones is starting to spread in Japan (JCIO 2022). In 2019, a private company in Japan conducted a field study on the accident sharing system consists of small radio devices with GNSS (Global Navigation Satellite System). The device also has salt water sensor so that the device recognizes the falling of water and share the location information automatically. The device was attached to the lifejackets (Japan Cabinet office 2019). In 2020, private companies in Japan conducted a demonstration experiment by attaching transmitters with buttons to boats and people, assuming the discovery, rescue, and search of victims of boats that were unable to return to port due to an accident (Kyocera 2020). In general, this information sharing system is intended to share the information on the "accident" to detect the location of the "person" who is overboard into the ocean.

4.2 People Flow Data System

Disasters are classified into sudden disasters such as earthquakes and delayed disasters such as population migration according to the lead time until the disaster occurs. Disasters are classified into local disasters and wide-area disasters according to the geographical spread of the affected area. A large-scale disaster such as a tsunami requires a sudden response over a wide area, which is the most difficult operation for relief agencies (Apte 2010). In addition, it is extremely difficult to identify where the disaster area is and where the victims are evacuated in the initial stage of emergency disaster relief activities. For this reason, in Japan, efforts are being made to utilize people flow data to grasp the evacuation situation in the event of a sudden disaster.

In 2010s, application of people flow data gradually started in Japan. Funakoshi and Hatayama (2017) utilized the location data of mobile phone to identify shelters that were not designated by local authorities in the 2016 Kumamoto Earthquakes.

Udagawa (2021) used people flow data to estimate the number of people in Kumamoto City during the pre-evacuation due to the typhoon. Kuroda et al. (2022) utilized people flow data to estimate the recovery process of sales levels and transport network performances aftermath of the heavy rain in July 2018. In July 2020, Japanese Red Cross Kumamoto Hospital utilized the people flow data in the flood affected areas in decision-making for the deployment of an assessment team.

In November 2022, the Japan's first demonstration experiment on real-time monitoring of tsunami evacuation status (Fig. 2) using human flow data and AI acquired by a smartphone app in the Tsunami disaster prevention drill was conducted in Nemuro City of Hokkaido, Japan (Red Social Innovation 2023). In this way, the application of the people flow data has been in progress in the areas of disaster management. Compared to the information sharing system in the sea, people flow data collected from mobile phones are utilized to grasp the real-time movement of the people in the disaster affected areas to identify the isolated places, the places where refugees are gathering and local authorities hardly recognize them, and monitor the evacuation behavior in the tsunami prevention drills in the community.

4.3 An Example of Search and Rescue by Drones

As the progress of the development and widespread uses of drones, drones are becoming a new solution for the search and rescue tools in the accidents and disasters in the sea. Soshino et al, (2020) reported the design and development of the image sharing system by using mobile phones and drones. The images taken by drones

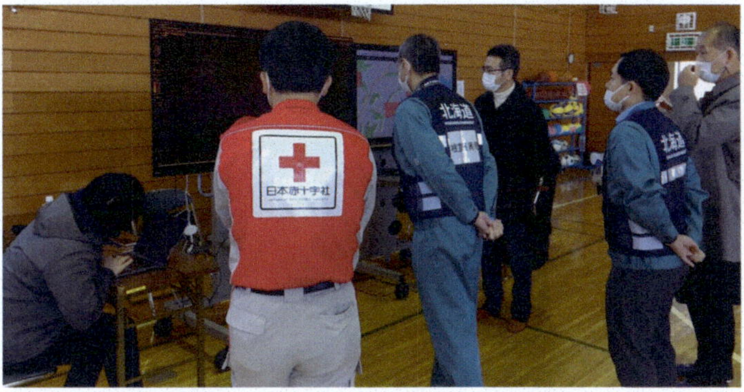

Fig. 2 Real-time tsunami monitoring experiment in the tsunami evacuation drill

and smartphones are uploaded on a cloud-based map in a real time and shared with rescue workers and the emergency response headquarters. The system was developed by a venture company with a Non-Profit Organization called "Emergency medical and Disaster coping Automated drones support system utilization promotion Council (hereafter EDAC)."

In 2018, EDAC organized an emergency tsunami assessment experiment by using the system together with the fire department of Saiki City of Oita prefecture, Japan. In the assessment experiment, the images taken by drones (Fig. 3) and smartphones of the fire department were shared with the staff at the emergency response headquarters in a real time (Fig. 4). In this way, mobile phones are effectively used to prepare for emergency response in future tsunami disasters.

Fig. 3 Drone operation by a tsunami assessment team

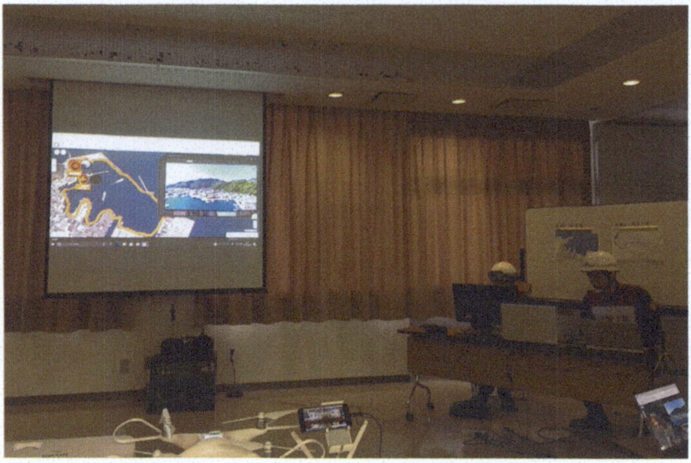

Fig. 4 The images taken by a drone and smartphones are shared at the headquarters

5 Personal Floatation Devices to Support Mobile Phone Applications in Emergencies

According to the Iwate Fisheries Technology Center (2011), when the GEJET occurred, the seawater temperature was very low (five to seven degrees Celsius). If the entire human body is immersed in seawater at such low temperatures, its core temperature will decrease to a critical level within two hours (Wilkerson et al. 1986). Even if the tsunami exposed people with lifejackets can float on the surface of the water and avoid drowning, they are at imminent risk of hypothermia. In order to save their lives, it is necessary to acquire information of their location immediately through their mobile phones and rescue them quickly before their core temperature drops to a severe hypothermia level. In this sense, it is an important precondition in the utilization of mobile phones to detect the locations of tsunami drifted people that mobile phones are not kept under the surface of the water. Wada (2006) successfully shared the location of the drifted person in the ocean by a mobile phone attached on a lifejacket. However, this study aimed to detect the location of the person who fell into the sea by accidents. In addition, the past studies on the effectiveness of lifejackets in recreational swimming (Leavy et al. 2015) and fishing (Jasper et al. 2017), (Barwood et al. 2014) while little study discussed the effectiveness of the lifejackets in tsunami.

While little study discussed the effectiveness of the lifejackets, or, even the causes of drownings in tsunami were not studied, Kurisu et al. (2018) conducted a series of experiments to analyze the cause of drowning during a tsunami at the Large Hydro Geo Flume in the Port and Airport Research Institute, Kanagawa Prefecture, Japan from 27 to 30 June in 2016.

In every experiment, the dummy's specific gravity was adjusted at one point zero five which is almost equal to a human's specific gravity and its weight in the air was maintained at about 48 kg.

In the Large Hydro Geo Flume, dummies with the lifejackets and the dummies without the lifejackets were exposed to the imitated tsunami waves. The heads of all the dummies not wearing lifejackets were entrapped in a vortex and drawn underwater immediately (Fig. 5). In contrast, the heads of all the dummies wearing standard lifejackets remained on the surface and were not dragged underwater (Fig. 6). These findings were obtained through video images, which were taken from the side window of a large flume. When a tsunami's extremely powerful vortex swallows' people under the water, and the rate of their breathing increase dramatically, people have no other choice but to inhale water. Even skilled swimmers without lifejackets would not be able to resurface and to remain afloat. This will greatly increase the risk of drowning. This severe whirling of tsunami waves is likely the main factors that cause the overwhelming majority of tsunami victims to drown. In this experiments that employed widely popular standard lifejackets which assures 15.5 lbs. (7.031 kg), the dummies with lifejackets were not dragged underwater. They remained afloat and the heads were higher than the water level. Thus, it can be concluded that when people are engulfed in the tsunami waves, lifejackets will provide them with a higher chance of survival, because they will remain on the surface of tsunami wave. In other words, a lifejacket is a critical piece of equipment for surviving tsunamis. Furthermore, in the experiments, the dummy with a buoyancy of 4 kg kept floating on the surface of the water and was not dragged underwater (Fig. 7). Therefore, in critical situations, when a tsunami wave is already approaching and there are no Personal Floatation Devices (hereafter PFDs) around, it is advisable to place empty plastic bottles between the skin and clothing, hold on to garbage cans, or wear helmets as substitutions for lifejackets and other protective gear. These actions represent the second-best tsunami survival technique.

In this way, to share the location data of the tsunami exposed people from their mobile phones, it is imperative for the people to wear the proper lifejackets to prevent the people from drowning by the vortex caused by tsunami. Even simple PFDs can support or become an infrastructure for the application of advanced technologies and system to prevent the tsunami exposed people from hypothermia.

6 Discussion

This chapter discussed the application of familiar tools such as mobile phones and PFDs to reduce human loss caused by tsunami exposure. In the past tsunami disasters in Japan such as the GEJET, a lot of inhabitants and first responders faced the challenges in communication by using mobile phones in the disaster affected areas. On the other hand, the review on the past tsunami exposure cases found that the uses of lifejackets or floating objects were effective while the decrease in body temperature caused by the prolonged water exposure was identified as a challenge. In addition, a

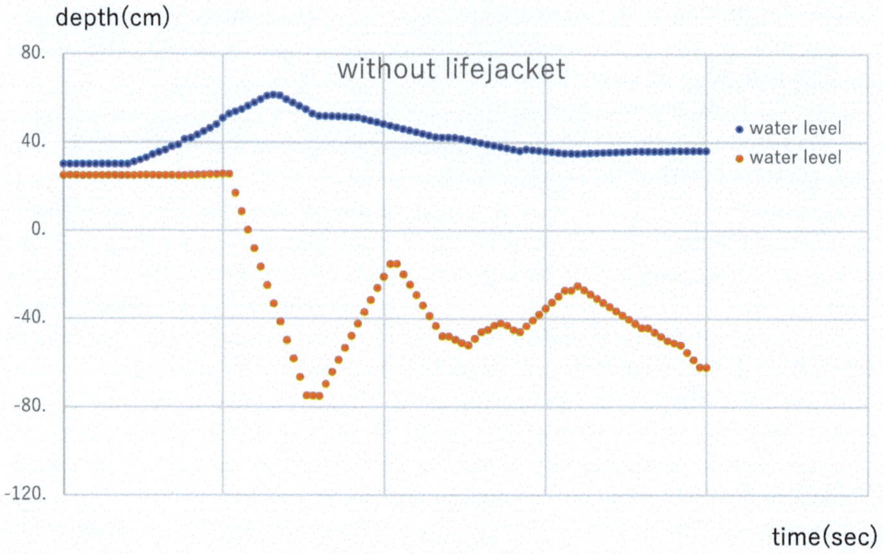

Fig. 5 The movement of the dummies without lifejackets exposed in the imitated tsunami

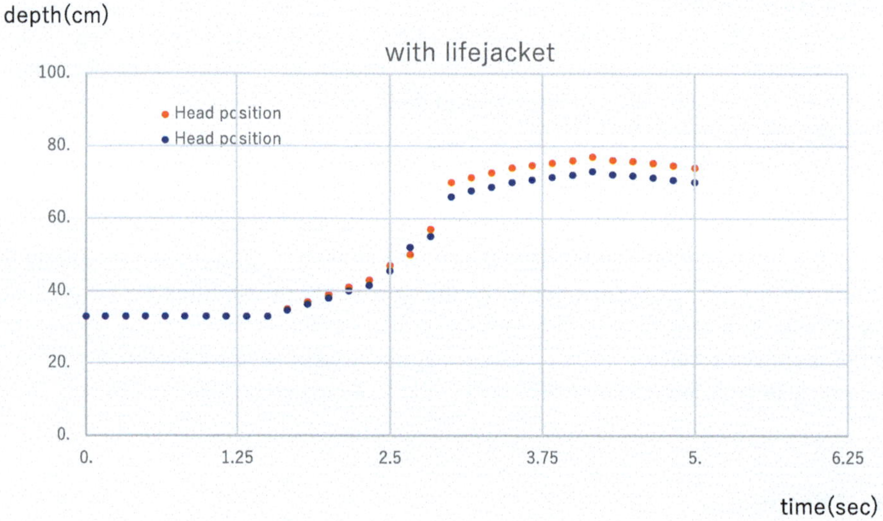

Fig. 6 The movement of the dummies with lifejackets exposed in the imitated tsunami

simulated tsunami demonstration using a dummy wearing a lifejacket pointed out the importance of lifejackets, not only in preventing users from drowning by ensuring buoyancy, but also in sharing location information with mobile phones. In this way, in order to prevent human casualties due to tsunamis, these two tools, the mobile phones and the PFDs, need to be used correctly by their users immediately after a

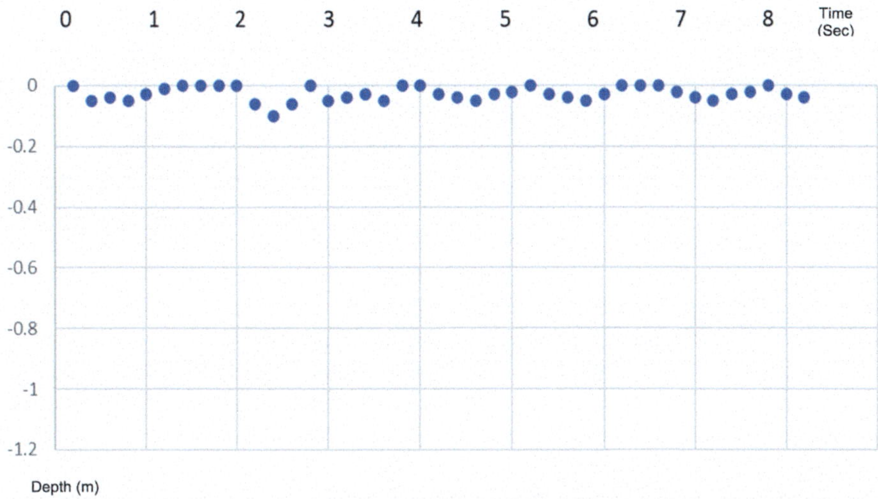

Fig. 7 The movement of the dummies with the four kg buoyancy exposed to the imitated tsunami

disaster occurs. Furthermore, if a person is actually exposed to a tsunami, it will be extremely difficult to move to land on their own.

To reduce human loss in tsunami, the public search and rescue services are crucial. Therefore, here, (1) improvements of the tools and (2) framework for utilizing these tools in emergencies are discussed.

Firstly, in terms of improving the PFD itself, it is necessary to prevent drowning and hypothermia by supporting buoyancy and body temperature maintenance. Particularly in the winter, in addition to buoyancy, it is necessary to have a function that prevents hypothermia and keeps the body dry. In this case, a lifejacket alone is not sufficient; rather, a set of waterproof cold-weather clothing that covers the entire body and a lifejacket, or waterproof cold-weather clothing may function as a PFD. Furthermore, research on the factors that inhibit the use of lifejackets in connection with past water accidents has shown that (1) lack of awareness of the wearability of lifejackets, (2) procurement and storage costs, and (3) discomfort when worn. These factors should be considered in designing PFDs by manufacturing companies. In particular, design and development of affordable, compact, and inflatable lifejackets are important so that lifejackets can be widely purchased and stored in the society. In terms of the framework for uses of PFDs in emergencies, the efforts by the people and the local authorities in tsunami hazards areas are important toward the widespread uses of PFDs before a tsunami occurs. In fact, in the coastal areas of Japan, local authorities support the stockpiling of lifejackets or provide financial support to the people in the community to purchase lifejackets. In addition, disaster prevention education to popularize lifejackets is also important. Lifejackets are the common solutions for tsunami exposure and other water-related accidents and disasters. For this reason, the link between the education on preventing water accidents and education on tsunami exposure should be strengthened.

Secondly, regarding the use of mobile phones, there are two issues in daily use: (1) issues with mobile terminals and (2) utilization of location information data of mobile terminals. At first, there is an issue of design of smartphones. The design of smartphones should be easy to use so that even the elderly can use them. In the demonstration of real-time tsunami monitoring exercise in Nemuro-City of Hokkaido, Japan, it was difficult to encourage the participants to install the mobile apps that helps to share the location of the users since most of the participants were elderly. In addition to responding to the needs of the elderly, manufacturers of mobile phones are required to make efforts to ensure that the devices themselves are waterproof and maintain battery capacity under low temperatures. Then, it is important to use mobile phone location information everywhere in society, especially in local governments, which are bases for disaster response. The location data of smartphone holders can be used in various ways by local authorities, such as tour routes for tourists, analysis of medical care at medical facilities, health behavior of residents including the elderly, formulation of evacuation routes in disaster drills, and traffic analysis for alleviating traffic congestion. In this way, by using mobile technology in a sustainable manner as a solution to everyday social problems, it becomes possible to quickly utilize mobile technology in the event of a disaster. As Sakurai and Watson (2015) pointed out the challenges in the uses of high cost of pre-paring disaster proof systems and proposed the application of daily use of the systems, the people flow data should be daily used in local authorities for various purposes in a sustainable way. The challenge in the application of mobile phone data in emergencies is the connectivity of the mobile phones. Aizu (2011) pointed out that in the GEJET, the tsunami washed away most of the physical infrastructure such as roads, railways, and telephone key stations which resulted in creating the "information blackholes" for a month or even longer. Furthermore, to effectively utilize the mobile phone data in emergencies, the user of the mobile phone location data, such as the staff of local authorities and relief organizations should have practical knowledge based on the past disaster response using the peoples flow data, in other words, how to read the data in emergencies. For example, small number of people's gathering in a refugee center can be read in various ways. This can mean the damage in the disaster affected areas is not severe so a little number of people evacuated to the refugee center. On the other hand, this can also mean that a little people reached to the refugee center due to the limited access to the refugee center. Or, limited number of mobile phone data is available due to the damage of the key mobile stations. In this way, the people flow data in emergency situation can be very tricky. Toward the better utilization in emergencies, the continuous trials and reviews are imperative.

The limitation of this study is lack of the exercises by using dummies wearing PFDs with mobile phones in the real ocean. The field experiment should be conducted in the future to monitor the location data and the body temperature of the dummies. In addition, the search and rescue drills based on the mobile phone location data should be done for the smooth relief operation in emergencies.

This discussion is concluded by pointing to the direction in which the familiar tools such as PFDs and mobile phones, can be utilized during disasters.

The lead time from the occurrence of an earthquake to the arrival of a tsunami is limited. During this time, in order for PFDs and mobile phone location information to be utilized quickly, these tools need to be daily used in communities. Therefore, an approach is required to utilize the common solutions such as PFDs and Mobile Phone's location Data to solve the social challenges in peacetime. For example, Promoting the spread of PFDs during normal times will ultimately help prepare for tsunami disasters. Likewise, during normal times, mobile phone location information contributes to city development by local governments, such as understanding traffic congestion, analyzing the dynamics of tourists, and the usage status of transportation facilities. In this way, daily use of people flow data in public institutions, which serve as bases for disaster response, will help ensure seamless use of people flow data in the event of a sudden disaster. In this way, it is important to take a stance to increase the resilience of society as a whole by using technologies that are useful during tsunami exposure to solve social problems in peacetime.

7 Conclusion

This chapter discussed the application of affordable and familiar technologies such as lifejackets and mobile phones for preventing human loss in future tsunami disasters. By reviewing the past tsunami disasters in Japan, the several cases were found that lifejackets and other floating items helped the survival of tsunami exposed people while the risks of hypothermia were reported. Application of mobile phone's location data is considered as a possible solution to identify the location of the tsunami drifted people. However, in the past tsunami disasters, a lot of communication challenges of mobile phones were reported while the various efforts have been done to identify the location of the drifted people in the ocean by using the telecommunication devices and systems. In addition, the imitate tsunami experiment by using a dummy with a lifejacket identified the importance of lifejackets not to save lives of the tsunami drifted people and to keep the mobile phones above the water to share the location data with relief agencies. In this way, these two common tools presented their possibility to save lives of tsunami drifted people. However, when a sudden onset disaster, earthquake, occurs, the lead time from the occurrence of an earthquake and the arrival of tsunami is limited. For the immediate uses of these common solutions in emergencies, the daily uses of the same solutions in peacetime are crucial. Therefore, this chapter proposed the daily widespread uses of the common solutions, which are effective in saving lives in emergencies, to solve social challenges of communities in peacetime.

References

Apte A (2010) Humanitarian logistics: a new field of research and action, foundation and trends®. In Technol, Inf Oper Manag, 3(1):1–100, ISBN:978–60198–336–7

Aizu I (2011), The role of ICT during the disaster—A story of how Internet and other information and communication services could or could not help relief operations at the Great East Japan Earthquake, available at: http://www.ispp.jp/ispp-wp/wp-content/uploads/2011/09/EarthquakeICT0825.pdf (Accessed) 2 May 2023

Akita Anglers Association (1983), The Nihonka-chubu earthquake as witnessed by anglers Available at: https://tsunami-dl.jp/document/108 (Available only in Japanese) (Accessed 9 May 2023)

Barwood MJ, Long GM, Lunt H, Tipton MJ (2014) Inherent work suit buoyancy distribution: Effects on lifejacket self-righting performance. Aviat. Space Environ. Med.; 85(9):960–964. pmid:25197896

BBC, Mobile phone inventor made first call 50 years ago (2023): Available at: https://www.bbc.com/news/technology-65112048, (Accessed) 25 April 2023

Bengtsson L, Lu X, Thorson A, Garfield R, Screeb J (2011) Improved response to disas ters and outbreaks by tracking population movements with mobile phone network data: a post-earthquake geospatial study in Haiti, PLoS Med, 8(8). [PMCID: PMC3168873] [PubMed: 21918643]

Calabrese F, Massimo C, Piero L, Dario., Carlo R. (2011) Real-Time urban monitoring using cell phones: a case study in rome". IEEE Trans Intell Transp Syst 12(1):141–151

Deville P, Linard C, Martin S, Gilbert M, Stevens RF, Gaughan EA, Blondel DV, Tatem J, A, (2014) Dynamic population mapping using mobile phone data. Appl Phys Sci 111(45):15888–15893. https://doi.org/10.1073/pnas.1408439111

Dynes RR (2003) The lisbon earthquake in 1755: the first modern disaster, Preliminary Paper #333, University of delaware disaster research center, Available at: https://udspace.udel.edu/server/api/core/bitstreams/2a85deff-b638-4d0b-867a-e37176f5e61d/content, (Accessed on 7 May 2023)

Funakoshi K, Hatayama M (2017) A study on the method to identify shelters and its support measures, IPSJ SIG Technical Report, 2017-IS-139(3): 1–8

Furukawa H, Kudo D, Nakagawa A, Matsumura T, Abe Y, Konishi R, Kushimoto S (2014) Hypothermia in victims of the Great East Japan Earthquake: a survey in Miyagi prefecture. Disaster Med Public Health Prep 8(5):379–389

GSMA (2022) The State of Mobile Internet Connectivity 2022: Available at: https://www.gsma.com/r/wp-content/uploads/2022/12/The-State-of-Mobile-Internet-Connectivity-Report-2022.pdf?utm_source=website&utm_medium=download-button&utm_campaign=somic22, (Accessed)25 April 2023

IFRC (2013) Emergency appeal final report, asia: Earthquake and Tsunamis available at:https://reliefweb.int/report/indonesia/asia-earthquake-and-tsunamis-emergency-appeal-n-m04ea028-final-report (Accessed 7 May 2023)

Inose N (2015) 救出: 3・11気仙沼 公民館に取り残された446人 (Rescue: 3/11 Kesennuma 446 people stranded at community center). ISBN: 978-4309023588

Iwate Fisheries Technology Center (2011) "Daily updated data of sea water temperatures for Iwate prefecture," Iwate Prefectural Office (in Japanese), available at: http://www.suigi.pref.iwate.jp/teichi/daily/report/2011/03/10. (Accessed on 7 May 2023)

Japan Cabinet Office (2007) White Paper on Traffic Safety in Japan Available at: https://www8.cao.go.jp/koutu/taisaku/h19kou_haku/pdffiles/gt_6.pdf (Available only in Japanese)

Japan Cabinet Office (2019) The MSK life jacket supports the search and rescue by using the location information that is obtained by SLAS assessment Available at: https://qzss.go.jp/info/archive/msk_190819.html (Available only in Japanese) (Accessed 7 May 2023)

Japan Cabinet Office (2021), Disaster management in Japan, Available at: https://www.bousai.go.jp/1info/pdf/saigaipamphlet_je.pdf (Accessed 7 May 2023)

Japan Craft Inspection Organization (2022), Report of a basic survey on the feasibility of personal life-saving equipment using information and communication technology: available at: https://jci.go.jp/jci/pdf/chousa/r3_zyouhou_tsuushin.pdf (Available only in Japanese)

Jasper R, Stewart BA., Knight A (2017) Behaviours and attitudes of recreational fishers toward safety at a 'blackspot' for fishing fatalities in western Australia. Health Promot. J Austr, 28(2):156–159. pmid:28135568

Kudo D, Furukawa H, Nakagawa A, Abe Y, Washio T, Arafune T, Sato D, Yamanouchi S, Ochi S, Tominaga T, Kushimoto S (2014) (2014) Reliability of telecommunications systems following a major disaster: survey of secondary and tertiary emergency institutions in Miyagi Prefecture during the acute phase of the 2011 Great East Japan Earthquake. Prehosp Disaster Med 29(2):204–208

Kurisu A, Suga H, Prochazka Z, Suzuki K, Oguri K, Inoue T (2018) Potential technique for im proving the survival of victims of tsunamis. PLOS ONE. 13(5): e0197498. pmid:29791490

Kuroda N, Kajitani Y, Tatano H (2022) Estimation of resilience factor on industrial sector under transport network deteriorations, J. JSNDS 41 (Special Edition):173–185

Kyocera Communication Systems Co., Ltd (2020) Press Release, 琵琶湖での水難事故の救助迅速化に有効性を確 認 (Confirmed effectiveness in speeding up rescue of water accidents in Lake Biwa). https://www.kccs.co.jp/news/release/2020/0123/. Accessed 1 Aug 2024

Leavy JE, Crawford G, Portsmouth L, Jancey J, Leaversuch F, Nimmo L, Hunt K (2015) Rec reational drowning prevention interventions for adults, 1990–2012: A review. J. Community Health; 40(4):725–735. pmid:25618578

Miyako City Central Community Center (2012) 3.11 大津波体験語り継ぎエピソード集 (3.11 A collection of episodes passed down from the great tsunami experience). https://miyakoarchive.irides.tohoku.ac.jp/nenpyo/jishin_tsunami/higashinihon/. Accessed 1 Aug 2024

NHK (2022) WEB Special, Aegis Ship "Miraculous Rescue" Secret Story ~ Japan Maritime Self-Defense Force 11th Year Testimony. Available at https://www3.nhk.or.jp/news/html/20220304/k10013511811000.html, (Available only in Japanese), (Accessed on 7 May 2023)

Peary MDB, Shaw R, Takeuchi Y (2012) Utilization of Social Media in the East Japan Earth quake and Tsunami and its Effectiveness. J Nat Dis Sci 34(1):3–18

Ratti C, Frenchman D, Pulselli MR, Williams S (2006) Mobile Landscapes: using location data from cell phones for urban analysis. Environ Plann B Plann Des 33:727–748. https://doi.org/10.1068/b32047

Red Social Innovation (2023), How to provide mobile real time monitoring on tsunami evacuation, available at: https://red-social-innovation.com/en/solution/how-to-provide-mobile-real-time-monitoring-on-tsunami-evacuation/, (Accessed on 1 May 2023)

Saito H (2020) As a wife and mother, I was swept away by the tsunami on 3.11. The day I failed and survived, yahoo news, available at: https://news.yahoo.co.jp/byline/saitohidetoshi/20200310-00166753, (Available only in Japanese), (Accessed on 7 May 2023)

Sakurai M, Watson TR, (2015) Securing communication channels in severe disaster Situations-Lessons from a Japanese earthquake, practitioner cases and Practitioner-Centered Research, Proceedings of the ISCRAM 2015 Conference:288–299

Seto S, Imamura F (2020) Classification of tsunami deaths by modifying ICD-10 categories in the 2011 Tohoku earthquake tsunami—A case study in Miyagi prefecture. Int J Disaster Risk Reduct 2020:171043

Shibata Y, Uchida N, Shiratori N (2014), Analysis of and proposal for a disaster information network from experience of the great east Japan Earthquake, IEEE communications magazine

Soshino S, Oohata T, Inada Y (2020) Sustainable and resilient community development by drones, Mobile Technologies for Delivering Healthcare in Remote, Rural or Developing Regions, edited by Ray KP, Nakashima N, Ahmed A, Soshino Y, chapter 15, ISBN: 9781839530470

Stiros CS (2001) The AD 365 Crete earthquake and possible seismic clustering during the fourth to sixth centuries AD in the Eastern Mediterranean: a review of historical and archaeological data, J Struct Geol, 23 (2–3): 545–562, ISSN 0191–8141

Udagawa S (2021) Estimating the number of evacuees during the alert period using People-Counting Data -Case Study of the 2020 Typhoon No.10-, Saigai Joho (Disaster Information) 19(2):133–143

UN OCHA (2021) From digital promise to frontline practice: new and emerging technologies in humanitarian action: Available at: https://www.unocha.org/sites/unocha/files/OCHA%20Technology%20Report.pdf, (Accessed) 25 April 2023

Ushiyama M, Kanata M, Imamura F (2004) A Study on Mitigation in Human Damage by disaster information for Tsunami Disaster, J JSNDS 23(3):433–442

Wada M (2006) Development and evaluation of rescue systems for Man-Overboard, IPSJ SIG Technical Report

Wilkerson JA, Bangs CC, Hayward JS, (1986) Hypothermia frostbite and other cold injuries, first edition. Seattle, Washington, U.S.A.: The Mountaineers

Wilson R, Zu Erbach-Schoenberg E, Albert M, Power D, Tudge S, Gonzalez M, Guthrie S, Chamberlain H, Brooks C, Hughes C, Pitonakova L, Buckee C, Lu X, Wetter E, Tatem A, Bengtsson L (2016) Rapid and near real-time assessments of population displacement using mobile phone data following disasters: The 2015 Nepal Earthquake, LoS Curr. 2016 Feb 24;8:ecurrents.dis.d073fbece328e4c39087bc086d694b5c. https://doi.org/10.1371/currents.dis.d073fbece328e4c39087bc086d694b5c. PMID: 26981327; PMCID: PMC4779046

Emerging Technologies to Avoid Preventable Death in Emergencies: A Case Study on Hypothermia Prevention in a Refugee Center in Winter

Masahiro Nemoto and Yasuhiro Soshino

1 Introduction

The 2011 Great East Japan Earthquake and Tsunami (hereafter GEJET) caused catastrophic damage in the wide geographic areas in the northern coastal areas of Japan (Mimura et al. 2011). In addition to the historical size and damage of the earthquake and tsunami, GEJET showed a face of the tsunami disaster in winter. After the occurrence of the GEJET, in Japan, there is a growing concern about the "National Catastrophe" such as Tokyo Inland Earthquake and Nankai Trough Earthquake and Tsunami (Japan Cabinet Office 2021). To prepare for these catastrophic disasters, a lot of efforts are made by the government, local authorities, Non-Profit Organizations, the private sector, and local inhabitants. In addition to these future catastrophes, there is another concern about the occurrence of massive earthquake and tsunami disasters along the Japan Trench and Chishima Trench, which are expected to strike northern coastal areas on the Pacific Ocean side of Japan. The challenges of this disaster are: (1) the catastrophic damage in the wide geographical areas from the coastal side of Hokkaido and Tohoku region, (2) the long power outage, and (3) the typical winter health risks such as hypothermia and carbon monoxide poisoning. To prepare for this earthquake and tsunami, another challenge is the lack of historical documents on the earthquake and tsunami before the eighteenth century in Hokkaido, where only the oral traditions of the Ainu people are recorded (Takashimizu 2005) (Niizato

M. Nemoto (✉)
Japanese Red Cross Hokkaido College of Nursing, Kitami, Japan
e-mail: nemoto@rchokkaido-cn.ac.jp

M. Nemoto · Y. Soshino
Disaster Management Research Institute, Japanese Red Cross College of Nursing, Shibuya, Japan
e-mail: y-soshino@jrcdmri.redcross.ac.jp

Y. Soshino
Japanese Red Cross Kumamoto Hospital, Kumamoto, Japan

P. K. Ray et al. (eds.), *Technology Innovation for Sustainable Development of Healthcare and Disaster Management*, Disaster Risk Reduction,
https://doi.org/10.1007/978-981-97-2049-1_6

et al. 2006). In different parts of the world, tsunami disasters were reported in the cold regions. In 1958, an earthquake with a magnitude of seven point eight caused the mega-landslide into a narrow harbor that resulted in the 524-m-high tsunami. Recently, in Greenland, the landslides caused the 100-m-high tsunami, which struck a village of Nuugaatsaq. This mega-tsunami washed eleven houses and killed four people (Schiermeier 2017). On March 4, 1952, a magnitude 8.2 earthquake occurred off the coast of Tokachi, Hokkaido. This earthquake caused a drift ice tsunami that caused severe damage in Hamanaka town, Hokkaido (Kioka et al. 2008). Following this earthquake, 4th of November 1952, the Kamchatka earthquake and tsunami caused serious damage to the Kamchatka Peninsula and the Kuril Islands. The exact number of death is not clear but estimated more than one hundred (Gusiakov 2000).

Historically, tsunami disasters in cold regions are not frequently reported in the world. In addition to the natural hazards such as earthquakes and tsunamis, the low temperature of the weather condition and seawater increase the health risks of the tsunami-affected people. In the 2011 GEJET, a lot of cases of hypothermia were reported in various places such as the tsunami-struck areas, the houses of the disaster-affected people, and the refugee centers. In the 2011 GEJET, hypothermia patients were identified in coastal areas, which means hypothermia risk increased due to exposure to tsunami water and a cold living condition at refugee centers (Kamata et al. 2022).

Hypothermia is generally defined as an unintentional drop of core body temperature below 35 °C. Hypothermia can be caused by exposure to a cold environment. In addition to impaired thermoregulation and other factors, decreased heat production and increased heat loss are the main factors that increase susceptibility to hypothermia (Wong et al. 2004). From the human body, heat is transferred via conduction, convection, and radiation. Heat is also lost via evaporative mechanism (Romanovsky 2018). Preventing or delaying further heat loss, and rewarming as soon as possible are crucial in the treatment of hypothermia out of hospital. Mild hypothermia with uninjured patients can be managed on site without transport to the hospital. Passive rewarming by moving the patients from the cold environment, providing warm food and drinks, and promoting movements usually sufficient (Paal et al. 2022). In this way, to prevent hypothermia aftermath of tsunami disasters, it is important to minimize the risk factors according to the basic mechanism of human heat loss for example, changing the dry clothes of the people directly exposed to tsunami, warming up the air in shelters by using jet heater, preventing the negative effect by the wind in shelters, and installing the beds instead of lying on the cold floor in shelters. While a lot of studies pointed out that pre-hospital treatment for mild hypothermia patients can be managed on site, most of the studies on hypothermia aftermath of tsunami disasters focus on the treatment in the hospital, or the location where the hypothermia cases were found. On the other hand, there is little study on how to set up the living conditions in shelters to prevent hypothermia aftermath of tsunami disasters in winter. While Paal et al. (2022) mention that hypothermia has been a disease of disasters such as avalanches, earthquakes, and tsunamis, very limited studies such as Oshiro et al. (2022) discussed hypothermia in natural disasters. To reduce the burden of the local hospital in the tsunami-affected areas, it is crucial to prevent hypothermia cases as many as possible. Prevention in the prehospital setting reduces resource utilization and improves mortality (Arthurs et al. 2006). In terms of living conditions at refugee centers, Soshino et al. (2017) discussed the application of international medical relief equipment such as tents, generators, water and shower equipment to improve the living conditions of refugee centers in Japan, while their study did not cover the prevention of hypothermia.

Therefore, this study aims to examine the practical solutions to prevent hypothermia in shelters in the aftermath of tsunamis in winter. In the Hokkaido region, where only trains link to the mainland of Japan, it is necessary to be self-sustainable till the external relief teams arrive and establish shelters by sharing familiar tools. This study aims to examine the effectiveness and identify the challenges to alleviate hypothermia by conducting shelter management drills in winter.

The structure of this study is as follows. Following the introduction, Sect. 2 explains the background and methods of this study. The results of the practical shelter management drills are shown in Sect. 3. Section 4 describes the effects and the challenges of the examined measures in this study. The Discussion is described in Sects. 5 and 6 concludes this study.

2 Background and Methods

2.1 Background

Since 2010, the author (Nemoto) has been organizing the refugee center management exercises at the Japanese Red Cross Hokkaido College of Nursing (hereafter "the JRCH College"). At the end of September 2010, the first exercise assuming the disruption of the heating system due to the power outage in winter commenced in Kitami City, where the JRCH college is located and the temperature in winter decreased to minus 20 ºC. It was 5 ºC outside and 17 ºC in a gymnasium where approximately 20 nursing students stayed on the plastic sheets with a blanket per person, dried biscuits, some salts, and water at the Japanese standard refugee center. Few students were able to sleep and take biscuits and water. The temperature of the floor, the noises of the plastic sheets that occurred when the participants rolled over, and the air stream in the gymnasium prevented the participants from sleeping well. Since this first exercise, it took the JRCH college three years to organize the exercise in winter while the college experienced the challenges in heating the whole space in the gymnasium with portable stoves that local authorities officially introduced as relief items from 2011 to 2013. During this period, the JRCH college also learned to shut the uncomfortable air stream in the gymnasium by installing tents. Finally, in 2014, the exercise was held in January and the outside temperature was minus 17 ºC. The various items were tested to create comfortable living conditions by covering the floor with mats, installing plastic sheets to separate the spaces, heating by three jet-heaters that resulted in the temperature on the floor kept at approximately 12 ºC.

However, approximately 60% of the participants were not able to sleep enough due to the cold floor. In addition, carbon dioxide concentration reached approximately 10,000 ppm, which is largely beyond the national standards (below 1,000 ppm). Furthermore, this exercise also identified the risks of getting sweat in the sleeping bags and unbearable noises caused by the utilization of the emergency aluminum blankets with no moisture permeability. In 2016, the exercise made a good result by installing cardboard beds, an instant partition that separates the large space for 70 people in the gymnasium within 30 min. The 2016 exercise also identified the risks of the temporary toilets installed outside in severe winter. In this way, since 2010, the JRCH college has accumulated knowledge and experiences in shelter management in winter. For the proper management of refugee centers in winter, ensuring a sleeping environment, proper nutrition, and sanitation are required. Furthermore, to prepare for future tsunami disasters, prevention of hypothermia in refugee centers is imperative.

2.2 *Overview of the Field Test*

To examine the effectiveness of solutions to prevent hypothermia, this study reviews past exercises organized by the JRCH college from 2010 to 2023. The latest exercise was conducted from 21st and 22nd of January 2023. In the latest exercise, approximately one hundred participants joined. Table 1 presents the overview of this study. According to the basic mechanism of heat loss (Evaporation, Conduction, Convection, and Radiation) and rewarming to prevent hypothermia, each countermeasure was tested.

This study divides the factors in hypothermia prevention into two large groups; Heat Loss and Heating. Heat loss is caused by evaporation, conduction, convection, and radiation. In this study, a simple tented sauna system, a sauna car, and a kitchen bus (hereafter "Kitchen Cruiser" were mobilized as the solutions to evaporation. The Kitchen Cruiser provided hot soup to prevent dehydration and a hot meal to

Table 1 The Relation between the solutions and factors in hypothermia prevention

	Heat loss				Heating
	evaporation	Conduction	Convection	Radiation	
Solutions	(3.1) Sauna (Moderate Heating with Humidity) Kitchen (Hot Soup)	(3.2) Cardboard Beds	(3.3) Shelter System	(3.4) Heating System Sleeping Bags	(3.6) Electric Blankets Kitchen Cruiser Footbath Hot water bottle
		Uses of portable toilets (3.5)			

heat the participants. Cardboard beds were used against conduction, a shelter system was established to reduce the effect of convection, heating system and sleeping bags prevented the participants from heat loss by radiation. To help the participants to warm up themselves, electric blankets, footbaths, and hot water bottles were used. In addition, the Kitchen Cruiser provided the participants with hot meals. To reduce the heat loss in excretion behaviors, portable toilets were used inside the gymnasium.

3 Results

3.1 Solutions to Evaporation in Heat Loss

3.1.1 Sauna

The main purposes of bathing are to ensure proper sanitation, warmth, and relaxation. However, in emergencies, due to the damage to the water and energy supply system, ensuring bathing facility becomes a challenge. The international disaster response teams of the Finnish Red Cross equip the sauna system to establish the bathing facility even in the disaster-affected areas. Compared to the shower system that needs plenty of hot water, after sauna, relatively limited water is needed which is suitable for emergency uses. In this way, the sauna is the efficient solution for ensuring proper sanitation with limited energy and water. In addition to the sanitation purpose, bathing is done to warm the human body. However, in severe winter conditions, sweating from bathing easily deprives human body temperature. Hence, it would be ideal to establish a place for warming without sweating in the refugee center in severe winter. Therefore, in this study, the simple sauna system with a tent and a stove was tested under the outside temperature was minus 20 °C.

This simple sauna system consists of a tent and a stove inside the tent (Figs. 1, 2). The stove with sauna stones is heated by firewood.

Fig. 1 A simple Sauna room

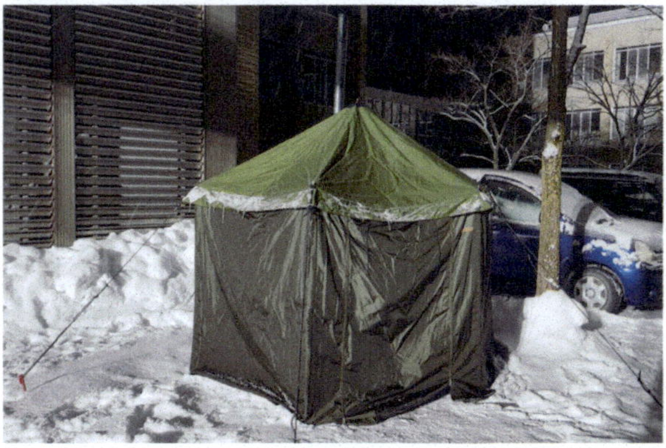

Fig. 2 Putting the water on sauna stones(löyly)

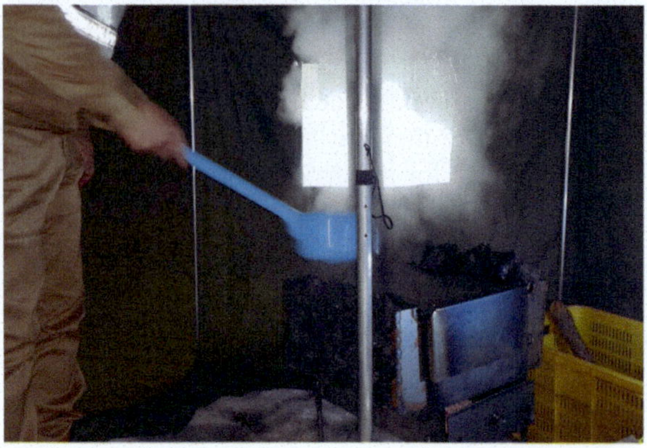

In the beginning, the temperature inside the tent was approximately 0 degree Celsius, however, after putting the water on the sauna stones (Fig. 3), which is called "löyly" in Finland, the temperature increased to 35 °C, and the humidity increased to one hundred percent. This is not suitable as a normal sauna; however, this environment is suitable for warming the body with winter clothes on. This simple sauna system is suitable for warming the refugees without sweating and taking off their clothes.

Fig. 3 Sleeping on the plastic sheets without heating

3.2 Solutions for Conduction in Heat Loss

3.2.1 Cardboard Beds

To prevent hypothermia cases in refugee centers, the flooring material is important by reducing heat conduction. However, for a long period, thin plastic sheets, called "Blue Sheets" in Japanese, were used as flooring material in refugee centers in Japan (Fig. 3). Unfortunately, these thin plastic sheets have no means of reducing heat conduction to prevent hypothermia if the sheets were used as flooring material.

Therefore, this study tested cardboard beds instead of using plastic sheets for flooring material. Cardboard is a useful material that can be rapidly mass-produced and suitable for manufacturing various items. The height of the bed is 35 cm from the floor and the cardboard consists of five layers. In this way, the cardboard material shields the cold air from the floor of the refugee center, and the board of the cardboard bed retains the body temperature of the evacuees themselves, helping them maintain their body temperature while sleeping. Hence, these layers of cardboard beds reduce heat conduction to prevent hypothermia in refugee centers in winter. Mizutani and Nemoto (2017) examined temperature changes in the use of cardboard beds and identified the effectiveness of the beds in reducing heat conduction. In the drill, nearly one hundred cardboard beds were installed by the participants and the participants slept one night on the beds (Fig. 4). The feedback from the participants was positive.

Fig. 4 Sleeping on the cardboard beds

Fig. 5 Shelter system to separate the large space in a gymnasium

3.3 Solutions for Convection in Heat Loss

3.3.1 Application of Shelter System

In Japan, public facilities such as schools, parks, and stadiums are used as public refugee centers. Among them, school gymnasiums are common facilities in establishing refugee centers. However, gymnasiums are designed for playing basketball and volleyball, which means gymnasiums have one single large space. The height of gymnasiums, which is more than 11 m on average in Japan, are suitable for playing ballgames, however, this large space creates convection which is one of the main risks for hypothermia.

Therefore, in the drill, a shelter system which consists of large curtains with ropes and support pillars was used to limit ceiling space by covering the space with large curtains (Fig. 5). In addition, small tents were also set up to separate the large space. Furthermore, a partition system that was made of cardboard was also installed to alleviate the effect of convection in a gymnasium.

3.4 Solution to Radiation in Heat Loss

3.4.1 Heating System

To minimize the heat radiation from the human body, utilization of heating systems in refugee centers is crucial in severe winter environments. In Japan, most local authorities use portable fuel heaters in refugee centers. However, this type of heater increases the risk of carbon monoxide. In fact, in the Great East Japan Earthquake and Tsunami, in Yamagata prefecture where the power outage continued for approximately 24 h, 51 patients were referred to hospitals due to the disaster-associated carbon monoxide poisonings. Among them, 35% of the patients used electric generators, 16% of the patients used electric generators and oil stoves, and 45% of the patients used charcoal briquettes (Iseki et al. 2013). Therefore, this study examined a heat-exchange jet heater (Fig. 6). The electricity for operating the jet heater was supplied by a Plug-in Hybrid Vehicle (Fig. 7). Heat-exchange jet heaters are used as the thermal technology for emergency preparedness in Finland and Italy. The carbon dioxide produced by these heaters is vented through a chimney, which provides fresh outdoor air to interior spaces. When this heater was utilized in a large gymnasium, it delivered warm and clean air. Therefore, the concentration of carbon dioxide decreased. The heat-exchange jet heaters "ventilated" the refugee center, which was densely populated with evacuees and rapidly accruing carbon dioxide from their exhalations.

Fig. 6 Heat-exchange jet heater

Fig. 7 Power supply by a plug-in hybrid vehicle

3.4.2 Sleeping Bags with Permeability

To prevent and treat hypothermia patients, the golden rule is to take off their wet clothes. Wet clothes rapidly deprive the body temperature, which easily leads to hypothermia not only in emergencies, but in various scenes such as marathons, trekking, camping, and other outdoor activities. In the 2011 Great East Japan Earthquake and Tsunami, among hypothermia patients received in 72 hospitals in Miyagi prefecture within 72 h of the earthquake, 72.5% of the hypothermia cases were caused by direct exposure to tsunami (Furukawa et al. 2014).

In addition to the direct tsunami exposure, items for sleeping such as aluminum-made emergency sheets have become another risk of hypothermia under severe winter conditions. In the past experiment (Nemoto and Oyama 2014), which was conducted at under minus ten degrees Celsius, organized by the Author (Nemoto), the aluminum-made emergency sheet was distributed to the participants who stayed in a vehicle with its engine off. In this test, the moisture from the human body froze. Therefore, to prevent heat loss by evaporation, utilization of permeable fabric is important in emergency response under severe cold conditions.

In this way, in the public shelters of the coastal areas, the pre-positioning of clothes and sleeping bags to prevent evaporation is one of the main risk factors for hypothermia.

Therefore, in this study, sleeping bags made of permeable fabric were examined during the emergency shelter management drills. Sleeping bags are used for treating hypothermia patients in rescue activities under severe cold conditions. In addition, sleeping bags were used to prevent heat conduction by sweating during sleep in refugee centers and vehicles under severe cold conditions. The sleeping bags, which are examined in the drills were made of Tyvek Silver made by Asahi-DuPont. Tyvek Silver is a fabric that retains heat and dissipates moisture coated with aluminum. In the drill, the author (Soshino) slept for six hours using this sleeping bag with winter clothes on in a gymnasium (Fig. 8) where the temperature of the room was approximately eight degrees Celsius to find no problem while sleeping.

Fig. 8 Sleeping bags with permeability

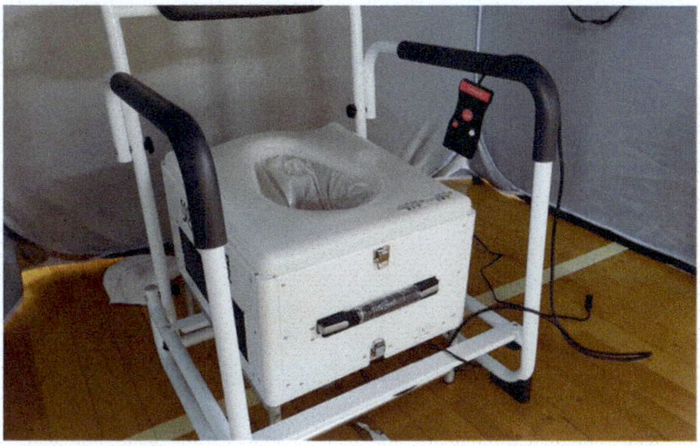

Fig. 9 An indoor wrapping toilet

3.5 *Sanitation and Hypothermia Prevention*

In Japan, toilets are risky places where a large part of hypothermia cases occur even in peacetime. In toilets, heat conduction to cold lavatory sheets and radiation to cold air can occur. In addition, in disaster situations, portable temporary toilets were often installed outside of the refugee centers. In this case, in severe winter, refugees walk to the outside toilets even though it snows, which means exposure to convection. In addition, such toilets have no heating function, which supports heat conduction and convection in toilets. Hence, as in peacetime, special attention should be paid to toilets to prevent hypothermia in emergencies.

Therefore, in the practical drill, wrapping toilets were installed inside of gymnasium (Fig. 9). The wrapping toilet uses heat to seal the upper and lower extremities of rolled plastic sheets without odor. In the practical drill, these toilets were installed in the usual toilet spaces. In addition to wrapping toilets, temporary toilets (Fig. 10) and trailer type toilets (Fig. 11) were installed. The trailer type toilets were very comfortable to use, however, the participants of the drill had to walk to the toilets under snow. In addition, the inside of the temporary toilets is frozen which has risks for hypothermia (Fig. 12). This experience emphasized the difficulty in balancing hypothermia prevention and providing a comfortable sanitation environment in refugee centers in winter.

3.6 *Solutions to Personal Heating*

For hypothermia prevention, heating is one of the key approaches. In the practical drill, we conducted various approaches to support personal thermogenesis and warming.

Fig. 10 Temporary toilets

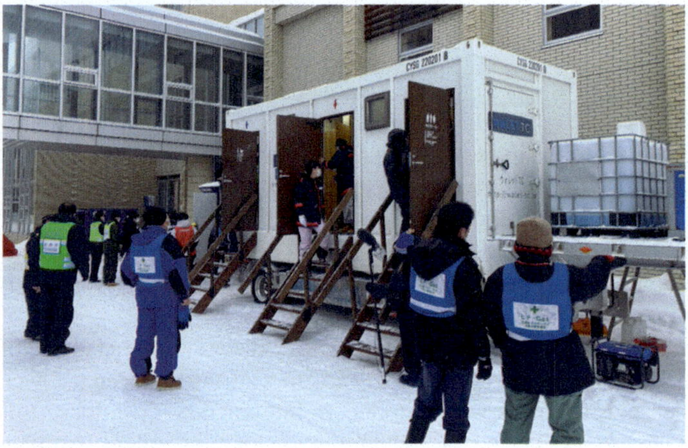

Fig. 11 Trailer type toilets

3.6.1 Electric Blankets

Even though cardboard beds reduce heat conduction, it is not sufficient to prevent hypothermia. Electric blankets were spread on cardboard beds (Fig. 13) to support the comfortable sleep of participants. Electric blankets were promising solutions in emergencies since each blanket consumes only 40–70 watts, this means that a single 2.2 kW emergency generator can provide heat for 30–40 people during an emergency. Instead of using a generator, nowadays, the application of vehicle with an external power supply function can be considered. All the participants who used the electric blankets confirmed their effectiveness in the refugee center under severe winter conditions.

Fig. 12 Inside of a temporary toilet

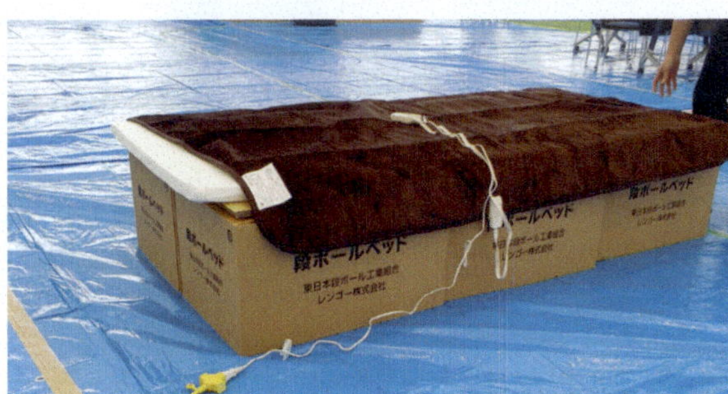

Fig. 13 Electric blanket on a cardboard bed

3.6.2 Food and Drinks for Emergencies to Prevent Hypothermia

To support thermogenesis and warming in refugee centers in winter, providing hot meals and drinks plays a significant role. In the practical drill, a large food truck called "Kitchen Cruiser" was deployed to provide hot meals to prevent hypothermia (Fig. 14). This vehicle is a bus, which is converted into a mobile kitchen to open a restaurant at any place in Hokkaido, Japan. The kitchen system consists of two gas burners, a fryer, a microwave, a soup range, and three sizable sinks (Fig. 15). In the drill, while the outside temperature decreased to minus 12 °C, the inside of

Fig. 14 The kitchen cruiser

Fig. 15 Inside of the kitchen cruiser

the kitchen remains more than 15 °C. By using this vehicle, four hundred hot meals including hot soup to prevent dehydration (Fig. 16) were served within four hours.

A lot of participants pointed out the effectiveness of the vehicle and the importance of providing hot meals in refugee centers. The utility of a Kitchen Cruiser lies in its capacity to serve large quantities of tasty and comforting food in a hygienic manner while keeping the food hot (or cold, depending on the season). Due to the diversity of disaster victims' circumstances, it is necessary to provide food that takes into consideration their illnesses. Professionals such as chefs, dietitians, caregivers, and social workers are essential to the efficient operation of these kitchens and are expected to play an active role as disaster support personnel.

3.6.3 Footbath

Footbath is popular in Japan for easy-to-use and heating effects in peacetime. In emergencies, footbaths can contribute to preventing hypothermia and promote blood

Fig. 16 Distribution of the hot meals and soup

Fig. 17 Making hot water by melting snow

flow in the whole body. In the practical drill, we tried footbath by using a gas stove for melting snow (Fig. 17) and large boxes made of cardboard with large plastic bags (Fig. 18). Each participant wore another plastic bag on foot to prevent themselves from getting wet. The amount of hot water required for one person was approximately three liters. 80% of the participants who took a footbath reported that the footbath was effective for heating themselves.

3.6.4 Hot Water Bottle

Hot water bottles are commonly used in Japan and other part of the world. Without electricity, the bottles can heat the users when sleeping. In the practical drill, hot

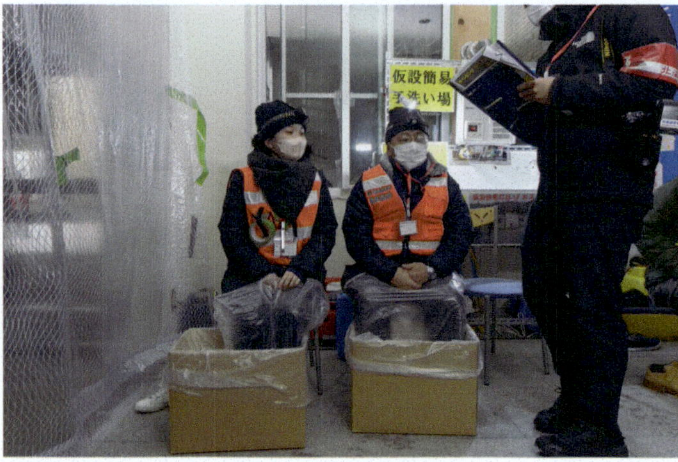

Fig. 18 Footbath made of cardboard boxes and plastic bags

water bottles (Fig. 19) were distributed to participants to support their sleep. Same as footbath, the hot water was made by using a gas stove to melt snow. The hot water bottles were used to warm the sleeping bags or clothes of participants to prevent low-temperature burns. Almost all participants confirmed the effectiveness of the use of hot water bottles in refugee centers under cold conditions. Since hot water bottles can be prepared without electricity, they are effective solutions even in power outages caused by disasters.

Fig. 19 Preparing for a hot water bottle

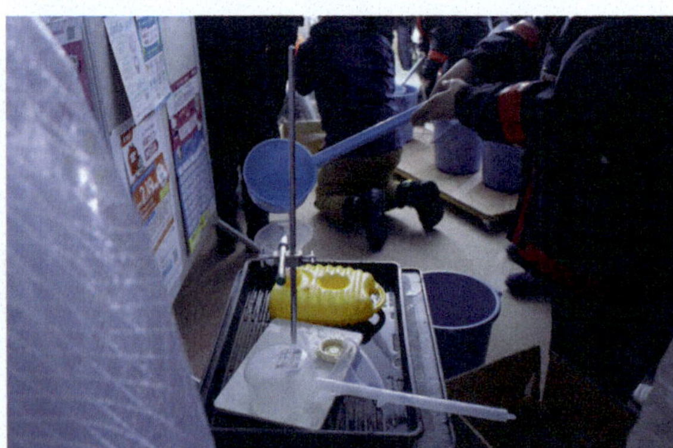

4 Effectiveness and Challenges of the Measures

Table 2 summarizes the effectiveness and challenges in all the tested solutions to the four main causes of hypothermia: evaporation, conduction, convection, and radiation in the refugee centers.

4.1 Solutions for Prevention of Heat Loss

First, to cope with evaporation, the sauna was effective by increasing the temperature and humidity around the body to prevent the transpiration of body temperature and moisture. On the other hand, the application of the sauna needs the appropriate temperature control to prevent sweating which causes heat loss. Second, the kitchen cruiser is expected to contribute to prevent dehydration by serving hot soup while the served meals also help to maintain the body temperature through the dietary thermogenesis. The challenges in the deployment of kitchen cruisers are the establishment of the deployment system and procedures in disasters in wide geographical areas. In addition, the kitchen cruisers are required to take measures against the freezing of doors and snow accumulation around the vehicle. Furthermore, the vehicle has difficulty in alleviating the noises and exhaust gas when operating engines and generators which is annoying for the people in the refugee centres in emergencies. Third, Cardboard beds were effective in preventing the conduction by keeping the space between the body and the cold floor in the gymnasium. The challenges in installing the cardboard beds are (1) ensuring the space within the limited areas in refugee centers, (2) fire preventions, and (3) pre-consent of the stakeholders in the refugee centers for the smooth installation of the beds in emergencies. Then, the shelter system was effective in isolating the people in the gymnasium from the cold outside air and the wind inside of the gymnasium. The shelter system was also useful in enhancing the effect of the heating. However, the shelter system has the challenges in fire prevention, daily uses in peacetime, and stock space in peacetime. In addition, while shelter systems reduce indoor winds due to convection, they are accompanied by an increase in carbon dioxide concentrations. Next, the heating system was effective to alleviate the heat loss by radiation since the heating system reduces the gap between the body temperature and the temperature around the body. However, the application of heating systems has challenges in ensuring the fuel and electricity. In addition, the heating system increases the possibility of causing carbon monoxide poisoning cases if the exhaust gas continues to be released indoors. Furthermore, it is not easy to warm up the large space of a gymnasium by using a limited number of heaters. Then, sleeping bags were effective in alleviating the radiation while they had challenges in the gaps between the blankets which were daily used in peacetime. In addition, the establishment of the stock management system as relief items is necessary to be used in emergencies. Next, portable indoor toilets were effective in preventing heat loss caused by going to the outdoor toilets and radiation from the

Table 2 Effectiveness and challenges of each solution as a countermeasure for hypothermia caused by heat loss

Purpose	Approaches	Solutions	Effectiveness	Challenges
Prevention of heat loss	Evaporation	Sauna	Prevention of heat loss from the body surface by increasing ambient temperature and humidity	– Temperature control to avoid sweating – Secure heat source – Prevention of carbon monoxide poisoning (when burning firewood in a tent)
		Kitchen cruiser	– Prevention of dehydration – Promotion of thermogenesis by diet	– Establishment of a system for dispatching chefs in the event of a wide-area disaster – Maintaining business as usual – Anti-freeze measures for vehicle doors – Countermeasures against snow accumulation around vehicles – Noise and exhaust gas generated during engine operation
	Conduction	Cardboard beds	Prevention of heat conduction to cold floors	– Stockpile system in wide areas – Securing installation space – Fire prevention – Consent of related parties regarding installation
	Convection	Shelter system	– Isolation from outside air wind suppression – Improved heating effect	– CO2 charging – Stockpiling during normal times Daily Uses
	Radiation	Heating system	Maintain body temperature by reducing the difference from outside temperature	– Fuel stockpiling – Prevention of carbon monoxide poisoning (when burning indoors) – Securing electricity – Heating large spaces

(continued)

Table 2 (continued)

Purpose	Approaches	Solutions	Effectiveness	Challenges
		Sleeping bags	Appropriate transpiration and body temperature maintenance with aluminum vapor-deposited sleeping bag	– Gap with daily life – Generalization and dissemination as disaster prevention supplies
	All	Portable toilets in an indoor	– Prevention of exposure to outside air – Prevention of body heat loss due to cold toilets – Prevention of disaster-related diseases by refraining from using the toilet	– Gap between normal quality of life Measures for evacuees using cars – Collection of excrement and establishment of final treatment process

Table 3 Effectiveness and challenges of each solution as a countermeasure against hypothermia by heating

Purpose	Approaches	Solutions	Effectiveness	Challenges
Heating	Heating	Electric blankets	Maintain body temperature by heating	– Prevention of low-temperature burns – Secure power supply – Prevention of ticks
		Kitchen cruiser	Promotion of thermogenesis by hot meal	– Establishment of a system for dispatching chefs in the event of a wide-area disaster – Maintaining business as usual – Anti-freeze measures for vehicle doors – Countermeasures against snow accumulation around vehicles – Noise and exhaust gas generated during engine operation
		Footbath	Maintain body temperature by heating	– Securing raw water – Securing heating means- – ·Maintaining hygiene
		Hot water bottle	Maintain body temperature by heating	– Securing raw water – Securing heating means – Prevention of low-temperature burns

body to the cold toilet bowl. Installing indoor toilets prevents victims from withholding food and hydration in order to avoid going to outdoor toilets. This reduces the risk of hypothermia and other illnesses.

However, the portable toilets cause the gap between the daily lives and the lives in emergencies which makes people in refugee centres hesitate to drink and eat. In addition, establishing the proper management system to dispose and treat the excreta would be also challenging.

4.2 Solutions for Warming the Body

To keep the body temperature under the cold air condition in the refugee center, the electric blankets were useful while careful attention should be paid to prevent low-temperature burn cases. In addition, ensuring electricity will become a challenge in emergencies. Furthermore, it is necessary to be careful to prevent health damage caused by ticks and other insects. Then, hot meals provided by the kitchen cruiser are effective in supporting heat production through the intake of hot meals. Next, taking footbaths and use of hot water bottles would be effective in keeping the body temperature as long as the body will not get wet, however, ensuring the water and energy for boiling the water in emergencies is expected to be the challenge in emergencies. In taking footbath, ensuring proper sanitation is also a challenge. In addition, in the use of hot water bottles, careful attention should be paid to prevent low burn cases same as electric blankets.

5 Discussion

This study aimed to examine the practical solutions to prevent hypothermia in the refugee center in severe winter. Hypothermia results from an imbalance between heat loss and heat production. Therefore, this study examined various solutions against the main facilitating factors of hypothermia: Evaporation, Conduction, Convection, and Radiation to alleviate the heat loss. In addition, to support heat production, various solutions were tested. The tested solutions were effective in alleviating heat loss and promoting heat production. On the other hand, this study identified the challenges in ensuring electricity, fuels and in preventing carbon monoxide poisoning and low-temperature burn cases if the tested solutions are used in real emergency situations. If a disaster occurs in winter when river water freezes, it would be challenging to ensure water in disaster-affected areas since piped water also freezes due to the lack of electricity and low temperature. This water problem may be a typical challenge that is not expected to happen in emergencies in Japan.

The past studies on the living conditions in shelters focused on particular topics such as minimum required personal space at shelters (Kawano et al. 2016), sleep quality at shelters (Ogata et al. 2020), and providing shelter to nursing home evacuees (Laditka et al 2008). Basically, there is limited studies on the improvement of the living conditions in shelters by using practical solutions while Soshino et al. (2017) reported support to improve the refugee centers by using various field hospital equipment. In addition, there are several reports on the hypothermia cases in the Great East Japan Earthquake and Tsunami, however, few studies presented practical solutions to prevent hypothermia in the refugee centers. This study strategically tried to examine various solutions such as food, space, air-conditioning, flooring, energy, and sanitation according to the basic factors to prevent hypothermia by trying to balance the heat production and heat loss even in emergencies. In this way, the uniqueness of this study is that it targeted the prevention of hypothermia in refugee centers during severe winter and tried to test the familiar solutions for alleviating the heat loss and promoting the heat production.

The limitation of this study is that some of the solutions tested in this study have not been deployed in real disaster response in severe winter. In the real disaster situations, at present, it is not realistic to immediately deploy kitchen cruisers, which were tested in this study, to a lot of refugee centers. In addition, in past disasters in Japan, even gymnasiums to be used as refugee centers were not able to accommodate refugees due to the damage caused by earthquakes. In such cases, a lot of people were forced to evacuate by using vehicles or staying in their damaged homes without water supply and power supply. In particular, ensuring proper sanitation for the refugees who stay in their vehicle instead of staying in the refugee centers. Therefore, in addition to the countermeasures to hypothermia cases in refugee centers, the study under the different settings will be necessary. Furthermore, this study targeted to prevent the hypothermia cases in the refugee centers. However, the hypothermia prevention among the relief workers and volunteers should be considered.

From a global perspective, this study discussed tsunami preparedness in cold regions, where few cases were reported in the past. Mega-earthquake and tsunami disaster alongside Japan and Chishima Trenches is a unique disaster since this disaster strikes in highly populated coastal areas of northern Japan. Therefore, as Oshiro et al. (2022) discussed hypothermia cases in various disasters occurred in various countries, the knowledge and experience on tsunami disasters in winter should be shared with other countries such as Canada, the United States, Chile, New Zealand, where tsunami disasters may occur under the harsh winter conditions.

The solutions that this study examined are relatively simple, affordable, and popularized in other countries. In this sense, the solutions in this study can be the common solutions to prepare for tsunami disasters in severe winter.

6 Conclusion

This study aimed to examine the practical and familiar solutions to prevent hypothermia in the refugee center during emergencies in severe winter. The solutions were proposed aiming to alleviate heat loss and promote heat production for ensuring the heat balance to prevent hypothermia in the refugee centers. The field tests were conducted in a gymnasium under severe winter conditions. To alleviate the heat loss, which consists of four main promotion factors: Evaporation, Conduction, Convection, and Radiation, the solutions were deployed according to these four factors. In addition, the solutions to promote heat production were also tested. Both approaches presented the effectiveness in preventing hypothermia cases while the challenges in ensuring electricity and water were identified. In addition, this study pointed out the risks of low-temperature burn and carbon monoxide poisoning to support heat production in the refugee centers. Further studies on the real shelter management in severe winter are needed. In addition, this study pointed out the need for future studies on the accommodation management for relief workers and volunteers deployed to disaster in severe winter conditions. While there are limited studies on shelter management in severe winter, this study presented practical and familiar solutions to prevent hypothermia in emergency shelters. Finally, this study is concluded by pointing out the importance of sharing knowledge on hypothermia prevention in disaster situations under severe winter conditions.

References

Arthurs Z, Cuadrado D, Beekley A, Grathwohl K, Perkins J, Rush R, Sebesta J (2006) The impact of hypothermia on trauma care at the 31st combat support hospital. Am J Surg 191:610–614

Furukawa H, Kudo D, Nakagawa A, Matsumura T, Abe Y, Konishi R, Kushimoto S (2014) Hypothermia in victims of the Great East Japan Earthquake: a survey in Miyagi prefecture. Disast Med Public Health Prep 8(5):379–389

Gusiakov V.K. (2000) Two great kamchatka Tsunamis, 1737 and 1952 [Online]: available at: https://bibliotecadigital.ciren.cl/handle/20.500.13082/28986 (Accessed: 24 April 2023)

Iseki K, Hayashida A, Shikama Y, Goto K, Tase C (2013) An outbreak of carbon monoxide poisoning in Yamagata prefecture following the Great East Japan Earthquake. Asia Pac J Med Toxicol 2(2):37–41

Japan Cabinet Office (2021) White paper on disaster management. https://www.bousai.go.jp/en/documentation/white_paper/2021.html, Accessed 19 April 2023

Kamata H, Seto S, Suppasri A, Sasaki H, Egawa S, Imamura F (2022) A study on hypothermia and associated countermeasures in tsunami disasters: a case study of Miyagi Prefecture during the 2011 great East Japan earthquake. Int J Disast Risk Reduct 81:103253

Kawano T, Nishiyama K, Morita H, Yamamura O, Hiraide A, Hasegawa K (2016) Association between shelter crowding and incidence of sleep disturbance among disaster evacuees: a retrospective medical chart review study. BMJ Open 6:e009711. https://doi.org/10.1136/bmjopen-2015-009711

Kioka S, Mori M, Yamamoto Y, Takeuchi T (2008) Medium scale test and numerical simulation on ice impact to a structure due to tsunami during sea ice season. In: Proceedings of coastal engineering, JSCE, vol 55. Japan Society of Civil Engineers, pp 851–855

Laditka BS, Laditka NJ, Xirasagar S, Comman BC, Davis BC, Richter EVJ (2008) Providing shelter to nursing home evacuees in disasters: lessons from Hurricane Katrina, research and practice. Am J Public Health 98(7):1288–1293

Mimura N, Yasuhara K, Kawagoe S, Yokoki H, Kazama S (2011) Damage from the Great East Japan Earthquake and Tsunami – a quick report. Mitig Adapt Strat Glob Change 16:803–818

Mizutani Y, Nemoto M (2017) Evaluation of thermal effects by cardboard-bed at winter evacuation shelter. Hokkaido No Seppyo (snow and Ice in Hokkaido) 36:101–104

Nemoto M, Oyama T (2014) Safety verification of emergency supplies for staying overnight in automobile in heavy snow storm. Hokkaido No Seppyou (snow and Ice in Hokkaido) 33:145–148

Niizato T, Shigeno K, Tahashimizu Y (2006) Ainu oral traditions and historical records on earthquakes in Hokkaido Japan. Histor Earthquakes 21:121–136

Ogata H, Kayaba M, Kaneko M, Ogawa K, Kiyono K (2020) Evaluation of sleep quality in a disaster evacuee environment. Int J Environ Res Public Health 17:4252

Oshiro K, Tanioka Y, Schweizer J, Zafren K, Brugger H, Paal P (2022) Prevention of hypothermia in the aftermath of natural disasters in areas at risk of avalanches, earthquakes, tsunamis and floods. Int J Environ Res Public Health 19(1098)

Paal P, Pasquier M, Darocha T, Lechner R, Kosiski S, Wallner B, Zafren K, Brugger H (2022) Accidental hypothermia: 2021 update. Int J Environ Res Public Health 19(1)

Romanovsky A (2018) Thermoregulation: from basic neuroscience to clinical neurology part1. In: Handbook of clinical neurology

Schiermeier Q (2017) Huge landslide triggered rare Greenland mega-tsunami. Nature. https://doi.org/10.1038/nature.2017.22374

Wong CS, Chen KC, Wang TL (2004) Clinical aspects of hypothermia. Ann Disast Med 2:2

Soshino Y, Kuroda A, Miyata A (2017) Review of the domestic utilization of pre-positioned international medical relief equipment in Japan. J Natur Disast Sci 38(2):167–178

Takashimizu Y (2005) Ainu oral traditions and historical records on tsunami in Hokkaido Prefecture Japan. Histor Earthquakes 20:183–199

Complexities Involved

Techniques and Process Development for Quick Response and Sustainable Recovery

Sankar Halder, Padmanava Sen, and Somnath Mukherji

1 Introduction

Disaster management in coastal regions should take into account the local context because every region has its specific geographical and societal challenges. For example, the challenges in coastal regions, hills, earthquakes, or volcanic zones are different in nature. Based on the geophysical context, the intensity of the natural calamity could vary. In order to survive in these regions, human inhabitants have to be able to adapt to the natural challenges and they also have to make sure that their livelihood and other basic needs of life are secure and resilient.

When we talk about resiliency, a community needs to be prepared in such a way that the disaster damage is minimal and in case of significant damage, the recovery is fast with minimal/zero human loss of lives. Livelihood options, education, and housing need to be resilient for the community to bounce back. That is why we can subcategorize the disaster management as follows:

- Preparedness
- Immediate response
- Long-term recovery and rehabilitation

S. Halder
MUKTI, Raidighi, India
e-mail: sankar.halder@muktiweb.org

P. Sen (✉)
Asha for Education, Covina, USA
e-mail: padmanava.sen@barkhauseninstitut.org

Barkhausen Institut, Dresden, Germany

S. Mukherji
Association for India's Development, College Park, USA
e-mail: somnath@aidindia.org

P. K. Ray et al. (eds.), *Technology Innovation for Sustainable Development of Healthcare and Disaster Management*, Disaster Risk Reduction,
https://doi.org/10.1007/978-981-97-2049-1_7

It is like a virtuous circle. A well-prepared community will make sure the damage is less with fast recovery. Such a community will be resilient enough to respond immediately after the disaster without creating panic.

To stay well and resilient, a community should be holistic (Paton and Johnston 2017) and not only focus on human lives. It should consider nature, including animals, plants, and the environment. Any ideal community cannot sustain itself in silo ignoring its biodiversity and other ecological factors. That is why, if we consider Indian coastal regions such as Sundarbans, we must consider the response and recovery of wild animals, mangroves, river embankments, plants, domestic animals, etc.

The challenges that we need to consider are summarized.

1. Agriculture

The rural areas considered in this section are primarily agriculture-dependent and hence, the short-term and long-term impact of a natural disaster are considered in detail as its impact on the livelihood is very high. The impact of such a disaster in urban areas will be different and outside the scope of this chapter. Damage to agricultural lands and lack of knowledge about preparatory and revival methods have caused mass exodus and migration.

2. Food

As evident, food production, transport, and consumption play an important role and thus, have been considered in this chapter.

3. Education

While education sometimes takes the backseat during the natural disaster, but the impact on it takes a toll on the long-term aspects of the well-being of learners.

4. Health

Health is one of the most important factors during the relief and rehabilitation with a special focus on children, women health, and people with comorbidities.

5. Housing

A cyclone/disaster-resistant house and efficient management of the architecture will be the key for the disaster-prone communities.

6. River Embankment

While not every natural disaster-prone area has rivers but, in this chapter, the effects of rivers in an estuarine community will be considered.

The above six components will be covered in detail throughout Sects. 2.1–2.6. A few other components will be discussed in Sect. 2.7 Section 3 will cover a few examples and use cases to conclude this chapter.

2 Different Aspects of Cyclone Resilient Livelihood

It is important to define a livelihood that is resilient to natural disasters, covering preparedness, response, and long-term recovery. In that sense, livelihood covers not only the basic necessities of life, but it goes beyond them—holistic aspects of lives. In this section, the aspects that will be covered are agriculture, food security, education, health, housing, and river embankment protection in a natural disaster-prone area.

Preparedness

This is the most important aspect to save lives as well as property damage caused by any possible natural disaster, prior to the disaster. Preparedness means building resilience in the community as it ensures quicker relief and rehabilitation after a disaster. For preparedness, understanding the possible calamities is needed. For example, a community in the mountains needs to be aware of landslide possibilities after rains whereas a coastal community needs to prepare for cyclones, flood, tsunamis, water logging, etc. Knowing the vulnerability/weakness and strength will be required for readiness. More details about each aspect of livelihood will be covered below in terms of preparedness, specifically in an estuarine area where some areas may be below sea level.

Immediate Response

This 2nd phase usually receives the most media attention and support from the government and other organizations. While the main focus is on health and saving human lives, all aspects of livelihood need to be considered as described in the following subsections.

Long-Term Recovery and Rehabilitation

This 3rd phase is closely associated with the first phase, i.e., preparedness. While the support from government and other supporting organizations may run out in this phase, it remains the least highlighted but crucial in stabilizing the community for the longer term.

The roles of technology and organizations are embedded in these three phases of six major components as well as a dedicated Sect. 3.

In the following sections, more details will be given for each component. The sections are written considering any natural disaster-prone coastal area, however, in some cases, specific factors that are important for a specific region in India (Sundarbans) will be mentioned considering the association and expertise of the authors. Sundarbans is the largest delta in the world. Over many centuries, the rivers Ganges, Brahmaputra, and Meghna have brought silt (small particles) along the 2500 km route across India and Bangladesh to the Bay of Bengal. Over the years, these sediments deposited on the seabed and formed delta (islands). Gradually, the area got covered by natural salt-tolerant forests, called mangroves (Rudra 2017; Mukhopadhyay 2016). The Indian portion of the delta starts around 50 km south of Kolkata and currently falls in the districts of North and South 24 Paraganas in the state of West Bengal. The rest of the Delta is in the south of Bangladesh.

2.1 Agriculture

Preparedness

The main components will be

- Land shaping—If the heights of the lands are raised where crops are grown, they will be safe under the conditions of water inundation.
- Preparation of the soil—Agents needed to recover salinized land should be stored within the community. For example, sesbania (Wikipedia, Sesbania 2023) can be used for both nitrogen fixation and green manure, reducing the salinity of the land. But, the seeds of sesbania need to be available.
- Salt-tolerant seeds—The incursion of the salt water into farmlands leaves behind salt in the land. Salt-tolerant paddy varieties (Qin et al. 2020) can grow well in saline conditions. Also, if the existing seeds are not salt tolerant for some cost measures, the farmers need to be prepared with salt tolerant seeds during disaster seasons.

Immediate Response

The following components and technologies need to be considered.

- Drainage of saline water that entered agricultural fields.
- Certain agricultural practices like leaching (excess irrigation to drain salt from the soil)
- Seedling preparation can be done in a land that is not impacted and can be transferred to the impacted lands using salt-tolerant seeds as mentioned earlier in this subsection.

Long-Term Recovery and Rehabilitation

The following components and technologies need to be considered.

- Keeping water drainage systems intact—We must make sure that the agricultural field is managed with a proper drainage system at any time during the monsoon or flood calamity. It is essential to make the land shape suitable for flushing out the saline water or excess rainwater.
- Using organic manures and assessment of soil health—To bring back soil health from high salinity to suitable pH factor. Normally, after flooding of the agricultural fields, the salinity becomes very high. To recover that, we need to use organic farming methodologies utilising locally available resources, especially biomass (e.g., vermicompost, normal compost from bio compostable matters) and green

manures (e.g., sesbania) before each crop cycle on a regular basis. Bio and green manures can bring down the salinity and make land suitable for agriculture.

- Using salt-tolerant varieties of seeds—Coastal regions in Sundarbans are normally good for paddy and tropical vegetables. It is essential to cultivate salt-tolerant varieties of paddy and vegetables. It is also important not to use any hybrid seeds that usually require more controlled weather conditions. Local people are advised to produce their own salt-tolerant seeds and reserve them for future use.
- Continuation of paddy seedling preparations on higher grounds—If the salinity is continuing for a longer period for a specific area, seedlings can be prepared on a relatively higher ground and relatively better soil conditions as mentioned in the preparedness section as well. The seedlings are usually more sensitive to the salinity of the soil but matured seedlings can tolerate higher salinity/more adverse soil conditions.
- Sustainable land-usage practices—Farmers must adapt to the rotational crop production methodologies. Same crop cannot be cultivated repeatedly on the same land. Normally, the plants get nutrients out of NPK (nitrogen, phosphorus, and potassium) in the ratios suitable for the plant. As an example, paddy crops require more nitrogen, repeating cultivation of paddy will lead to nitrogen-depleted land. Hence, rotation between paddy crops and crops that require less nitrogen and replenish the soil (e.g., pulses) will ensure a better balance of soil nutrition and an adequate amount of bacteria.

2.2 *Food Security*

Preparedness

The main components will be:

- Storage of dry food items and drinking water—For immediate use after a disaster, adequate quantities of dry food (e.g., puffed rice, flattened rice for Sundarbans communities) need to be stored and kept ready because running a kitchen may not be possible immediately after a disaster. At the same time, adequate drinking water for a few days should be kept in reserve. Drinking water sources get contaminated or inaccessible.
- Storage of cattle foods—At least a few days, a stock of fodder for the cattle needs to be stored.
- Storage of produce at higher grounds—Normally, rural coastal areas people do not go to markets daily. After a good harvest, they store the produce to use for consumption for the entire year. It is essential to save their produce at a higher ground.
- Securing live stocks—Live stocks (chicken, goats, and fishes) are their main protein source as well as provide an additional source of income. Securing them will be key to recovery after stored dry food items are consumed.

Immediate Response

The following components and technologies need to be considered.

- Use of dry food items for a first few days
- Prepare a temporary community kitchen or make their individual kitchens functional. Community kitchen operations may need help from local governments and other organizations. These organizations need to provide the cooking ingredients as needed.
- Food for infants and people with comorbidities need to be provided.
- Food for cattle and livestock need to be provided from the storage or need to be brought into the impacted area.

Long-Term Recovery and Rehabilitation

The following components and technologies need to be considered.

- Community food sharing—It is important to establish a food storage facility that can be maintained by the community. During disasters, that will be used to fill the gaps and other times, it will provide food security. Time to time, the food has to be distributed or shared to avoid any wastage.
- Reducing the dependency on imported foods—The community should be self-reliant on basic food items thus reducing the dependency on money during disaster relief as well as the transportation from other areas.

2.3 Education

Preparedness

The main components will be:

- Education materials—After a disaster, it usually takes 3–6 months to bring back normalcy. During this time, most families give importance to food, clothes, and shelter while education of children is generally ignored. This causes dropouts from schools and irreversible damage to learning. Hence, saving the education materials is as important as saving basic necessities.
- Importance to board exam students—The students from higher classes, appearing in state-level board exams (e.g., Class 10, 12) are specifically vulnerable to such disasters, as the dates of the exams are not moved for them. So, it is important to prepare for a fallback option to move them to relatives' houses away from disaster-prone areas as alternative study/preparation points.
- Storage of digital devices and batteries—It is important to store files and folders in clouds (e.g., google drive, one drive) rather than in desktops/laptops that may get damaged by flooding. This is not only for personal files but relevant to schools, colleges, and alternative educational centers.

Immediate Response

In the first phase, the main goal is to save as much educational materials as well possible. Apart from that, the books and aiding educational devices (e.g., tablets) need to be saved. For higher class students, especially students appearing in board exams, immediate distribution of books will be needed as well. Also, educational gaps may create a major bottleneck in future, hence, even during immediate response, education should not be ignored.

Long-Term Recovery and Rehabilitation

The main components will be:

- Education on disaster resilience—This should be given to everyone in coastal regions including students. Also, the sustainable development and environment educational aspects should be taught to reduce the environmental impact of life choices (e.g., deforestation, climate change, global warming, sea level rise). This should also include risk assessment knowledge to understand the warning signs for upcoming disasters.
- Health education including first aid—Disasters normally cause health hazards. It is essential for people, including children, to have preliminary first-aid care knowledge to treat minor injuries.
- Psychological support—Any natural disaster is a psychologically traumatic event and people need consistent mental health support to deal with such events.

2.4 Health

Preparedness

The main components will be:

- Medicines/water purifiers—Expecting water-borne diseases after disaster, water purifiers and essential medicines against fever, diarrhea, skin diseases need to be stored. Additionally, people with comorbidities should always be prepared with storage of at least one month of medicines.
- Sanitary napkins need to be available to all women in a community affected by disaster.

Immediate Response

The main components will be:

- Assurance of clean drinking water
- Preparation of avoiding contagious diseases, spreading through communities

Long-Term Recovery and Rehabilitation

The main components will be:

- Health care center—There should be preparedness and equipment to combat water-borne diseases. The health care center staff should have reasonable knowledge of such injuries and diseases and should be accessible.
- Connection to a doctors' network with mobile phone/e-consultation
- Adequate vehicles—Will be needed to transfer people with comorbidities as well as pregnant women.
- Availability of drone(s) for use and technicians to operate them.
- Training and readiness of community health care workers—They should be trained regularly in the community health center, specially to handle the disaster situation.
- The elevated tubewell—In these communities, tube wells are good sources of drinking water, and they should be placed on higher ground (at least 15 ft above sea level). The tap of these sources must be at higher levels.

2.5 Housing

Preparedness

The main components will be:

- House structure—The houses should be built with a cyclone-resilient design (World Bank 2016; Habitat 2018) that can withstand high winds. The foundation should be strong enough to withstand flood water for multiple days.
- Fixing vulnerable/weak structures—Housing preparations will include the knowledge of vulnerable areas of a house and livestock should not be stored under that. For the same reason, bedrooms should not be vulnerable.
- Deciding the items to bring—It is important to separate the important items and store them in the best/strongest part of the house. When disaster strikes, that part can be used for shelter and also, when that part breaks down, it will be easier to take the important items while leaving the house.

Immediate Response

The main components will be:

- Assessment of damage to the house and immediate repair for structural strengthening
- Help/support from supporting organizations and local governments (insurance is mostly not applicable in the regions covered in this chapter)

Long-Term Recovery and Rehabilitation

The main components will be:

- Floor/height—The building should either be on a high ground or should have multiple floors. The foundation should be strong and elevated.

- Design—The architectural design should be able to handle high winds. The design should also consider the wind direction and an expert should be consulted for that.
- Materials—The materials should be sustainable and readily available locally.
- Sanitation facilities—The design of sanitation systems should be made carefully to avoid floating feces after flooding.
- Building a community flood center—There should be adequate flood-tolerant elevated relief centers where people can stay long term who cannot stay in their own houses during disaster. Existing school buildings can also be used in certain cases. The local government should be proactive in this.

2.6 River Embankment Protection

Preparedness

In the estuarine area, a well-protected and strong river embankment is the lifeline of the people. There should be preparation in the community to store items/materials (e.g., bamboos, palm trees, wood logs) that can be used for strengthening the embankment of the river during disaster. These items should be stored close to the riverbank. This needs to be done before and after the disaster, not during the disaster. There should be regular checks to find the weak points of the embankment and to strengthen the weaker sections.

Immediate Response

The main components will be:

- Repair of any breach in the embankment and stop any leakage of saline water into fields
- It needs to be observed during both high and low tides.
- Government bodies need to be prepared for damage repair after assessment.

Long-Term Recovery and Rehabilitation

The main components will be:

- Importance of embankment—The Netherlands has built embankment systems that protect them from sea water and floods even while living below sea level. However, many flood-prone areas in developing countries could not build such an embankment system due to a lack of technology, budget, or awareness.
- The role of silt and sand in the riverbed has to be realized (Purkait 2022). There are natural and human-made reasons to mitigate for. In (Purkait 2022), certain specific solutions have been extensively discussed, namely, backup embankments, dredging to clear the sediments, roles of mangrove and vetiver grass, repurpose of the sediment, use of the canals, etc.

2.7 Human and Animal Rights

Other than the six factors described in previous subsections, human and animal rights have a role to play in a cyclone resilient community.

Human–Animal Conflict

At the time of calamity or disaster, normally animals sense the sudden changes before humans. As a result, there is agitation and chaos among the animals (wild and domestic). It is the community's responsibility to take care of the wild and domestic animals before and after the disasters. That requires good preparation and will minimize the conflict between humans and animals. It will also be important not to kill and eat wild animals during disaster times. Awareness needs to be spread so that proper handover mechanisms of wild animals are in place. The coordination with government shelters will be required to enable this during disaster times.

Human Trafficking and Basic Rights Issue Due to Societal Differences

At the time of crisis in individual families/communities, basic rights are compromised. There is an increase in domestic violence against children and women. There is also an increase in human trafficking (mostly children and women) as well as sex and organ trafficking when the families go through financial turmoil. Hence, the preparations to prevent these very important during and after a disaster. The community needs to be vigilant, and it should be sensible enough to provide the basic necessities and psychological support in relief and rehabilitation situations. Government and non-government organizations need to be proactive in building connections with underprivileged communities to hold basic human rights.

3 Role of Organizations, Governments, and Technology

The government and organizational support, namely, community-based organizations (CBO) and civil society organizations (CSO) are covered in Sect. 3.1. As the impact and role of technology is growing in the last two decades, they are mentioned in Sect. 3.2.

3.1 Role of Community-Based Organizations (CBO)/Civil Society Organizations (CSO), and Local Governments

Preparedness

The CBO/CSO and local governments need to make sure that the residents take care of the above precautions in the six aspects mentioned. This requires regular awareness training and knowledge transfer among representatives in local communities. Local governments also need to plan higher grounds for gathering people during floods and alert them accordingly. The different bodies need to circulate the dos and don'ts as referred by the government, e.g., (National Disaster Management Authority of India, Cyclone Do's and Don'ts 2023).

The following steps can be followed in the following order.

(1) Plan the logistics and start cooperations between government bodies as well as non-government organizations (CBO/CSO)
(2) Figure out the logistics in case of requirements of basic necessities (depending on the impacted population size)
(3) Inform the residents about the upcoming disaster using all possible communication/ broadcasting mode
(4) Prepare the community gathering places along with the basic necessities (water purifiers, food)
(5) People need to be taken to the safe zones in due time
(6) Provide basic facilities based on the groups/needs
(7) Precautions in case of special medical situations (e.g., cyclone/flood along with COVID-19 pandemic)

Immediate Response

The specific roles will be:

- Community kitchen, providing ingredients.
- Community medical camps, providing medicines.
- Immediate clothes, women hygiene, baby food distribution
- They have to confirm the safe return of residents.

Long-Term Recovery and Rehabilitation

The roles are broken down based on the different organizations.

For the government, the roles will be:

- Government should set up long-term policies to address large-scale disasters that impact a sizable population.
- There should be research funding to set up early warning systems and make local people aware of the possibility and frequency of such disasters. The geological information systems (GIS) should be available.
- Government should build adequate infrastructure to deal with disaster as well as help in the restoration of the community infrastructure (e.g., schools, hospitals)
- Government should work in a collaborative governance model to make a policy framework in order to deal with such a situation, including all stakeholders (residents, CBO/CSO, research institutes)

For CBO/CSO, the roles will be:

- They should aid government initiatives and should be engaged in active advocacy. Also, in certain communities, the presence of local organizations can be crucial in spreading awareness.
- They would play a role in the capacity building of local people for housing, livelihood, and other basic means of life.

For International communities, the roles will be.

- To create a compensational model, if possible, for people affected by global warming and climate change, which might be causing frequent disasters in coastal regions (e.g., sensitive estuarine areas like Sundarbans)

3.2 Role of Technology

Preparedness

In the above subsections, certain roles of technology (e.g., in education) are covered in the preparedness phase. However, the role of tracking disasters and timings during the preparation phase has transformed in recent years with the help of technology. Sites like (Windy Community, Homepage 2023) have been very useful.

Immediate Response

The specific roles will be:

- Restoring the power systems as well as charging of mobile phones
- Using drones for loss and damage assessment
- Digitally accessible educational material

Long-Term Recovery and Rehabilitation

The specific roles will be (Figs. 1 and 2)

- Restoring the power systems as well as charging of mobile phones
- The early warning system development is important, for example, (Windy Community, Homepage 2023). These sites can be used for real-time movement of the cyclone and the accurate prediction of landfall. This gives a concrete window for evacuation and protection. Wind speeds and directions can be understood accurately so that protection, evacuation, and relief work can be carried out with confidence.
- In the coastal areas, it is very important to understand the combination of how the timing of cyclones interact with the tides. In areas like Sundarbans, it becomes all the more important because the integrity of the embankments surrounding the islands is of prime importance for the safety and long-term livelihood of the community. Sites like (Tideschart, World-Tides 2023) give accurate prediction of the timing and height of the tides. If landfall of a storm coincides with high tide, we can expect higher storm surges. For example, Cyclone Amphan in May 2020

(a) Satellite map

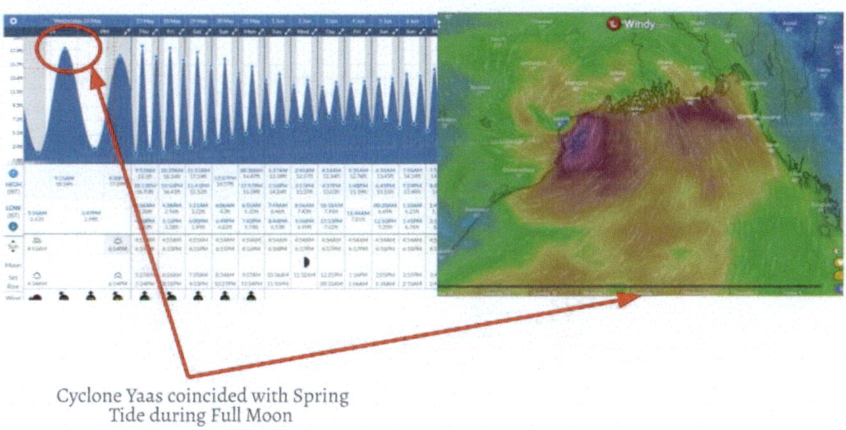

(b) A closer look

Fig. 1 The snapshots from windy.com (Windy Community, Homepage 2023)

in Sundarbans did not coincide with the high tide and hence caused less water damage, but Cyclone Yaas of May 2021 in the same region coincided with the high tide and caused massive water damage.

- Publicly available satellite imagery can be used very effectively for understanding areas of flooding and the extent of every cyclone/storm. Further, if the historical imagery and data are compared over several cyclones/storms, it would be possible to detect patterns of repeated flooding and develop a vulnerability map of an area. Of course, other socio-economic inputs have to be considered to map vulnerability.

(a) Before flood

(b) After flood

Fig. 2 The satellite images from Sentinel, before and after Cyclone Amphan in Sundarbans to assess the extent of flood damages (Sentinel Hub, EO Browser 2023)

We had used satellite imagery from Sentinel to understand the areas of flooding and plan accordingly for relief operations. (Sentinel Hub, EO Browser 2023). We ensured the satellite image reading with ground-truthing to remove potential sources of error.

- Use of altitude-based maps, open street maps, or data from the meteorological department to understand the sensitive and flood-prone areas
- With the help of meteorological maps and available flood/wind models, the disaster modeling and simulations need to be performed in advance to prepare and to build cyclone-resilient infrastructure. The research group involved from the government should be proactive in this.

- Availability of drones and training of operators among local people will be of importance.
- Communication and coordination channels through mobile phones should be setup proactively.

4 Use Case, Results, and Evolving Solutions

The disaster relief and rehabilitation in a cyclone and flood-prone area like Sundarbans have been part of several previous research works as well (Sen 2019; Samanta et al. 2017; Dasgupta and Shaw 2015). As we were referring to Sundarbans in the earlier subsections, we would like to bring a live use case on the methodologies implemented by Mukti (Mukti, Homepage 2023), Asha for Education (Asha for Education, Homepage 2023), and Association for India's development/AID (Association for India's Development, Homepage 2023).

Mukti (a Bengali word that means Liberation) is a socioeconomic development organization, working for holistic development in Sundarban coastal regions since 2003. Since then, Mukti has been working for Sundarbans people in the Delta region of South 24 Parganas, WB, India with the programs of health, water and sanitation, education, agriculture, livelihood, environment, and rights. Mukti is working with thirty island communities that are in catastrophes of climatic hazards resulting in penury, vulnerability, and violence where the Sundarbans grassroots are subjected to such menaces. It has addressed disasters (cyclones) like Aila (2009), Bulbul (2019), Amphan (2020) and Yaas (2021). The organization worked on disaster management holistically taking help from the local community and other like-minded organizations like AID and Asha for Education. After cyclone Aila, Mukti has used all 5 methods together described in Sect. 2.1 (recovery phase) to convert saline land into arable land within a year that would have taken 3–5 years without this intervention. This benefitted 1000 hectares of land and 4000 farmer families.

Mukti has adopted a HEALERAID concept toward the societal challenges where H stands for health, E stands for education, A stands for agriculture, L stands for livelihood, E stands for environment, R stands for rights, A for awareness, I for integrated development, and D for disaster recovery. MUKTI works as the HEALERAID in the communities, to heal and reduce vulnerability of the people of the last miles through its various projects under six focused programs. More details are given in (Halder and Sen 2022; Mukti, Homepage 2023).

AID is a volunteer-driven organization working with grassroots groups across India in sustainable agriculture, health equity, gender, and environmental justice. AID has been working with Mukti since Cyclone Aila struck Sundarbans in 2009, engaging with immediate relief and long-term recovery by focusing on climate-resilient agriculture, rebuilding ecological integrity, and gender justice. Mukti, a long-term partner of AID was recognized by FICCI (Business Chambers of Commerce) in 2021 (CSRBOX, Mukti page 2023).

Asha for Education (AfE) is a secular organization dedicated to change in India by focusing on basic education in the belief that education is a critical requisite for socioeconomic change. Since Tsunami, AfE has extensively worked with organizations dealing with natural disasters to prevent child trafficking, dropouts from schools, early marriage for girl students, migration, and child labor. Mukti has been one of the prominent partners in this cause. More details are available in this webpage (Asha for Education, Mukti page 2023) Apart from the immediate response to a disaster, AfE worked with Mukti closely for educational rehabilitation. Some initiatives taken right after cyclone Amphan in 2020 are documented in this report (Asha for Education, Mukti page Amphan report 2023). The partnership is still going strong, and the legacy is being continued as a sustainable partnership. Even this chapter is co-written by persons of these three organizations.

5 Conclusion

To conclude, disaster relief and rehabilitation is a complex process that needs a generation to create awareness and to chalk out the three phases—preparedness, immediate response, and long-term recovery. The role of correct methodologies and technical aids has become more and more important to reduce the loss of lives, property damage, and the duration of recovery. It is still surprising to see the differences in response to disasters across the world. Pragmatic solutions that consider the local community needs and culture can make the process more efficient and create a new cyclone-resilient generation.

References

Asha for Education, Mukti page Amphan report, https://ashadocserver.s3.amazonaws.com/671_DisasterResponsebyMUKTI.pdf Accessed 29 Oct 2023

Asha for Education, Homepage, https://ashanet.org/ Accessed 29 Oct 2023

Asha for Education, Mukti page, https://ashanet.org/project/?pid=671 Accessed 29 Oct 2023

Association for India's Development, Homepage, https://aidindia.org/ Accessed 29 Oct 2023

CSRBOX, Mukti page, https://csrbox.org/India_organization_West-Bengal-Mukti_26288 Accessed 29 Oct 2023

DasGupta R, Shaw R (2015) An indicator based approach to assess coastal communities' resilience against climate related disasters in Indian Sundarbans. J Coast Conserv 19:85–101. https://doi.org/10.1007/s11852-014-0369-1

Habitat for Humanity (2018) Guidelines for building a cyclone-resilient home

Halder S, Sen P (2022) Sustainable societal development through holistic education programs. In: Ray P, Shaw R (eds) Technology entrepreneurship and sustainable development. Disaster risk reduction. Springer, Singapore. https://doi.org/10.1007/978-981-19-2053-0_7

Mukhopadhyay A (2016) Living with disasters: communities and development in the Indian Sundarbans. Cambridge University Press, Cambridge. https://doi.org/10.1017/CBO9781316227572

Mukti, Homepage, www.muktiweb.org Accessed 29 October 2023

National Disaster Management Authority of India, Cyclone Do's and Don'ts, https://ndma.gov.in/Natural-Hazards/Cyclone/Dos-Donts Accessed 29 October 2023

Paton D, Johnston D (2017) Disaster resilience: an integrated approach. Charles C Thomas Publisher

Purkait SK (2022) Sundarbans-society, environment and development

Qin H, Li Y, Huang R (2020) Advances and challenges in the breeding of salt-tolerant rice. Int J Mol Sci 21(21):8385. https://doi.org/10.3390/ijms21218385.PMID:33182265;PMCID:PMC7664944

Rudra K (2017) Rivers of the Ganga-Brahmaputra-Meghna delta: a fluvial account of Bengal (Geography of the physical environment)

Samanta B, Das S, Hazra S (2017) Micro level vulnerability assessment of a community living in mousuni island in the Indian sundarban: an integrated study employing geoinformatics. In: Environment and earth observation: case studies in India, pp 195–213

Sen H (2019). The Sundarbans: a disaster-prone eco-region. Springer, Berlin. https://doi.org/10.1007/978-3-030-00680-8

Sentinel Hub, EO Browser, https://apps.sentinel-hub.com/eo-browser/ Accessed 29 Oct 2023

Tideschart, World-Tides, https://www.tideschart.com/ Accessed 29 Oct 2023

Wikipedia, Sesbania, https://en.wikipedia.org/wiki/Sesbania Accessed 29 Oct 2023

Windy Community, Homepage, https://www.windy.com/ Accessed 29 October 2023

World Bank (2016) Building for cyclone resistance: a guide for architects and engineers https://documents1.worldbank.org/curated/en/326581468337788007/pdf/ACS15966-WP-PUBLIC-BRR-report-002.pdf

Toward a Housing Design that Can Withstand Natural Disasters: A Study in Sundarbans

Ankita Kothiyal and Sankar Halder

1 Introduction

Cyclones are considered as one of the most devastating natural hazards across the world considering the gravity of severity, destruction created, and frequency of occurrence. The Sundarbans is a unique ecosystem straddling the coastal areas of India and Bangladesh. Spanning approximately 10,000 km^2, it is recognized as the largest contiguous mangrove forest in the world. The Indian portion of Sundarbans, located in the state of West Bengal, comprises an area of about 4,262 km^2. This constitutes approximately 40% of the total area of the Sundarbans. In 1989, the Indian part of the Sundarbans was designated as a UNESCO World Heritage Site. The vibrant communities of the region rely on the forest and waterways for their livelihood.

The Sundarbans region is uniquely positioned at the confluence of the Ganges, Brahmaputra, and Meghna Rivers. This geographical positioning, coupled with its environmental characteristics, makes the region highly susceptible to riverine and coastal flooding. Moreover, as a low-lying, densely populated delta region, the Sundarbans are extremely vulnerable to sea-level rise and storm surges. The communities here are continuously at risk due to the inundation of settlements, damage to homes and infrastructure, loss of lives, and threats to the unique ecosystem of the region.

More than just mere statistics, these calamities represent shattered lives and broken dreams. Each cyclone, each flood, leaves in its wake a multitude of climate refugees—individuals and families forced to abandon their homes and livelihoods, their heritage, and histories, as they grapple with the devastating impacts of an ever-worsening

A. Kothiyal (✉) · S. Halder
MUKTI, Kolkata, West Bengal, India
e-mail: ankita.kothiyal@muktiweb.org

S. Halder
e-mail: sankar.halder@muktiweb.org

P. K. Ray et al. (eds.), *Technology Innovation for Sustainable Development of Healthcare and Disaster Management*, Disaster Risk Reduction,
https://doi.org/10.1007/978-981-97-2049-1_8

climate crisis. The scars of these disasters are etched not only on the landscape but also on the hearts of those who once called these endangered lands home. We are at a juncture where the raw reality of climate change is no longer a distant threat, but an immediate and existential crisis. The urgent need for action echoes in the broken embankments, the eroded coastlines, and most poignantly, in the eyes of those displaced. If ever there was a moment to confront our climate reality and rally for transformative action, it is now.

This chapter is organized into distinct sections, each delving into a crucial facet of the pressing issue at hand. We begin with an exploration of the profound impact of cyclones on the local population, highlighting the stories of shattered lives and broken dreams left in the wake of these calamities. Following this, we delve into a comprehensive understanding of cyclones, their behavior, and the unique challenges they present, particularly in the Sundarbans.

Moving forward, we explore the historical context of cyclones in the Sundarbans and how climate change is projected to alter their patterns. We then investigated the vulnerability of traditional housing in the region, examining the typical housing styles and construction methods that have left communities exposed to the destructive forces of nature.

Shifting our focus to potential solutions, we discussed the importance of Cyclone-Flood resilient housing as a sustainable countermeasure. We also draw insights from similar initiatives implemented globally and emphasize the vital role played by MUKTI, a socioeconomic development organization, in addressing this pressing problem.

Our objectives behind the construction of Cyclone-Flood resilient houses outlines the goals guiding the construction of resilient housing by a discussion of the design considerations and key principles that guide their construction. Moreover, we explore strategies for repairing or rebuilding damaged homes to facilitate community recovery and resilience.

This chapter culminates with an urgency of action and calling upon all relevant stakeholders to prioritize and allocate resources toward cyclone-resilient housing initiatives in the Sundarbans. Through this structured exploration, we aim to pave the way for a paradigm shift in disaster management, championing a more proactive and pre-emptive approach that safeguards communities and sustains their socioeconomic activities in the aftermath of cyclonic events. It is our hope that this chapter serves as a resounding call to action, offering a path to a more secure and prosperous future for the inhabitants of the Sundarbans.

2 The Impact of Cyclones on the Local Population

Cyclones in the Sundarbans are often catastrophic, causing a tragic loss of life. Despite advances in early warning systems and meteorological predictions, the ferocity of cyclones often leads to significant fatalities and injuries, primarily due to high wind speeds, destruction, flooding, and resulting drownings The infrastructural damage caused by these cyclones is substantial. Homes, schools, healthcare facilities, roads, and bridges often bear the brunt of these storms.

Cyclones can have devastating effects on local livelihoods. Many residents in the Sundarbans depend on agriculture and fishing. The displacement caused by cyclones can lead to temporary or even permanent migration, as people are forced out of their damaged or destroyed homes. This displacement can disrupt social networks, strain resources in host communities, and cause considerable psychological stress.

In terms of health, they often cause outbreaks of waterborne and vector-borne diseases due to the destruction of sanitation infrastructure and flooding. Similarly, they result in extended disruptions to children's education. The unique mangrove ecosystem of the Sundarbans is also negatively impacted by cyclones. The vicious cycle of disaster and recovery hampers the efforts of communities to build resilience and improve their living conditions.

3 Understanding Cyclones

Cyclones, the atmospheric phenomena known by different names worldwide. In the Atlantic and Northeast Pacific, these powerful storms are referred to as hurricanes. The term "typhoon" is used for cyclones in the Northwest Pacific, particularly around Japan and China. In the Southwest Pacific and Indian Ocean, the generic term "cyclone" is used, encompassing cyclonic storms in both regions. The Southwestern Indian Ocean commonly uses the term "cyclone" for these weather systems. The Cyclones originate exclusively over warm equatorial ocean waters. The formation process begins as warm, moisture-laden air ascends from the ocean's surface. This upward movement results in the creation of a cyclonic system. As the air moves away from the ocean, a region of lower air pressure forms beneath it. This low-pressure zone induces the movement of air from adjacent areas characterized by higher atmospheric pressure. The influx of air toward the low-pressure area initiates a warming process, causing the air to ascend once again.

3.1 Cyclones in the Sundarbans

West Bengal's South 24 Parganas district, within which the larger share of the Sundarbans is located, is impacted by cyclones, the most frequent among Indian districts, found a recent study. Sundarbans region, which is in the Northern Hemisphere, cyclones usually have an anti-clockwise direction. The return period of cyclonic storms in the district was 1.67 years on a scale of 1.5–60 years, according to the analysis by the India Meteorological Department (IMD), Pune. Shorter return periods indicate more frequent cyclones (Basu 2022).

The Sundarbans, being located on the Bay of Bengal, is highly susceptible to cyclones. South 24 Parganas was the second-most vulnerable to all cyclones and ranked third in terms of severe cyclonic storms. These cyclones typically form during the pre-monsoon (April-June) and post-monsoon (October-December) periods when sea surface temperatures are high. They can bring high-speed winds, heavy rainfall, and storm surges, causing widespread damage and loss of life. The region's vulnerability to cyclones is expected to increase due to climate change, with projections indicating possible increases in the frequency and intensity of the most severe cyclones.

In general, the region experiences about one to two cyclones per year, although this can vary. Climate scientists are also studying the potential effects of climate change on cyclone patterns, with some suggesting that while the overall number of cyclones might not increase, the intensity of the most severe storms may increase.

3.2 Projected Changes in Cyclone Patterns Due to Climate Change

Over the last 60 years, the Indian Sundarbans have witnessed a loss of 210 km^2 of land due to rising sea levels. By the year 2050, the area is expected to face a notable rise in temperature, more frequent storms, and increased occurrences of flooding. Climate-related disasters are posing a reported risk to at least one-fourth of the population residing in and around the Sundarbans, amounting to eight million people.

Islands like Lohachara and Suparibhanga have completely disappeared, with Ghoramara now facing imminent extinction due to rapid coastal erosion. The Bay of Bengal's water levels have risen at a rate of 4.4–6.3 mm/year, double the global average. This accelerated rise, coupled with the continuous sinking of the Bengal coastline, exacerbates the crisis. At present, villages are situated at altitudes ranging from 1.5 to 3 m above the average sea level. They are susceptible to flooding during lunar tides, which can reach five to six meters. The vulnerable mud embankments offer limited defense against these high water levels, posing an immediate threat of submergence for the residents (Dhara 2020).

3.3 The Vulnerability of Traditional Housing- Typical Housing Styles and Construction Methods

The community residing in the Sundarbans region is among the most severely affected by the impacts of climate change. Repeated instances of devastating cyclones and floods have forced residents to rebuild their houses as many as nine to ten times, given the considerable damage incurred by these natural disasters. The traditional housing pattern, locally known as "Kuccha House," is prevalent in the Sundarbans, and it is particularly vulnerable due to the soft nature of the soil in the region. The type of soil plays a crucial role in house construction. The Ganga, Brahmaputra, and Meghna rivers deposit sediments from various areas, resulting in the formation of islands with soft clay soil, making them susceptible to damage during construction.

Houses in the Sundarbans are primarily constructed using locally available materials such as mud, bamboo, local mangrove wood, straw, and thatch. The walls are typically made from mud, or a bamboo framework filled with mud, while the roofs are thatched with straw or leaves. These houses are elevated off the ground using bamboo or wooden stilts or earthen plinths to protect them from regular flooding. While these houses are cost-effective and use sustainable materials, they are highly vulnerable to damage from cyclones and strong winds. Nowadays, some people have started ordering materials from the city, but the economic conditions of the residents do not permit the construction of concrete houses with high-quality materials.

Considering the frequent occurrence of natural disasters in the Sundarbans, such as cyclones and floods, Pucca houses are not a viable solution due to the substantial land pressure they create. Moreover, the cost of constructing Pucca houses is often prohibitive for the majority of Sundarbans residents, many of whom depend on agriculture or fishing for their livelihoods. Less than 1% of the population can afford such houses. Additionally, researchers have voiced concerns about the adverse environmental impact of cement on the delicate ecological balance of the Sundarbans.

After assessing the devastation caused by cyclones and floods in the region, MUKTI, a socioeconomic development organization with 18 years of experience in the Sundarbans, has devised a solution. They aim to design and construct cyclone and flood-resilient houses for the people of the Sundarbans that offer the safety and durability of Pucca houses but at a more affordable cost. These houses will be lightweight, exerting less pressure on the soil. Locally available materials like Golpata, Nipa Palm, mud, and bamboo are used for construction, minimizing the expenses associated with transporting materials such as sand, bricks, and boat charges.

4 Impacts of Cyclone

4.1 Cyclone Yass (2021)

A relatively strong and very damaging tropical cyclone that made landfall in Odisha and brought significant impacts to West Bengal during late May 2021 The storm brought heavy rainfall, high winds, and significant storm surge, leading to widespread flooding and damage. More than 4,500 villages were damaged. Various rural homes and agricultural lands were hard-hit.

Fig. 1 Impact of Super Cyclone Amphan (Mukti Stands by the Affected People During the Super Cyclone Amphan 2020) (image consent from MUKTI vide email dated 29.01.2024)

4.2 Cyclone Amphan (2020)

The high-speed winds and storm surges caused extensive damage to homes throughout the region Traditional houses were hit hard due to their vulnerability to strong winds and flooding. Many were destroyed or suffered severe damage, resulting in large-scale displacement of residents. 129 people were killed and 4.9 million were displaced from their homes (Fig. 1).

4.3 Cyclone Bulbul (2019)

Though not as severe as Amphan, Bulbul still caused considerable damage in the Sundarbans. The damage was again most severe for kutcha and semi-pucca houses. There were reports of houses losing their roofs, walls collapsing, and flooding causing damage to the structure and contents of the homes. A total of 517,535 homes and 1,489,924 hectares (3,681,680 acres) of crops were damaged or destroyed, with losses reaching Rs 238.11 billion (US$3.34 billion). (Source Wikipedia) The cost of repairs was a significant burden for many families, and some were unable to rebuild their homes to the same standard, resulting in reduced living conditions.

These case studies highlight the vulnerability of traditional and semi-pucca houses in the Sundarbans to cyclone damage. They also underline the importance of developing and promoting more cyclone-resistant housing designs for the region. While pucca houses tend to fare better, they are not immune to damage, particularly from storm surges and flooding, indicating that even these structures could benefit from design improvements to increase their resilience.

5 Importance of Cyclone-Flood-Resilient Housing

Cyclone- and flood-resilient housing plays an integral role in securing communities prone to extreme weather conditions. These structures are designed to withstand the destructive forces of high-speed winds, heavy rains, storm surges, and flooding.

In terms of safety and protection, these houses are capable of withstanding not only the high wind speeds associated with cyclones but also the water pressure and potential structural damage caused by floods. These houses offer residents a stable living environment, reducing the likelihood of displacement caused by natural disasters.

From an economic perspective, cyclone- and flood-resilient housing can lessen the financial strain of post-disaster recovery and rebuilding. By withstanding both types of events, these houses diminish the recurring costs associated with repairing or rebuilding traditional structures. Also, these houses often use locally sourced, ecofriendly materials, integrate energy-efficient features, and incorporate measures to manage excess water, contributing to a sustainable lifestyle and reduced environmental impact.

Low cost involves and long-term benefits of safety, resilience, and sustainability make cyclone- and flood-resilient housing a wise investment in regions like the Sundarbans, which face the dual threat of cyclones and floods.

6 Similar Initiatives Globally

Similar initiatives have been undertaken in other parts of the world, where communities face similar climatic challenges. Here are a few examples:

6.1 (a) Cabadsan Barangay in the Philippines

Typhoon Haiyan which is known as one of the strongest storms to make a landfall in Philippines has destroyed many houses. Acted intervened and helped some of the families in building their new resilient houses (Fig. 2).

6.2 (b) Cyclone Resilient Homes in Australia

Northern Australia is frequently hit by cyclones, leading to developments in cyclone-resistant housing. Architects and engineers have developed designs featuring reinforced concrete and steel structures, cyclone-rated windows and doors, and roofs with a low profile and high wind resistance (Fig. 3).

Fig. 2 Resilient house in Philippines (Purvis 2015) (image consent from Acted vide email dated 24.01.2024)

Fig. 3 Cyclone Resilient Home in Australia (Cyclone and Storm Tide Resilient Building Guidance for Queensland Homes 2024) (image consent from Queensland Reconstruction Authority vide email dated 31.07.2023)

7 What Role MUKTI Has Played to Solve This Problem?

Drawing from its extensive experience in dealing with natural calamities, MUKTI had forged a path toward resilience against nature's fury. The cornerstone of their novel approach lay in the creation of cyclone- and flood-resilient houses—a potent antidote to the perennial scourge of cyclones that plagued the coastal region of Sundarbans, West Bengal. The Sundarbans region, particularly susceptible to the whims of cyclonic destruction, bore witness to a devastating economic impact that rippled through the country, representing an annual loss of nearly 2% of the nation's GDP.

Illuminated by the harsh lessons of the past, MUKTI decided to construct low-cost cyclone & flood resilient houses on a prototype basis, aiming to protect the underprivileged population of Sundarbans during natural disasters and generate an alternative livelihood for them. The creation of a truly robust, cyclone-resistant structure was no small feat; it required a thorough understanding of multiple factors such as wind velocity, building geometry, selection of building materials, soil characteristics, and even the aerodynamics of airflow around the buildings.

Fully cognizant of these complexities, MUKTI engaged the expertise of a seasoned architect. Their task was to craft a low-cost prototype of a cyclone- and flood-resilient house, a project with the noble aim of shielding the most vulnerable sections of society from nature's capricious wrath. Through this pioneering endeavor, MUKTI aspired to empower these communities, bestowing upon them not just a haven from the storm, but the tools to weather it and emerge stronger.

7.1 Objectives Behind Construction of Cyclone and Flood Resilient House

- To ensure that every individual residing in the vulnerable villages has access to a safe home during disastrous events. The project seeks to design and construct houses resilient to cyclones, providing a secure roof over villagers' heads when nature unleashes its fury.
- To safeguard the most vulnerable sections of the community from cyclonic destruction. By constructing robust, cyclone-resistant houses, it intends to significantly minimize the loss of life and property in these high-risk zones.
- Implementation of sustainable and ecofriendly building techniques. The intent is to minimize the environmental impact of construction, thereby promoting harmony with nature.
- To deliver a cost-effective housing solution without compromising on safety or resilience. By leveraging local materials and innovative design, it seeks to ensure affordability for the economically disadvantaged sections of the community.
- To involve the local community in its execution. This inclusion seeks to empower the villagers, enhancing their understanding of the importance of cyclone-resistant construction, and inspiring them to maintain and propagate these structures in the future.
- To create more job opportunities by providing training and capacity-building opportunities to the locals in the construction of these homes. This initiative not only improves local livelihoods but also ensures the sustainability and upkeep of the cyclone-resilient houses.
- To design a cyclone- and flood-resilient housing model that can be replicated and scaled up across all cyclone-prone regions.

8 Design Consideration

Cyclones form over warm tropical oceans and seas, where the air is calm. These favorable conditions can persist for extended periods around the world during late summer. While many cyclones dissipate over the open sea, those that make landfall can have devastating consequences. They often result in severe damage to buildings, leaving countless individuals without homes and causing injuries and loss of life (Fig. 4).

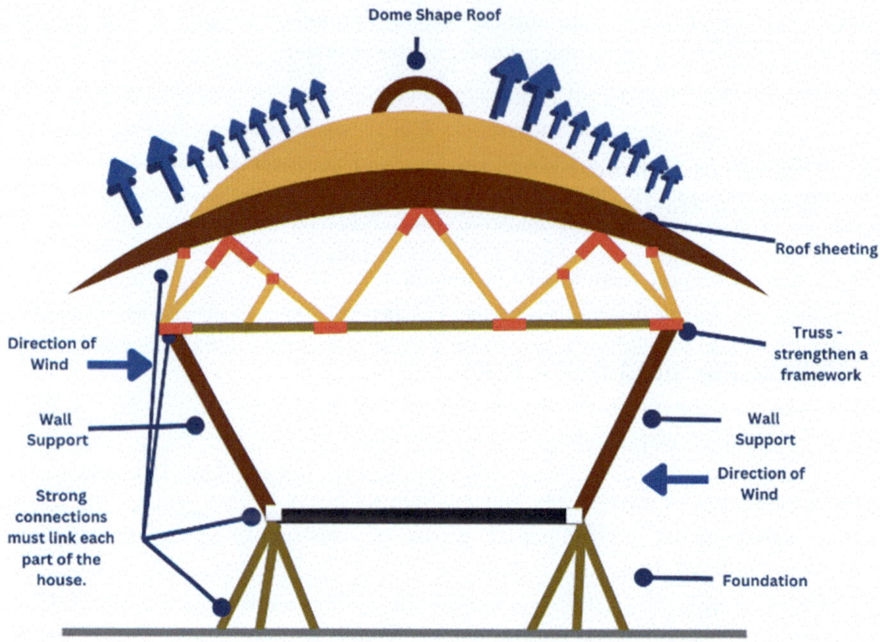

Figure. 4. The image illustrates the concept of a continuous chain of structural elements and connections that transmit the forces from the roof cladding to the ground. Each link in the chain is vital as it only takes one weak connection for cyclonic winds to damage a home

Fig. 5 Cyclone-Flood Resilient House (Mukti 2022a) (image consent from MUKTI vide email dated 29.01.2024)

9 Key Principals Followed in Construction

While designing the resilient houses, the organization kept in mind three things, i.e., affordability, adoptability, and disaster resilience, and aimed to create an appeal for sustainable practices at the same time.

1. **Pyramidal foundation on strip beam**—made of re-engineered prefab RCC posts thus taking the opportunity to improvise upon the existing cottage industry of prefab RCC posts of cross section 5"x5". The broad base of the foundation tied with a strip beam also helps to evenly distribute the load and decrease localized pressure.
2. **Stilt system of splayed RCC columns casted in-situ**—waste metal corrugated sheets were used for shuttering, which lead to the rounded fluted form of the exposed.
3. **Ferrocement staircase**—the splayed column provoked an opportunity to intertwine a staircase within which was made in-situ entirely (Fig.5).
4. **Shallow brick dome floor slabs** contained within a grid of tension tie beams. This was an important technique to show a low-cost method to make floor slabs at the upper level at one-third cost of the conventional RCC slab.
5. **Bamboo roof—**the Bengal "do-chala" shape of the roof was a good solution for a strong aerodynamic Bamboo roof. The structure is designed to take a wind speed of 180km/hr. The bamboo was sourced locally and treated. All the joineries employed lashing techniques with ropes made of "tyre doris" (fish-net material). The masons employed were local boatmakers and bamboo scaffolding experts.

   *The design of the roof—height and span ratio—based on the lateral thrust calculation from wind load, was a dominating factor to define the floor plate size and hence define the beam and foundation grid.
6. **Thatching**—thatch panels were made on the ground and then put up on the roof saving. Seventy percent of the labor work was traditionally done on the roof. The panels were placed very close to each other leading to an overall thickness of about 1.5'. The thatch was also sourced from a particular native species of rice which gives longer and sturdier strands [7].

Successful parameters	
Specifications of house	Features of house
300Sqft of area	Low cost and affordable
Made up from locally available material	Need very less space for construction
Tested in 180 km/hr of high wave	Cyclone resilient
12ft of height which makes it safer in case of high waves	Flood resilient
Tested at the time of Cyclone Yaas	Creating employment opportunities for local
A high-end design team is involved in construction	Can be used as a Home Stay

TYPE A^

- 12ft height
- Withstand 180km/hr of wave.
- Interlock ash bricks for wall
- Tiles roof (local mud tiles)
- Supporting bamboo structure for roof
- Cell roof for first floor
- Two floored

TYPE B^

- 12ft height
- Withstand 180km/hr of wave.
- Made of fully Bamboo
- Mud used for wall.
- Metal sheet for roof, Thatched underneath
- Two floored

TYPE C^

- 12ft height
- Withstand 180km/hr of wave.
- Tripod Structure to withstand high speed wind from any direction.
- Shallow dome
- Japanese Umbrella technology
- Thatched rooftop

^ Images depict the work done by MUKTI in the Sundarbans region of West Bengal, India (Mukti 2022a) image consent from MUKTI vide email dated 29.01.2024).

10 Repairing or Rebuilding Damaged Homes

MUKTI provided local people assistance in rebuilding their homes. Till now, 100 such cyclone-and flood resilience have been rebuilt. Hundreds of people in remote areas of the Sundarbans are dreaming of building their own houses that will not be damaged by the cyclone. With the support of their partners, MUKTI is transforming these dreams into reality. These sturdy dwellings are poised to serve not just as safe homes, but also as homestays, opening an alternative avenue of income for the locals.

MUKTI supports the community in three distinct ways:

- Assisting those who lack government support and are in poor financial conditions. Beneficiary contributions in this scenario encompass locally available materials, items already present in their homes, and labor. Essentially, the costs are offset by what the beneficiaries can manage themselves.
- Helping those who have completed the roof casting, making their homes cyclone resilient. However, they still need to construct an additional floor to make their houses flood resistant. In these cases, the beneficiaries are expected to provide labor.
- Providing aid to those who have neither cyclone-tolerant roof casting nor flood-tolerant infrastructure. In this context, beneficiary contributions involve locally sourced materials and labor

11 Conclusion

In the face of an increasingly unpredictable climate, cyclone-tolerant housing emerges as a critical element for the future resilience of the Sundarbans. These robust structures offer a lifeline of safety to the residents of this cyclone and flood-prone region, mitigating loss of life, reducing economic hardship, and fostering a sense of stability in communities persistently under threat. The recent disaster in the Turkey–Syria earthquake has led people to think about design technology as the poor construction of buildings has led to massive destruction. This becomes a call to action for policymakers, builders, and residents alike where each can contribute to build a better and safer environment for its people. The Joshi math tragedy in Uttarakhand has proven that it's important to involve residents in the process, contributing their invaluable local knowledge and experience to help tailor these solutions to their specific needs and circumstances. Throughout this chapter, we have explored the profound impact of cyclones on the local population, understanding cyclones' behavior and challenges they present, the historical context of cyclones in the Sundarbans, and the vulnerability of traditional housing. We have also delved into the significance of Cyclone-Flood resilient housing as a sustainable countermeasure and highlighted the crucial role played by organizations like MUKTI in addressing this pressing problem.

As we conclude this chapter, the need for cyclone-tolerant housing is not limited to the Sundarbans alone. Many other regions facing similar challenges could benefit from these innovative solutions. As we continue to contend with the impacts of climate change, embracing such locally driven, adaptive measures becomes not just beneficial but essential.

"For future prospects, it is imperative that we actively promote and advocate for the adoption of sustainable housing solutions. This not only helps in reducing the environmental impact but also enhances safety and resilience in the face of ever-increasing climate challenges". By raising awareness, providing incentives, and facilitating access to cyclone-tolerant housing, we can empower communities to make informed, sustainable choices for a safer and more secure future.

In summary, this chapter serves as a resounding call to action, offering a path to a more secure and prosperous future for the inhabitants of the Sundarbans and other vulnerable regions worldwide. Through collective efforts, we can pave the way for a paradigm shift in disaster management, championing a more proactive and pre-emptive approach that safeguards communities and sustains their socioeconomic activities in the aftermath of cyclonic events.

References

Basu J (2022) Sundarbans is cyclone capital of India: IMD report. https://www.downtoearth.org.in/news/naturaldisasters/sundarbans-is-cyclone-capital-of-india-imd-report-81244 (n.d.) (Assessed on 22nd Sept 2023)

Cyclone and Storm Tide Resilient Building Guidance for Queensland Homes (p19) published by the Queensland Reconstruction Authority (https://www.qra.qld.gov.au/resilient-homes), with images courtesy of the James Cook University Cyclone Testing Station (https://www.jcu.edu.au/cyclone-testing-station). (Assessed on 29 Jan 2024)

Dhara C (2020) West Bengal's climate change conundrum part III. https://www.firstpost.com/india/west-bengals-climate-change-conundrum-part-iii-extraordinarily-rapid-sea-level-rise-in-sundarbans-turns-families-into-refugees-6120781.html (Assessed on 29th January 2024)

Mukti (2022) Glimpses of model village, Purba Sridharpur. https://muktiweb.org/gallery/album/glimpses-of-model-village-purba-sridharpur, (Assessed on 29 Jan 2024)

Mukti (2022) Thatching work for cyclone tolerant houses is going on at Mukti Gram, Available at https://muktiweb.org. (Assessed: 29 Jan 2024)

Mukti stands by the affected people during the Super Cyclone Amphan (2020) https://muktiweb.org/mukti-old/mukti-stands-by-the-affected-people-during-the-super-cyclone-amphan/ (Assessed on 29 Jan 2024)

Purvis K (2015) Years of the floods: building homes to withstand the weather – in pictures. The Guardian. 21 April. https://www.theguardian.com/global-development-professionals-network/gallery/2015/apr/21/flood-cyclone-resistant-buildings-asia-gallery (Accessed on 29 Jan 2024)

Innovative Sanitation System for Disaster Victims and Social Sanitation Challenges

Yasuhiro Soshino and Akira Miyata

1 Introduction

In human history, the numbers of natural disasters have afflicted people. Even in the twenty-first century, natural disasters such as the 2001 Gujarat earthquake in India, the 2003 Bam earthquake in Iran, the 2004 Sumatra earthquake and tsunami disaster, the 2005 Pakistan earthquake, the 2007 Myanmar cyclone disaster, and the 2008 cyclone in Bangladesh, the 2010 Haiti earthquake, the 2011 Great East Japan Earthquake and Tsunami, the 2013 Philippines typhoon Haiyan, the 2015 Nepal earthquake, and the 2023 Turkey earthquake caused catastrophic human losses in the disaster affected areas. In addition, disasters cause devastating damage to the basic infrastructure (Okada et al. 2011) such as the water supply, the power supply, the fuel supply, and the drainage. This makes the people in the disaster-affected areas more and more difficult to survive and forces them to move to the refugee centers. In the past disasters in Japan, the disaster-affected people took refuge to public facilities such as schools, stadiums, parks, gymnasiums, car parks, and the office of the local authorities. However, the challenges in the living conditions at such refugee centers have been frequently reported including the past disasters. In the Great East Japan Earthquake and Tsunami, due to the harsh living conditions in the refugee centers, various health problems (Nohara 2012; Omama et al. 2013) were reported. The poor sanitation facilities in the refugee centers cause significant health problems. For example, the refugees hesitate to drink water due to the poor sanitation conditions in the toilet at the refugee center, which increase the risks of the dehydration and other cardiovascular diseases such as economy-class syndrome (Shibata et al. 2014). The

Y. Soshino (✉)
Disaster Management Research Institute, Japanese Red Cross College of Nursing, Tokyo, Japan
e-mail: y-soshino@jrcdmri.redcross.ac.jp

Y. Soshino · A. Miyata
Japanese Red Cross Kumamoto Hospital, Kumamoto, Japan

P. K. Ray et al. (eds.), *Technology Innovation for Sustainable Development of Healthcare and Disaster Management*, Disaster Risk Reduction,
https://doi.org/10.1007/978-981-97-2049-1_9

hesitation for drinking water also increases the risks of heatstroke. In addition to the health risks, sanitation in the refugee center is an important factor that may affect the evacuation behavior of the disaster-affected people. In the 2010 Chile earthquake and tsunami, due to the poor condition of the toilet at the evacuation centers, nine percent of the tsunami evacuees returned home in the lower places while the tsunami alert had been lifted in Iwate prefecture, Japan (Yoshida et al. 2011). After the 2015 Nepal Earthquake, before adequate sanitation was built in settlement areas, the people who lost their houses or who were afraid of returning homes had to consider open defecation (Hall et al. 2017).

While the past reports on the school facilities in Japan frequently pointed out the importance of the service continuity of the basic school infrastructure, the living conditions of the refugee centers were quite tough for the refugees. The harsh living conditions at refugee centers cause challenges in protecting the health and the human dignity of the refugees.

In addition to the refugee center, disasters disturb the service continuity of the medical facilities in the disaster-affected areas. The past disasters in Japan frequently caused the damage to the hospital infrastructure in the disaster-affected areas (Ochi et al. 2014). The instability of medical services prevents the disaster-affected people from receiving the proper medical care in emergencies. Due to the 2003 blackout in the northeastern United States, toilets in the disaster-affected hospitals were not able to be flushed which resulted in improper sanitation conditions (Klein et al. 2005). Due to the burst of the water pipes in the city's water line, the hospital in Massachusetts closed unused toilet facilities (Milsten 2000).

Furthermore, the living conditions of the relief workers are also important issues in humanitarian workers. To ensure the health condition of the relief workers are crucial. In the international medical relief activities, the self-sustainable operation of the medical teams is the principle of the relief operation. However, it is always difficult in ensuring the proper sanitation facilities in the disaster-affected areas. The localization, distribution, and hygiene of toilets are a major challenge in a field hospital in Haiti. During the first day after the establishment of the field hospital, no toilet was available for healthcare staff, patients, and their family members (Lichtenberger et al. 2010).

In addition to the natural hazards, the aging of infrastructure can be likened to "the silent disaster." In Japan, where the coverage rate of sewerage system is 80.6% in 2022 (JSWA 2022), the maintenance of the public infrastructure is becoming more and more difficult due to the limitation of the budget in local authorities which have been suffering from the rapid progress of the super-aging society and the population reduction. In other words, in Japan, the vulnerability of the basic infrastructure increases while the occurrences of the future catastrophes such as Nankai Trough Earthquake and Tsunami, Tokyo inland Earthquake, and Japan and Chishima Trench Earthquake and Tsunami are expected. In this sense, the simple and standalone borehole toilets, which used to be common before the spread of sewage systems in Japan, are the resilient solution since they function without water supply, drainage, and the sewage treatment system.

Technically speaking, in emergencies, the main technical reasons for the suspension of the usual flush toilets are caused by (1) lack of electricity, (2) lack of water supply, (3) lack of drainage, and (4) the damage to the sewage treatment plant in the community. As our houses, working places, and other buildings are linked to the power plant in the community with electric wires, the flush toilets are also linked to the water supply and treatment system with the water pipes and drainage. Once the links are cut off by natural hazards, a lot of flush toilets in the community are affected. In this sense, in terms of the damage control of the community, the application of off-grid flush toilets is one of the resilient solutions.

Therefore, this chapter presents the development of the innovative movable flush toilet. This innovative toilet functions without relying on existing water supply, drainage, and the sewage treatment plant. The movable toilets were deployed and utilized in three different floods operations in Japan. In addition, this chapter also describes the application of the movable flush toilets as a new solution for the social challenges in Japan. Following the introduction, Section two describes the design and development of the movable flush toilets. Section three presents the three case studies of the deployments of the toilets in emergencies and section four summarizes the application of the toilets in peacetime. Discussion is described in section five and section six concludes this chapter.

2 Background

2.1 The Smart Design Shelter Concept

In 2000, the Japanese Red Cross Kumamoto Hospital (hereafter Kumamoto Hospital) was appointed as an international medical relief hub hospital in the Japanese Red Cross Society. Since then, the Kumamoto Hospital started the design and development of field hospital equipment to be used for global disaster response operations. A field hospital can be compared to the convergence in technologies, in other words, the assembly of movable independent infrastructures such as generators, water supply system, tents, telecommunication, and other medical equipment. The improvement of such movable infrastructure directly affects the rapid transportation of the field hospital equipment, and medical services in the disaster-affected areas. The development of mobile flush toilets was commenced from 2008 since the hospital introduced commercial mobile flush toilets. After the 2011 Great East Japan Earthquake and Tsunami, the Kumamoto Hospital started the joint project on the application of lithium-ion batteries with Kyushu Electric Power. In this joint project, the hospital proposed the “Smart Design Shelter Concept,” which is the resilient network of movable off-grid infrastructures. This concept proposes the daily uses of movable infrastructure such as batteries, water-recycling shower, sanitation kits, and movable flush toilets in public facilities which are used as refugee centers in emergencies. In case of disasters, these movable infrastructures are transported and installed in

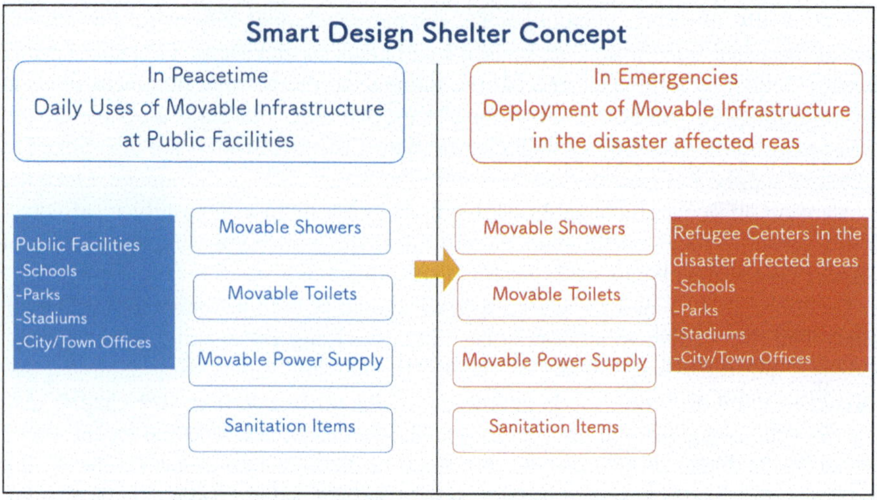

Fig. 1 The smart design shelter concept

disaster-affected areas (Fig. 1). To realize this concept, the Kumamoto Hospital started the design and development of the movable flush toilet.

2.2 Application of the Project Cycle Management Method

In designing the movable flush toilets, the authors applied some procedures in the Project Cycle Management (hereafter "PCM") Method (FASID 2004). The PCM method is developed by the Foundation for Advanced Studies on International Development (FASID) in Japan based on the logical framework which is developed by the United States Agency for International Development (USAID) and ZOPP method developed by the Gesellschaft fur Technische Zusammenarbeit (GTZ). The PCM method manages the whole cycle of a development project consists of planning, implementation, and evaluation by using a project management format called the Project Design Matrix. In designing the movable flush toilets, the authors used the Problem Analysis, the Object Analysis, and the Project Selection processes to identify the concept of the movable flush toilets.

In the planning process, the Problem Analysis process visualizes the logical relationship (causes and effects) between the existing challenges by drawing the problem tree (Fig. 2).

Following the Problem Analysis process, the Objective Analysis process, by drawing the objective tree (Fig. 3), clarifies the "means-ends" relationship between the desirable situation that would be achieved once the existing challenges have been solved. Then, in the project selection, multiple approaches are identified to select a project based on the results of the Objective Analysis (Fig. 4).

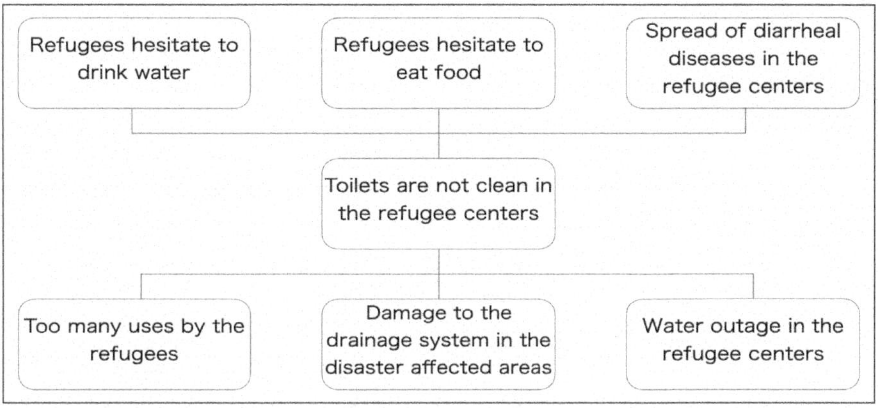

Fig. 2 The example of the problem analysis

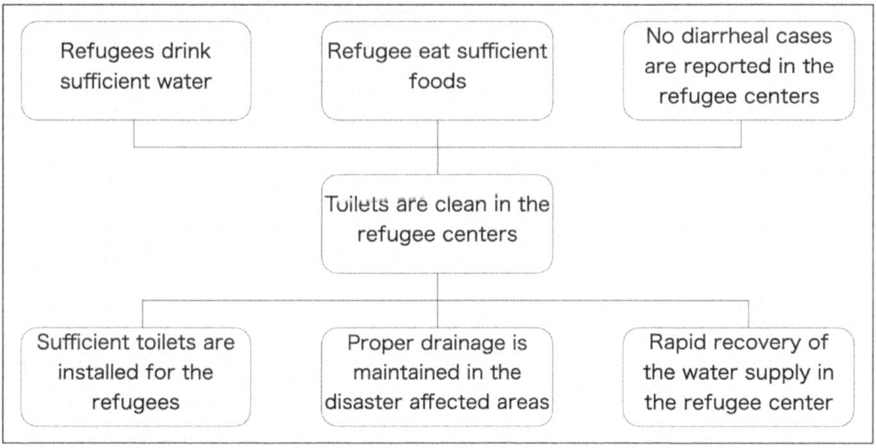

Fig. 3 The example of the objective analysis

2.3 Concept Extraction by Using the PCM Method

In designing the movable flush toilet, the authors used the PCM method to extract the concept of the movable flush toilet (Soshino et al. 2018). Firstly, problem analysis was conducted by drawing a problem tree on the sanitation challenges in refugee centers. From the focal problem "Improper Toilet at the refugee centers," the three direct causes were identified. Then, the causes of the three direct causes were analyzed. In this way, the existing challenges on sanitation in the refugee centers were shown in the problem tree (Fig. 5). Secondly, objective analysis was conducted by drawing the objective tree. In this process, the challenges posted in the problem tree have been rewritten as the direct objectives (Fig. 6). In the objective tree, each item formulates

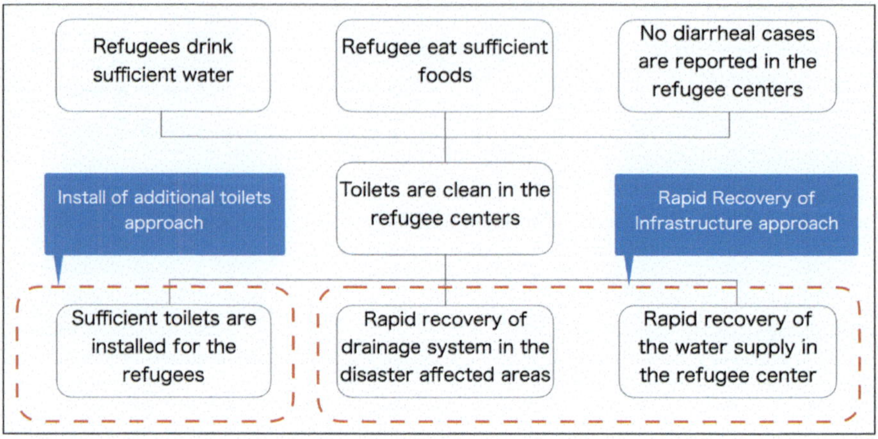

Fig. 4 The example of the selection of approaches

a group as a basis of the project, which is called an approach. In the approach selection, the approaches are selected by circling the groups in the objective tree and naming the selected approaches (Fig. 7). By using the PCM method, four approaches; "Standalone," "Monitoring," "Mobile," and "Design for Daily Use" are extracted. Based on this concept, the development of movable flush toilets started in 2013.

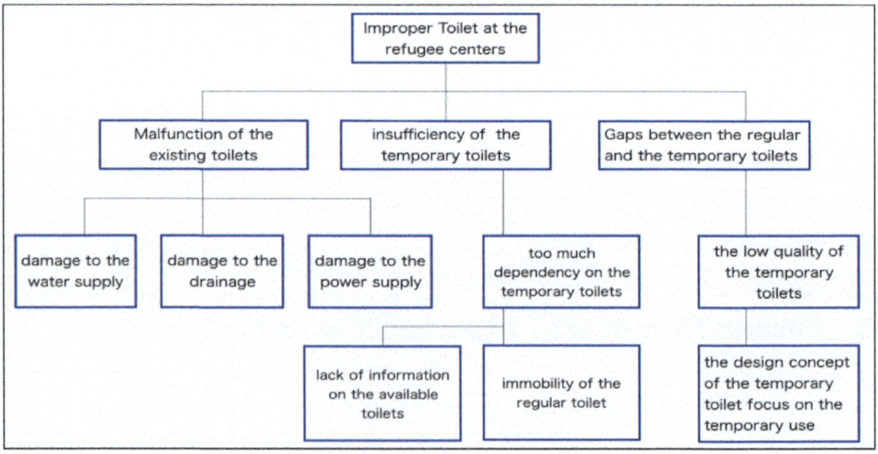

Fig. 5 The problem analysis in designing the toilet (Soshino et al. 2018)

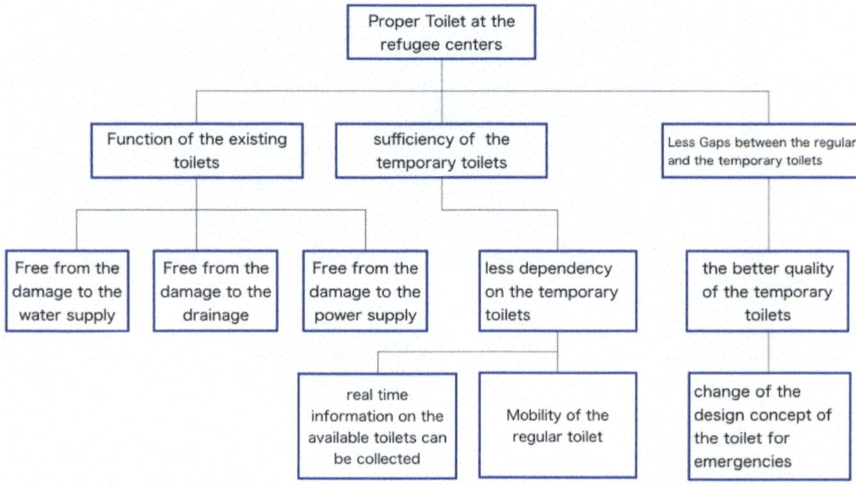

Fig. 6 The objective analysis in designing the toilet (Soshino et al. 2018)

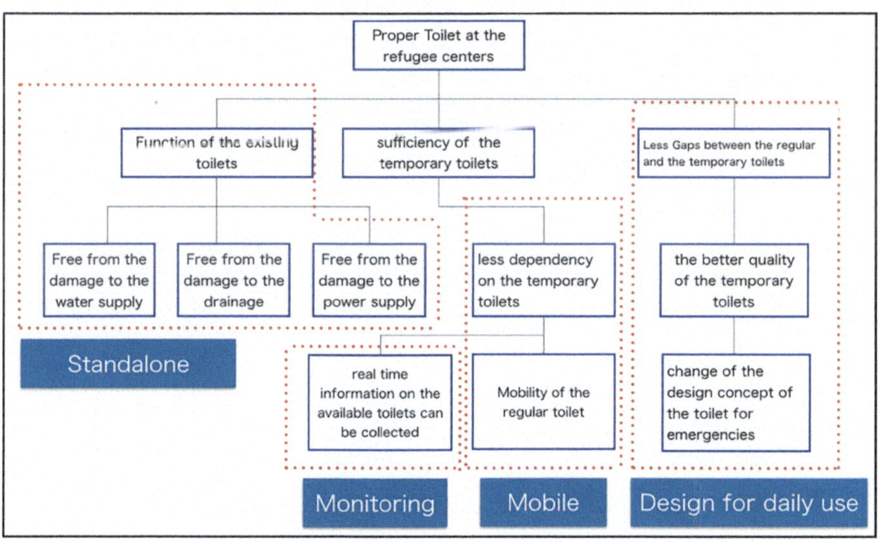

Fig. 7 Selection of approaches based on the objective tree (Soshino et al. 2018)

2.4 Development of Movable Flush Toilets

Figure 8 presents the overview of the basic system of the movable flush toilet. The toilet has two toilet rooms and one treatment room. Once the excreta is flushed with reserved water, the excreta is pulverized and reaches a raw water tank. From the raw water tank, the pulverized water is sent to the anaerobic treatment tanks. The water

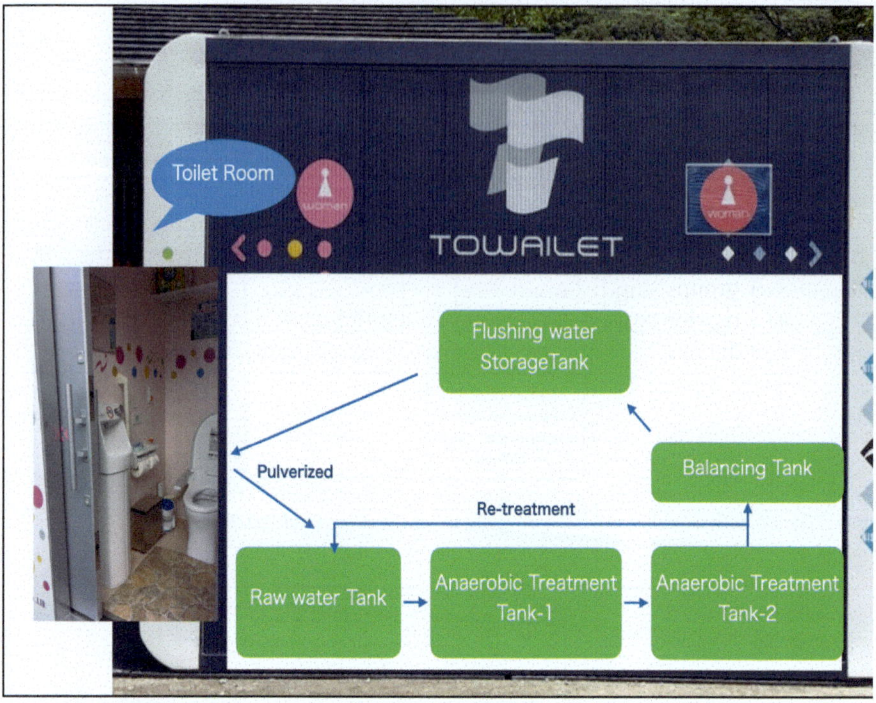

Fig. 8 The overview of the treatment flow of the movable flush toilet

is sent to the balancing tank and sent to the first aerobic treatment tank again, or sent to the flushing water storage tank. This process is cycled and finally, the water is sent to the flushing water tank to remove color and other impurities from water filter.

3 Deployment

3.1 The Northern Kyushu Heavy Rainfall in July 2017

3.1.1 Background

From the end of June and the beginning of July 2017, the active seasonal rain front stayed in the western part of Japan. Typhoon Nanmadol also caused heavy rains in western parts of Japan. The flow of the warm and moist air formed the liner rain belts which caused fierce rains in the same areas of Kyushu such as Asakura in Fukuoka prefecture and Hita in Oita prefecture. Due to the heavy rains, in Asakura, Hita and other areas of northern Kyushu, floods and landslides caused serious human losses and damage to infrastructure. In Oita and Fukuoka prefecture, 37 people were killed and a lot of houses were damaged or destroyed (FDMA 2018).

3.1.2 The Deployment of the Movable Flush Toilet

To support the volunteers who engaged in recovery activities, the prototype of the movable flush toilet was deployed and installed in Haki area where the floods caused serious damage in July 2017. In Haki, a lot of the volunteers were deployed to clean up the debris and mad in the flood-affected areas while limited public toilets were available. The toilet was transported by a medium size truck by using a crane for off-loading. Initially, one point five metric tons of water was put into the toilet to start the treatment procedures. The toilet was operated from September to November 2017 and 1,792 people used the toilet. During the operation, the quality of the recycled water was examined to find that the quality of the water qualified for the national standard.

3.2 The Western Japan Heavy Rainfall in July 2018

3.2.1 Background

In 2018, from July 5th to 7th, the seasonal rain front was stagnant near Honshu, and warm, moist air flowed toward this front, and the front continued to be active in the western part of Japan. In Hiroshima Prefecture, the occurrence of landslides was seen in various places, and In Okayama Prefecture, large-scale flood damage occurred mainly in the town of Mabi, Kurashiki City (FDMA 2019). On July 18, 2018, the Kumamoto Hospital and the staff from Nishimu conducted an assessment mission in the flooded areas of Mihara City in Hiroshima Prefecture and the town of Mabi in Okayama Prefecture to identify suitable place to support by installing a movable flush toilet. Among the several assessed places, the damage in the town of Mabi was severe. In the town of Mabi, due to the levee breach of the Oda River, nursing facilities, hospitals, and schools were flooded. The flood depth was estimated to reach a maximum of approximately five meters high. Due to the flood, the ground floor of the Mabi Memorial Hospital was severely flooded (Fig. 9). In addition, three hundred people such as hospitalized patients were temporarily isolated. The damage of the hospital infrastructure and equipment was quite severe and a lot of the hospital staff were engaged in the recovery effort. The electrical equipment in the hospital was flooded which resulted in the power outage in the hospital. Due to the damage of the hospital infrastructure, the hospital staff were using the temporary toilets which were installed on the premises of the hospital.

Fig. 9 The backyard of the flooded hospital in Mabi, Okayama prefecture

3.2.2 The Deployment of the Movable Flush Toilet

The assessment team decided to support the Mabi Memorial Hospital based on the discussion with the hospital staff. The mobile flush toilet was transported from Saga Prefectures and installed on 23 July 2018. A water tanker was deployed to pour 3 metric tons of water to start the treatment process of the toilet. A photovoltaic system consists of solar panels and batteries were installed to operate the toilet. The toilet installation work was completed in one day, and operation began the next day (Fig. 10). From 23 July to 26 August, 856 people used the toilet (Fig. 11).

During the operation, the feedback of the users were collected in a notebook in a female toilet room. Table 1 presents the feedback from the female users of the toilet in the Mabi Memorial Hospital.

Fig. 10 The movable flush toilet installed at the flooded hospital in Mabi

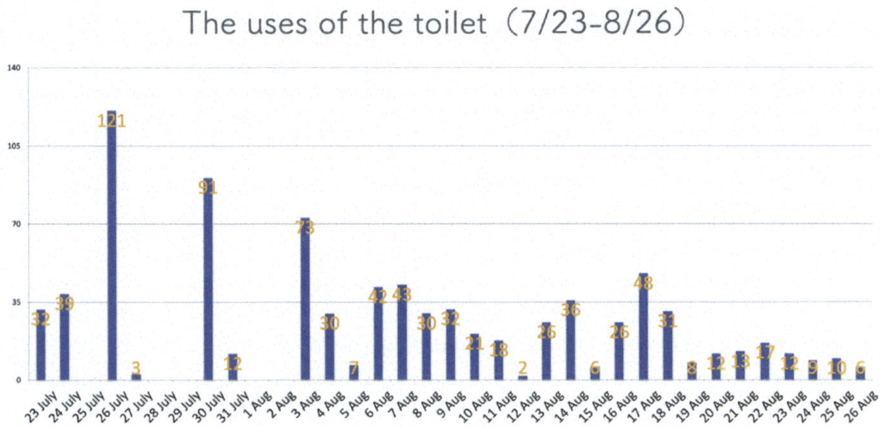

Fig. 11 The number of the users of the toilet

Table 1 The feedback from the female users of the movable flush toilet

Dates	Users	Comments
24 July 2018	A	(The toilet) is very clean, easy to use without stress. Thank you
26 July 2018	B	Never expected such a clean toilet (in the flooded areas). It is especially helpful during menstruation
27 July 2018	C	It is good that this toilet has a mirror while other (temporary toilets do not have mirrors. Compared to the temporary toilet, this toilet is clean and more comfortable. This toilet should be installed in various places (in the flooded areas), especially in refugee centers

3.3 *The 2020 Kumamoto Floods*

3.3.1 Background

In Japan, from the beginning of July 2020, the very active seasonal rain front extended from Kyushu region to Tohoku region in Japan. In Kyushu region, the historical rain was observed from 4 to 7 in July 2020. In various places in Kyushu region, the amount of precipitation for 4, 48, and 72 h has surpassed the highest value in recorded history (FDMA 2021). On July 3, in Hitoyoshi City of Kumamoto Prefecture, the radical elevation of the water level was seen in Kuma River and its tributaries. In result, floods and landslides struck in wide geographical areas in Kumamoto prefecture. In Kumamoto prefecture, 65 people were killed, two were missing and 51 people were injured. In addition, more than 7300 houses were damaged or destroyed and basic infrastructure such as roads, bridges, and railways were seriously damaged.

3.3.2 The Deployment of the Movable Flush Toilet

The Kumamoto Hospital deployed an assessment team to provide immediate assistance by using the movable infrastructure including the toilets. To prioritize the destination of the first assessment, the assessment team utilized the mobile phone location data provided by Agoop Corp. According to the mobile phone data, the concentration of the population was observed in a refugee center in Hitoyoshi City. Therefore, the assessment team decided to visit Hitoyoshi City and its surrounded areas. The central areas of Hitoyoshi City nearby Kuma River were severely flooded. Kuma village which was one of the most serious flooded areas (Fig. 12), the disaster response headquarters were established under the roof of the sports facility (Fig. 13). Even though the public toilet was functioning, emergency responders installed a portable packable toilet in a place for evacuees and inside the public toilet while little users were seen during the assessment. Based on the report by the assessment team, the movable flush toilet was installed next to the public toilet as a females' toilet (Fig. 14). The toilet was used by the female staff in medical teams, volunteers, and the staff of the local authorities including the Kuma Village office. According to the female staff at the Kuma Village Office, compared to the temporary toilets, the movable flush toilet was clean and comfortable to use (Shiga et al. 2020). Another movable flush toilet was installed in the debris collection site caused by the floods (Fig. 15). In Japan, debris and garbage caused by disasters are collected in wide spaces such as sport facilities with limited capacity of toilets. The movable toilet at the debris collection site was used by the workers in charge of the disposal of the debris. The movable flush toilet at the debris collection site was operated from September 2020 to February 2021 and 25,815 people in total used the toilet (Fig. 16).

Fig. 12 The road in the Kuma village which is one of the worst affected areas

Fig. 13 The emergency headquarters of Kuma Village Office

Fig.14 The movable flush toilet installed as a female's toilet

Fig. 15 The movable flush toilet installed at the debris collection site

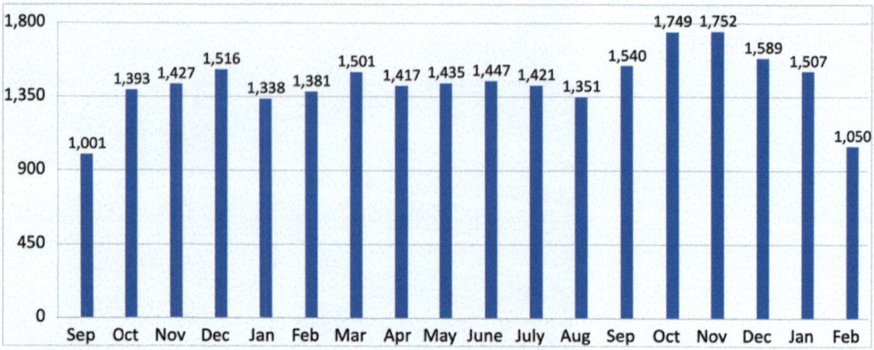

Fig. 16 The users of the toilet at the debris collection site

4 Application in Peacetime

4.1 Islands

In various islands in Japan where the population is decreasing, it is becoming difficult to install, renew, or maintain sewage treatment plants. For this reason, the excreta is transported from the island to the sewage treatment facility on the mainland. While the development of the sewage system on islands becomes more and more difficult, the local authorities in the islands are trying to attract tourists to stimulate the local economies. Therefore, the movable flush toilet with the sewage treatment system is one of the new solutions for the local authorities to install the sewage system in islands with the limited resources. By installing the movable flush toilets, the local authorities can reduce the construction costs of the toilet, and the transportation cost of the excreta from the island to the treatment center in the mainland. In 2019, Kamiamakusa City of Kumamoto prefecture decided to install a mobile flush toilet on the Yushima Island (Fig. 17). The population of the Yushima Island is approximately 300 people and the island can be accessed by a ferryboat within 30 min from the mainland. Nowadays, the Yushima Island is famous for a lot of cats that attract many visitors who like cats. The local authority of Kamiamakusa City chose the movable flush toilet since this new solution reduces the construction cost of the sewage system, transportation cost for the treatment of excreta. In this way, the installations of the movable flush toilets have been proceeded as a new solution for the infrastructure development in island areas of Japan.

Fig. 17 The movable flush toilet in the Island of Yushima, Kamiamakusa, Kumamoto, Japan

4.2 Campsites

Similar to the island areas, movable flush toilets are new solutions in camping sites where there are challenges in developing the sewage system since the camping sites are normally located in remote areas. In addition, the visitors concentrate on the particular seasons while there are limited visitors particularly in the winter season. In such circumstances, the periodical uses of the movable flush toilets are new way of infrastructure development. From 20 September to 30 October 2022, a field test of the movable flush toilet was conducted in a camp site at the Shikotsu Lake in Hokkaido (Fig. 18). During the field test period, more than eight hundred people used the toilet. Same as in island areas, the camp sites are located in remote areas from the sewage treatment centers. Therefore, to develop a new toilet at the camp sites, the construction cost naturally increases if the toilet is connected to the existing sewage system. On the other hand, the treatment cost is needed to transport the excreta from the camp site to the sewage treatment center. In addition, in order to protect the natural environment around the campsites, it is impossible to discharge untreated sewage. Therefore, the application of movable flush toilets is one of the new solutions to install the proper sewage system at the campsites.

5 Discussion

This chapter presented the emergency deployments and the daily application of the movable flush toilets. The application of the PCM method identified the concept of the movable flush toilets. While the movable flush toilets were effective for the rapid recovery in the disaster-affected areas, the toilets were also effective to reduce the cost of the infrastructure development in islands areas and campsites. In other words, the concept of the movable flush toilets, "Self-sustainable," "movable," "monitoring,"

Fig. 18 The movable flush toilet in the campsite in Hokkaido, Japan

and "for daily uses" are the common solutions for the social challenges in peacetime and in emergencies. In peacetime, the movable flush toilet contributes to reduce the construction cost in islands and campsites where the existing local sewage system is not available. In emergencies, the movable flush toilets offer the rapid recovery of the toilet services to fill the gap between the quality of lives before disaster and in emergencies. Regarding the disaster-related budgets of local governments, while the focus is on post-disaster recovery and reconstruction, limited budget is allocated for disaster preparedness in peacetime (Ministry of Economy, Trade and Industry 2023). Since the mobile flush toilets reduce the construction cost of the toilets for the visitors, local authorities can use the budget for tourism in peacetime. If disaster occurs, the local authorities in the disaster-affected areas can install the mobile flush toilet as a post-disaster recovery program to prepare for the future disasters. In this way, the mobile flush toilet contributes to build back better in the disaster-affected communities.

From the practical point of view in the humanitarian services, the movable flush toilet presented its potential to the disaster-affected people, in particular the females, by providing a level of quality that can be daily used in peacetime. In addition, the movable flush toilet also supported the staff in the first responders. In fact, the past global humanitarian operations faced the challenges in ensuring the proper management of sanitation issues in the field. After the 2010 Haiti earthquake, cholera spread from the UN camp while no cholera case had been reported before the earthquake in Haiti. Lantagne et al. (2014) pointed out that UN installations should introduce the on-site treatment system that can inactivate the treat fecal wastes before disposal to prevent the spread of infectious diarrheal diseases to the affected areas by aid workers. The movable flush toilet can treat the fecal wastes without relying on the local sewage system or releasing human waste into rivers. In this way, the movable flush toilet contributes to improve the sanitation conditions in the disaster-affected areas. Furthermore, the remote monitoring system in the movable flush toilets improve the efficiency and accountability in humanitarian services. In disaster response operations, a lot of sanitation facilities are installed to support the lives of the disaster-affected people. However, it has been very difficult to know how many times, when and by whom the sanitation facilities such as hand-washing devices, toilets, and showers are used in the disaster-affected areas. The remote monitoring system in the movable flush toilet reports the conditions of the toilet which ensures the accountability and efficiency in the disaster response operations.

The movable flush toilet can be installed without connecting with the existing water supply and drainage network. These days, the development and maintenance of the infrastructure have been more and more difficult in rural areas of Japan where the super-aging and the population decrease is ongoing. The movable flush toilet could be utilized instead of allocating a lot of budgets to build or keep the sewage systems to install a few toilets in rural areas of Japan. In this sense, the movable flush toilets are considered to be a new solution for the infrastructure development method that is different from the normal sewage system not only in Japan but in emerging countries. Urban areas in emerging countries are not equipped with sanitation infrastructure to keep pace with rapidly increasing populations. The movable flush toilets can be a future solution to substitute the regular flush toilets and sewage systems in emerging countries if the cost of the movable flush toilets is reduced in the future.

In addition to the cost issues, there are the challenges in the laws and regulations in Japan toward the widespread uses of the movable flush toilets. Japan's Building Standards Act stipulates that no toilets other than flush toilets connected to the sewage system may be installed within the treatment area specified by the Sewerage Act. In this way, the existing laws and systems were not developed with this new toilet in mind. As the development and improvement of movable flush toilets are proceeded, this will become a difficult challenge. In order to respond to this legal system, it is necessary to promote further joint demonstrations with related government agencies and local governments. In addition, toward the widespread uses in the world, the practical field tests in different countries should be conducted. Currently, the uses of the movable flush toilets have started in Japan and the toilet is developed mainly for domestic uses in Japan. However, in the uses in the different countries, different quality and function will be required. The performance of toilets designed and manufactured for the use in Japan could be excessive in other countries. Since the movable flush toilets have a potential to solve the global sanitation challenges, further field tests and new trials are needed for further improvement.

6 Conclusion

This chapter presented the design and development of the innovative movable flush toilet with an independent treatment system inside the toilet, which functions without relying on the existing water supply and the sewage treatment plant. In emergencies, the movable flush toilet was deployed for the flood relief operations to support the early recovery in the disaster-affected areas. In peacetime, the toilets have been installed at the islands and the campsites. The movable flush toilets are not conventional toilets connected to a series of systems consisting of a sewage system and a sewage treatment plant. In the islands and the campsites, the movable flush toilets contributed to the reduction of construction costs and excrement transportation costs. The widespread uses of the movable flush toilets that function without existing infrastructure could contribute to enhance the resiliency in the society to prepare for emergencies.

References

Fire and Disaster Management Agency (FDMA) (2018) Fire Service White Paper 2017 [Online]. https://www.fdma.go.jp/publication/hakusho/h29/items/special1.pdf. Accessed 31 May 2023
Fire and Disaster Management Agency (FDMA) (2019) Fire Service White Paper 2018 [Online]. https://www.fdma.go.jp/publication/hakusho/h30/items/special1.pdf. Accessed 31 May 2023
Fire and Disaster Management Agency (FDMA) (2021) Fire Service White Paper 2020 [Online]. https://www.fdma.go.jp/publication/hakusho/r2/items/special1.pdf. Accessed 31 May 2023
Foundation for Advanced Studies on International Development (FASID), PCM Project Cycle Management, Management Tool For Development Assistance, Participatory Planning, 6th edn.
Hall ML, Lee ACK, Cartwright C, Maharatta S, Karki J, Simkhada P (2017) The 2015 Nepal earthquake disaster: lessons learned one year on. In: Public health. Elsevier, pp 39–44
Japan Sewage Works Association (2022) Sewage treatment population penetration rate by prefecture. https://www.jswa.jp/wp2/wp-content/uploads/2022/11/R04_penetration-rate.pdf (Only in Japanese). Accessed 17 May 2023
Klein KR, Rosenthal MS, Klausner HA (2005) Blackout 2003: preparedness and lessons learned from perspectives of four hospitals. Prehospital Disast Med 20:5
Lantagne DS, Balakrish Nair G, Lanata CF, Cravioto A (2014) The cholera outbreak in Haiti: where and how did it begin? Curr Top Microbiol Immunol 379:145–164
Lichtenberger P, Miskin IN, Dickinson G, Schwaber MJ, Ankol OE, Zervos M, Munoz-Price LS (2010) Infection control in field hospitals after a natural disaster: lessons learned after the 2010 earthquake in Haiti. Infect Control Hosp Epidemiol 31(9):951–957
Milsten A (2000) Hospital responses to acute-onset disasters: a review. Prehosp Disaster Med 15(1):40–53
Ministry of Economy, Trade and Industry, Japan (2023) Current situation and issues related to service introduction to local governments [Online]. https://www.meti.go.jp/shingikai/economy/resilience_society/003.html (Only in Japanese). Accessed 17 May 2023
Nohara M (2012) Impact of the Great East Japan Earthquake and tsunami on health, medical care and public health systems in Iwate Prefecture, Japan, 2011. Western Pac Surveill Response J 2(4):24–30
Ochi S, Nakagawa A, Lewis J, Hodgson S, Murray V (2014) The great East Japan earthquake disaster: distribution of hospital damage in Miyagi Prefecture. Prehosp Disaster Med 29(3):245–253
Okada N, Ye T, Kajitani Y, Shi P, Tatano H (2011) The 2011 eastern Japan great earthquake disaster: overview and comments. Int J Disast Risk Sci 2:34–42
Omama S, Yoshida Y, Ogasawara K, Ogawa A, Ishibashi Y, Nakamura M, Sakata K (2013) Influence of the great East Japan earthquake and tsunami 2011 on occurrence of cerebrovascular diseases in Iwate, Japan. Stroke 44(6):1518–1524
Shibata M, Hanzawa K, Ueda S, Yambe T (2014) Deep venous thrombosis among disaster shelter inhabitants following the March 2011 earthquake and tsunami in Japan: a descriptive study. Phlebology 29:257–266
Shiga Y. Nishimura K. Matsumoto T (2020) Current status and issues of nursing support for disaster victims in the Covid-19 pandemic (Available only in Japanese) [Online]. https://plaza.umin.ac.jp/jrcsns/publication/report-202007/. Accessed 17 May 2023

Soshino Y, Kuroda A, Miyata A, Kurayama K (2018) Design and development of the red cross mobile flush toilet toward the smart design shelter. In: 2018 IEEE global humanitarian technology conference, San Jose, CA, USA, https://doi.org/10.1109/GHTC.2018.8601756

Yoshida K, Yuriko Matsubayashi Y, Ogasawara T, Sakai S (2011) On evacuation behavior in Iwate at 2010 Chile earthquake Tsunami. J Japan Soc Civil Eng, Ser. B2, Coast Eng 67(2):1256–1260

Benefits and Barriers of Using WhatsApp in Eye Health Communication in Deprived Settings in India

Chandrani Maitra, Jenny Rowley, and Esperanza Miyake

1 Introduction

Globally, it is estimated that 2.2 billion people have blindness or vision impairment; of which, at least a billion would have been preventable (WHO 2019, 2023). Ninety percent of the visually impaired population resides in the developing world and eighty percent of this blindness is preventable (WHO 2013). A multitude of socio-economic risk factors results in a high prevalence of visual impairment in the developing world (Jaggernath et al. 2014). Disability resulting from visual impairment results in lower employment and lower productivity. Consequentially, beyond the individual and the society, visual impairment has a far-reaching influence on global socio-economic development. India has almost a quarter of the global burden of vision impairment and this poses a major challenge for India's healthcare agenda (Nair 2015). The policy of the Government of India, in public–private partnerships with non-governmental organisations, is to provide eye care services through primary eye care settings; this includes raising awareness through health education (Misra et al. 2015). Primary Eye care services in India are delivered through two major models—fixed facility service provision (hospital-based) and mobile dispensing services (mobile dispensary and tele-ophthalmology).

Traditionally, media platforms such as radio and television, have been used as preferred communication channels by healthcare providers for achieving a behavioral change in the target population (Finkelstein et al. 2012). The access and appeal

C. Maitra (✉)
Centre for Social Change and Development, Manchester, United Kingdom
e-mail: chandranimaitra@gmail.com

J. Rowley
Manchester Metropolitan University, Manchester, UK

E. Miyake
University of Strathclyde Glasgow, Glasgow, UK

P. K. Ray et al. (eds.), *Technology Innovation for Sustainable Development of Healthcare and Disaster Management*, Disaster Risk Reduction,
https://doi.org/10.1007/978-981-97-2049-1_10

of social media in the developing world have increased exponentially with the availability of increasingly affordable mobile technology and cheaper internet service provision. This has resulted in wider access to target audiences and increased engagement, especially with the disadvantaged population (O-Adewuyi 2016). Social media has created a new form of communication platform by transforming monologue into dialogue (McMullan 2011). Bridging geographical, time and social barriers, and bypassing traditional media, social media has provided access to local communities in their own language. More specifically, educating and empowering people with health information is one area where social media has been exceptionally useful (Whitehead and Seaton 2016). The traditional vertical system of sharing health information from sources of scientific authority, passed down to the public through local authorities and clinicians, risks being made obsolete by emerging technologies that facilitate rapid horizontal and vertical information sharing (O-Adewuyi 2016). Currently, available online forums allow patients to air questions, grievances and suggestions and to provide mutual support. Social media can also be used in identifying areas in need of information, monitoring the response of the public to health issues and communicating appropriate health messages to targeted communities. Social media has been used in health education, disease surveillance, tracking, disease outbreak monitoring and providing communication in real time (O-Adewuyi 2016). However, research into the role of social media in healthcare has been limited and is restricted to Twitter, YouTube and Facebook (Moorhead et al. 2013; Lim 2016). Importantly, the role of social media remains largely unexplored in health communication in eye health research.

As is seen in many developing countries including India, geographical areas of extreme socioeconomic deprivation are often juxtaposed with areas of affluence. In these areas of abject poverty, with higher levels of prevalence of communicable diseases, a lack of knowledge and access to basic eye care services renders this population more vulnerable to eye ill-health and its consequent detrimental effects. Such communities are often ignored as they lie hidden in the shadow of the more affluent neighbouring areas. WhatsApp, a relatively new social media platform, is widely available, free to use, inexpensive and provides instant audio-visual, two-way messaging capabilities. WhatsApp is widely accepted in the Indian subcontinent and has a further reach in the general population, particularly in rural communities, over other social media platforms. Its presence and usage are particularly evident in the deprived sections of society, where the acquisition of cheap smartphones can provide a vital link between healthcare providers and the 'hard to reach' deprived population. However, the potential of WhatsApp, in eye health information dissemination remains largely unexplored.

Considering the potential of WhatsApp and the unmet needs of the deprived community, this study explored the use of WhatsApp in developing the community-based practice of promoting eye health in a deprived locality bordering a large metropolitan city in Eastern India (Maitra 2021).

Through the lens of a qualitative case study, this chapter has the stated objective of highlighting the burden of vision impairment, particularly in the context of India, and the role of social media, specifically WhatsApp-based health communication strategy, in increasing awareness of eye health amongst members of the deprived communities, and to synopsise its benefits and challenges. The chapter is organised into several sections: vision impairment and its impact, the role of social media in health communication, an overview of the case study and key findings leading to the discussion. The final sections discuss the benefits and barriers of WhatsApp-based health communication strategy and outline the directions for future research.

2 Vision Impairment

2.1 The Global Context

Vision impairment is said to be present when an eye condition affects the visual system and one or more of its vision functions (WHO 2019). Common eye conditions resulting in vision impairment and blindness include refractive errors, cataracts, glaucoma, corneal opacity, diabetic retinopathy and infections like trachoma. Multifactorial risk factors may lead to vision impairment and blindness which includes age-related changes, genetic makeup, lifestyle changes, risk-taking behaviours, a variety of infections and various underlying chronic health conditions like diabetes and hypertension (WHO 2019). Consequently, vision impairment varies by geographical region (countries and continents), socio-economic parity, age, gender, area of residence and ethnicity.

Whatever may be the reason behind it, vision impairment, through its consequent disability, has an impact that goes beyond the individual and the family. Vision impairment has serious consequences across the life course of the individual (Cupples et al. 2012). Early onset vision impairment may result in language and communication development disorders in preschool children (Mosca et al. 2015), affecting cognitive function in childhood (Bathelt et al. 2018) and the development of autistic patterning (Wrzesińska et al. 2017). Vision impairment affects the educational attainments of secondary school students (Chanfreau and Cebulla 2009) and affects perceived self-concept and self-esteem (Augestad 2017). Overall, there is a long-lasting effect of vision impairment on adaptive behaviour and quality of life of the children (Bathelt et al. 2019) and their parents (Lupón et al. 2018).

The overall cumulative effect of unaddressed vision impairment can be disproportionately large on social and economic well-being (Cumberland et al. 2016). Adults with vision impairment have less employment and productivity (Naidoo et al. 2019) and higher rates of mental illness including anxiety and depression (van der Aa et al. 2015). In the elderly population, vision impairment may result in social isolation (Hodge and Eccles 2013), walking difficulties (Swenor et al. 2014), a higher risk of falls and fractures (Hong et al. 2014), and a greater chance of being cared at nursing homes (Owsley et al. 2007). Vision impairment is also closely associated with significant cognitive decline (Bowen et al. 2016) and with harassment due to violence and abuse, including bullying and sexual violence (Brunes and Heir 2018).

Beyond the individual, the impact of vision impairment is evident in family members, friends and carers, who are often involved in providing crucial physical, emotional and social support (Bambara et al. 2009). Unsurprisingly, such high-intensity support is often associated with carer fatigue resulting in a higher risk of physical and mental health conditions amongst family members and carers (Dada et al. 2013).

At a societal level, vision impairment impacts employment, quality of life and caretaking requirements (WHO 2019). Vision impairment and blindness are ranked third for years lived with disability in the Global Burden of Disease (GBD) Study (GBD 2017 Disease and Injury Incidence and Prevalence Collaborators 2018). The annual global financial burden on the loss of productivity due to vision impairment is estimated to be US$ 244 billion and US$ 25.4 billion for uncorrected myopia and presbyopia alone respectively (Frick et al. 2015; Naidoo et al. 2019).

2.2 *The Indian Context*

There is a paucity of good-quality data on the state of eye health for the whole country, although recent global estimates indicate that India has the largest number of people with blindness and vision impairment in the world, at around 8.3 million (Stevens et al. 2013). Prevalence estimates show that over the age of 50 years, 3.6% were blind and 4.4% had severe visual impairment (Neena et al. 2008). Estimates also indicate that the disease burden is more in the rural population (8.2 vs. 7.1% urban), in women (9.5 vs. 6.5% males), and amongst the older (>70 years) population (20.6% compared to 1.3% in people aged 50–54 years) (Neena et al. 2008). Furthermore, the prevalence of blindness was found to be significantly higher amongst people from lower socio-economic backgrounds. It is also estimated that almost 88.2% of blindness is avoidable and that the major cause of blindness is due to untreated cataracts (81.9%) and uncorrected refractive errors (7.1%) (Neena et al. 2008). It is however important to highlight that these figures are over a decade old and there remains a lack of research reporting on current disease prevalence in India.

Despite significant economic advancements over the past few decades, pockets of extreme poverty exist in India (Gupta et al. 2014). It is estimated that approximately 22% of the population (270 million people) lives in poverty which reflects the largest concentration of poverty anywhere in the world. People from disadvantaged backgrounds carry a much greater risk of blindness and vision impairment (Ulldemolins et al. 2012). Although the reasons may be multifactorial, a lack of awareness, availability of local eye care services, accessibility and affordability of care, pose major challenges to alleviating this problem.

In response to the unmet needs of blindness, the Government of India has demonstrated a strong commitment by establishing a National Program for Control of Blindness (NPCB) in 1976, the first country to do so globally (Vemparala and Gupta 2017). This is a centrally sponsored programme with an ambitious goal of reducing blindness prevalence from 1.4 to 0.3% by the year 2020. Another initiative, ‘Vision

2020: The Right to Sight' programme was adopted in India in the year 2004 with the ambitious aim of blindness elimination by the year 2020 (VISION2020 India 2018). Vision 2020 aims to address blindness by addressing advocacy and strengthening existing eyecare and organisational development.

Barriers to accessing eye care services are related to a lack of availability, accessibility, affordability and acceptability of eye care services (WHO 2019). As the existence of a knowledge gap on eye care is accepted, NPCB, as one of its major objectives, aims to enhance community awareness of eye care by using print, electronic and social media (Vemparala and Gupta 2017). Specifically, NPCB supports awareness programmes that include the identification and motivation of potential beneficiaries, information sharing through media, and educating voluntary groups, teachers and community-based volunteers. This aspect of NPCB is particularly relevant to this study and acts as a mainstay on which the focus to increase community awareness of eye health rests.

3 Social Media and WhatsApp

Over the past decades, social media, an interactive Web 2.0 Internet-based application, has gained exponential popularity. Recent estimates suggest that amongst 4.54 billion registered internet users, approximately 3.8 billion actively use social media and the number of social network users rose from 0.97 billion (2010 estimate) to 2.96 billion (2020 estimate) (Statista 2019a). In the Indian context, 400 million Indians identify themselves as social media users, and almost all access social media on a mobile platform (Statista 2019b). Although Facebook remains the most common social media platform in India, the popularity of WhatsApp, the platform used in this study, has grown significantly in India over the last few years. WhatsApp usage has increased from 20 million users per month (2013 estimates) to 200 million in 2017 (Statista 2019c). Almost 30% of WhatsApp users report using it on a daily basis and estimates suggest that its penetration level is 28% of the population. Significantly, 39.1% of WhatsApp users in rural communities, spend between one and two hours on a daily basis (2018 estimates) thereby making it a popular social media in these communities.

4 Social Media Usage in Health Communication

Introduced in 2011, the US National Library of Medicine MeSH (Medical Subject Headings) thesaurus defines health communication as 'the transfer of information from experts in the medical and public health fields to patients and the public' and 'the study and use of communication strategies to inform and influence individual and community decisions that enhance health' (National Library of Medicine 2019). Health communication relates to all of the processes of communication that are evident in diverse healthcare contexts. Suitably devised behaviour-based health communication can significantly and positively impact health-related attitudes, beliefs and behaviours, thus rendering it an important tool in public health campaigns, and communicable disease prevention programmes.

Traditionally, radio, television and print media (newspapers, magazine articles) have been used as preferred communication channels by healthcare providers for achieving a behaviour change in the target population (Finkelstein et al. 2012). The traditional *vertical* system of sharing information from sources of scientific authority, which was passed down to the public through local authorities (municipal corporation, village heads, etc.) and clinicians, risks being made obsolete by social media that facilitate rapid *horizontal* information sharing (O-Adewuyi 2016). By transforming monologue into a dialogue, social media-based online forums have created a new form of communication platform, that allows patients to air questions, post grievances, make suggestions and provide mutual support (McMullan 2011).

The overall influence of social media-based health communication has been to educate and empower by the transfer of information from the healthcare provider to the target population and is perceived to have a greater potential to influence members of the public and decision-makers over traditional mediums (Côté and Darling 2018). Social media has been used in identifying information needs, monitoring public response to health issues and communicating appropriate health messages to targeted communities. Social media have also been used in health education, disease surveillance, tracking and monitoring of disease outbreaks and providing cost-effective communication in real time (Fung et al. 2015; O-Adewuyi 2016). Through the engagement of scientists and clinicians, and by promoting an exchange of ideas virtually, social media has also created the potential to reach the broader scientific community (Cevik et al. 2019).

Social media platforms have been used for various purposes that range from disseminating health information to the general public to collecting user opinions and attitudes about using social media in healthcare communication. Figure 1 provides a schematic diagram of the main themes of social media usage in health communication and the overarching flow of information between healthcare providers and end users.

4.1 Social Media Use in Eye Health Communication

Very little research has been undertaken in the field of social media usage in eye health. Mostly, this has been in the form of content analysis on the quality of information disseminated on eye health issues like retinopathy of prematurity (Şahin et al. 2019), cataract surgery (Bae and Baxter 2018), glaucoma (McGregor et al. 2014) and blepharospasm (Ehrlich et al. 2012). Amongst the handful of studies using social media in health communication on eye care, one has evaluated technology utilisation and patient preferences (Aleo et al. 2014), two others have examined social media as an epidemiological surveillance tool (Gesualdo et al. 2015; Deiner et al. 2018), whilst one other has used social media as an interventional tool (Sanguansak

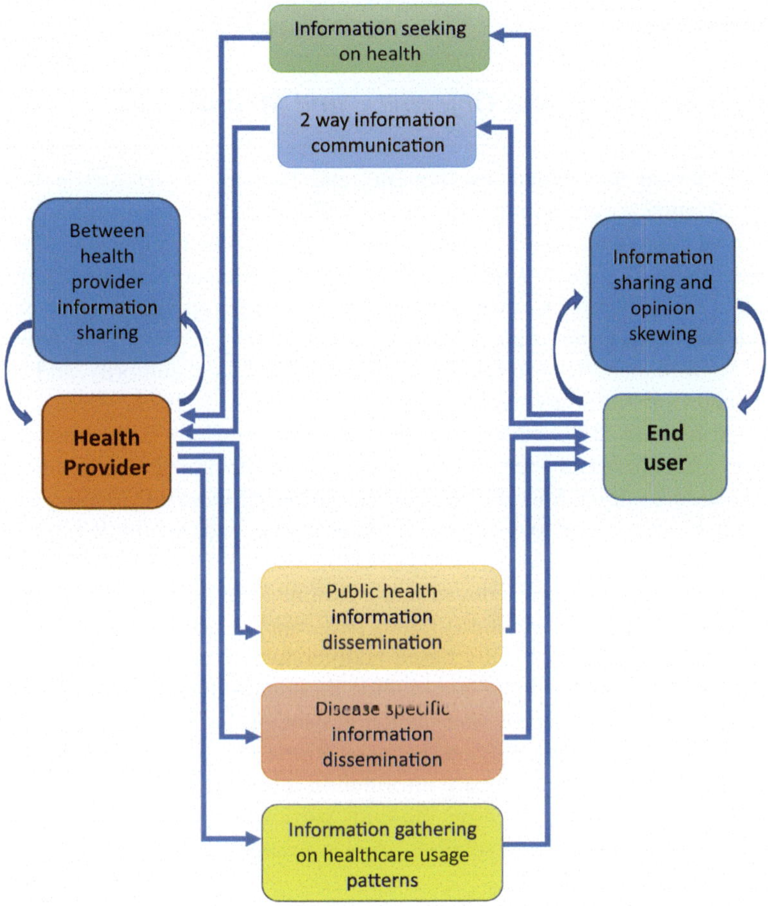

Fig. 1 Social media usage in health communication. Adapted with permission from Maitra (2021)

et al. 2017). It is apparent that there is a serious lack of research into social media engagement in eye health, and none has explored the role of WhatsApp as an intervention in health communication. This background sets the need to evaluate WhatsApp and other social media-based information dissemination strategies in eye health and forms the core principle of this research.

5 This Study

5.1 Overview, Aims and Objectives, and Methodology

This study aimed to evaluate the benefits and barriers of WhatsApp usage in supporting healthcare communication in eye health in deprived semi-urban settings. A qualitative case study approach was adopted to generate rich insights. The study was conducted on a group of women from deprived socioeconomic backgrounds, residents in the village of Chowbaga adjacent to the metropolitan city of Kolkata, the capital of the eastern Indian state of West Bengal (Fig. 2). Ethical approval was obtained from Manchester Metropolitan University, the local authority, and consent was obtained from the study participants.

WhatsApp was used as an interventional tool to create an information resource link on basic eye care between a tertiary city-based eye healthcare provider (Susrut) and the deprived community. Susrut Eye Hospitals (shortly Susrut) is a tertiary provider of comprehensive eye care to the city of Kolkata and beyond (Susrut Eye Foundation 2019).

Multiple WhatsApp-based 'educational' sessions were delivered by Susrut to 30 women from the deprived community (study participants), and this process was facilitated by specially selected locally based neoliterate women who were not part of the study. Chosen from the local community, these neo-literate women were well-placed to understand the needs and challenges facing the community, and, were selected for their approachability, and trained in the use of information usage on WhatsApp to disseminate healthcare information related to eye health. In an effort to facilitate two-way communication, the study participants were encouraged to post queries to Susrut during these sessions. A wide range of native language-based,

Political map of the province of West Bengal (courtesy mapsofindia.com)

Metropolitan map of Kolkata (courtesy bharatonline.com)

Area map of the location of study (courtesy mapsofindia.com)

Fig. 2 Map demonstrating the geographical location of the study

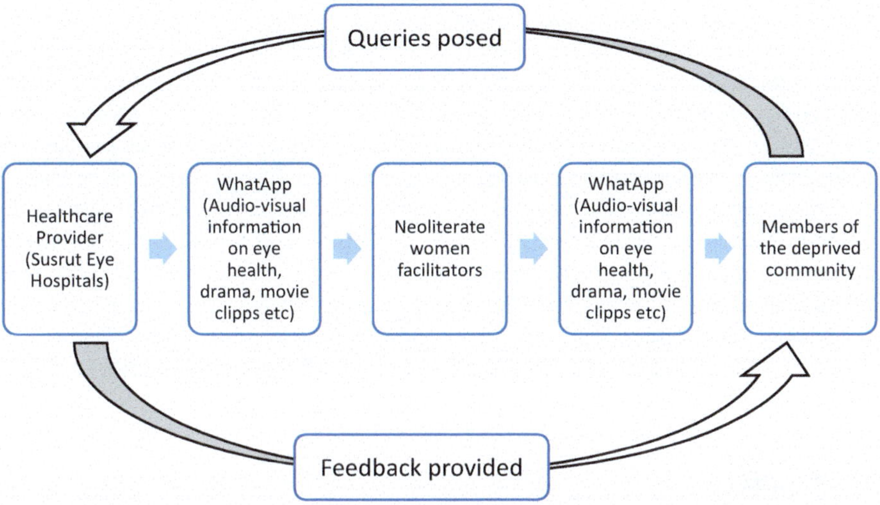

Fig. 3 Schematic diagram of the flow of information in the individual educational sessions. Adapted with permission (Maitra 2021)

easy-to-understand, relevant, and tailor-made eye health-related audio-visual information, was disseminated that covered a range of topics including, routine preventative eye care, common eye diseases, eye donation and eye-related emergencies. A variety of learning materials including pictures, videos, audio clips and remote-based direct interactive communication was shared across the WhatsApp platform. Quite commonly, storytelling was used as a pedagogic method to get the message across to the informal adult learners. Altogether, five educational sessions were conducted by Susrut over a period of five months. Figure 3 provides a schematic diagram of the flow of information in these individual sessions.

Prior to the commencement of the education sessions, participants' perspectives on the information gap in eye care were explored through a face-to-face structured interview. Post completion of the educational sessions, participant perspectives on this new way of health communication were explored through a face-to-face, focus group interview. A thematic analysis was performed to systematically interrogate data generated from these interviews to create meaningful themes.

5.2 *Characteristics of the Study Population*

Altogether, 30 women participated in this study. They were below the poverty level (BPL) in their gross family earnings, and resident mostly in rented dwellings that were thatched or tile-roofed houses or semi-permanent makeshift settlements. Education was low and unemployment was high. The male members of the community were mostly employed in low-pay temporary jobs, particularly in the construction sector which made them characteristically mobile. Although electricity was available at homes, drinking water was sourced from outside the house, overcrowding was common and the toilet was shared. Vehicle ownership was low and was mainly restricted to owning bicycles. Public transportation links and the road infrastructure were considerably underdeveloped. Smartphone ownership was low and access to information on any health topic was through television, radio and wall-posters. There was a high burden of eye diseases in this community mainly in the form of unaddressed refractive errors, untreated cataracts, infections and injury. Although, there was a village medical unit providing basic medical care, the nearest specialist eye unit was based in the city of Kolkata, almost 5 miles away (more than an hour of bus journey). The tertiary eye hospital in consideration for this study (Susrut Eye Hospitals) was located almost 8 miles away (almost 1.5 h on a bus journey).

5.3 *Findings*

5.3.1 Key Theme 1. Information Gap in Eye Health and Barriers to Access Information, as Perceived by the Study Participants

All participants agreed on the existence of a gap in eye health and barriers to accessing eye health-related information (Fig. 4). All participants reported that information on eye health was not readily available and that they had not been approached by healthcare organisations and governmental agencies. Some considered the negative influences of poverty, affordability, Illiteracy and inability to read or write effectively as a critical barrier. Others cited a lack of trust in healthcare providers as they perceived their information as biased with vested motives. Other issues with accessing information related to a lack of personal time, lack of privacy, the overwhelming burden of household chores, and the responsibility of attending to the needs of other (particularly elderly) members of the family. Others highlighted the barriers posed by long-prevailing sociocultural practices and taboos, a lack of women empowerment in the male-dominated environment, economic migration into poorly developed settlements, and unfamiliarity of surroundings.

5.3.2 Key Theme 2: Overall Benefits of WhatsApp-Based Health Communication on Eye Care in the Deprived Community

Acceptability of WhatsApp-based information dissemination amongst the participants was high (Fig. 5). Most regard this as the first intervention to bring easy-to-understand and relevant information to their ‘doorsteps’, to help understand the problems of poor eye health and break down the communication barriers. Most found this helpful in getting the right information without any misrepresentation and considered it instrumental in dispelling misinformation and mistrust spread by

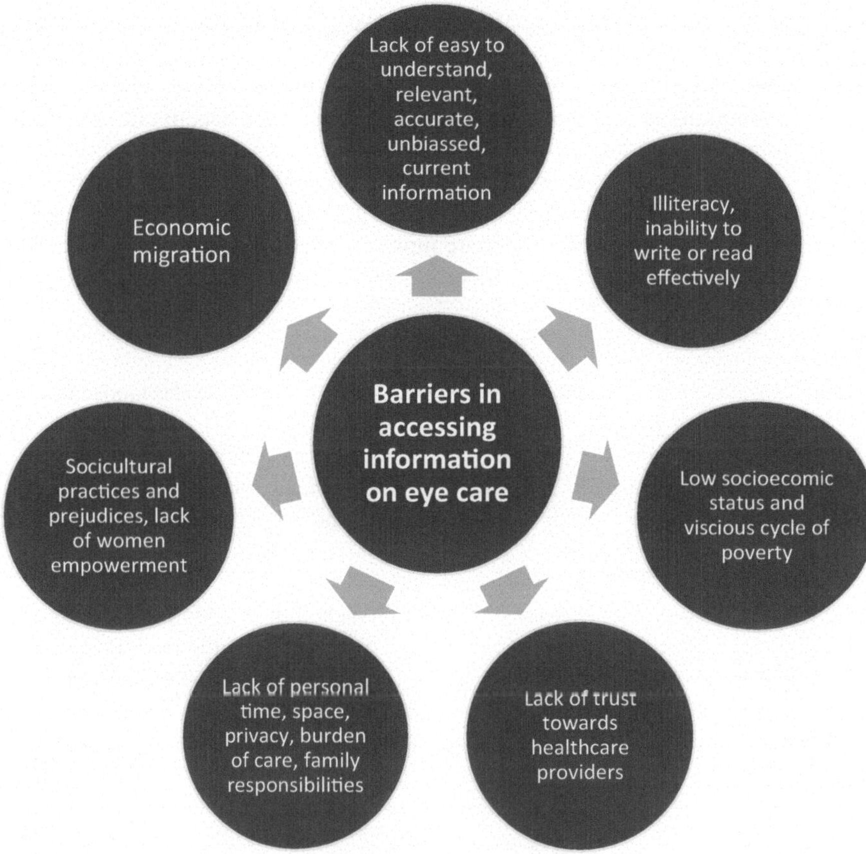

Fig. 4 Information gap and barriers in accessing eye care information (Deprived community member perspectives. Adapted with permission from Maitra (2021)

informal healthcare practitioners (village quacks) and other unscrupulous individuals with ulterior motives. This helped them rebuild their trust in tertiary healthcare providers. The real-time and rapid response from healthcare providers was helpful for some in accessing appropriate eye care. The capability of WhatsApp in sharing audio-visual information and the pedagogic style was preferred by the study participants, as it made the information relevant and easy to relate to. The choice of dramatic narration, and the use of native (Bengali) language, were particularly emphasised. The benefits of the two-way communication facility offered by WhatsApp over other traditional communication media, were highlighted, as it allowed questioning and voicing out concerns, whilst simultaneously receiving feedback information from a tertiary healthcare provider. The free access to information, which can be accessed at a time of their convenience, and a venue of choice, was of considerable appeal to the participants. WhatsApp interface was considered easy to use and of considerable appeal to participants whose first experience of interacting with the smartphone was during the learning sessions. This interactive experience boosted their confidence levels to use this technology to access more information on a wider range of issues. Beyond the benefits of increasing awareness of eye health, members of the deprived community were also keen to highlight the wider benefits of this study. Some admitted that participation in these educational sessions had enhanced their social status, as they are now considered resourceful members of their families and the wider community. The direction of the flow of information seems to have changed too as now participants were able to share more information rather than receive and some helped their extended family members and friends, resident geographically elsewhere.

Fig. 5 Overall benefits of WhatsApp-based health communication on eye care (perspectives of the members of the deprived community). Adapted with permission from Maitra (2021)

5.3.3 Key Theme 3: Benefits of WhatsApp in Increasing Awareness of Eye Problems in the Deprived Community

All participants reported being benefitted from increased awareness and knowledge about eye-related problems (Fig. 6). Accepting that this might be widely prevalent in their communities, a vast majority of participants reported being unaware of common eye problems and hazards that might lead to a loss of eyesight, and that simple precautionary measures, basic hygienic practices, and changes in lifestyle may preserve eyesight. They understood the role of good nutrition and antenatal care and addressed the widely prevalent social malady of drinking illicit liquor. These sessions have enabled them to access information on affordable eye care and the availability of governmental support schemes, thereby regaining their lost confidence in government-initiated programmes. It helped halt the spread of misinformation that often results in wrong practices with dire outcomes. Whilst, acknowledging the benefits of routine eye check-ups in detecting early detection of preventable eye diseases, participants reported increased awareness of matters related to eye donation, and their own positive liking towards this.

5.3.4 Key Theme 4: Barriers Anticipated by the Deprived Community in Implementation o WhatsApp for Eye Care-Related Health Communication

Despite the widespread acknowledgement of the benefits of WhatsApp-based eye-health information, concerns and reservations were raised by some of the participants (Fig. 7). The vast majority of the participants cited the affordability of a smartphone and the data charges as a matter of concern. Some cited this as a hindrance in sharing good eye care information amongst other members of the family or friends resident elsewhere. Most reported a lack of personal free time to access this information in a busy household as a major hindrance. Some of the participants expressed concerns with the sociocultural implications post-acquisition of the 'new' knowledge, in the context of a male-dominated and prejudiced society, especially on the issue of eye donation. Others highlighted the non-cooperative attitude of other traditional healthcare providers, towards this new form of communication.

6 Discussion

Despite a high prevalence of eye diseases, very little awareness has spread to people belonging to deprived communities, and the reasons are manifold. Amongst a host of factors like a lack of basic literacy, particularly digital literacy compounded by extreme economic disparities, deep-rooted sociocultural prejudices, economic migration, a lack of women empowerment, public apathy, and a prevailing culture of mistrust, may all independently, or collectively, has resulted in a wide knowledge

Fig. 6 Benefits of WhatsApp in increasing awareness of eye problems in the deprived community. Adapted with permission from Maitra (2021)

gap on eye care. Additionally, there is a lack of coordinated and concerted approach between allied key healthcare workers, a lack of leadership from the governmental health authorities, and apathy from health insurance providers. On the other hand, there is also a simultaneous lack of accessible and easy-to-understand, relevant, unbiased, and accurate information on eye health, especially from health agencies and healthcare providers. Amongst the members of the deprived community, there is a prevailing sense of hopelessness, frustration, fatigue and a feeling of being socially isolated from the mainstream population.

This study demonstrated a strong commitment amongst the members of the deprived community to engage with a WhatsApp-based health communication portal. All participants from the deprived community reported having benefitted in a significant manner from the WhatsApp-based health communication intervention carried out in this research. This intervention was regarded as the first of its kind to bring understandable, acceptable, trustworthy, and relevant information to their doorsteps

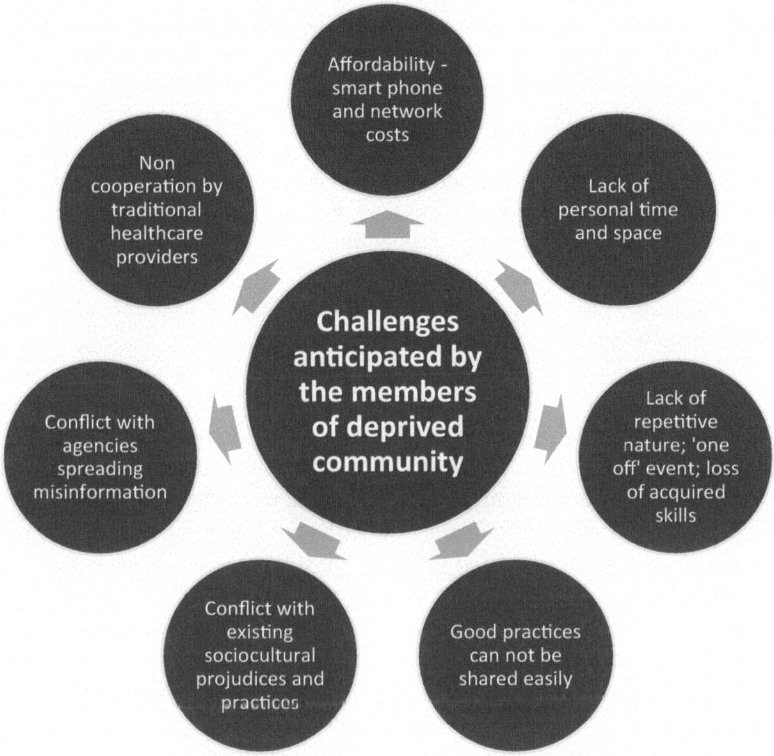

Fig. 7 Barriers anticipated by the deprived community in the implementation of WhatsApp in health communication on eye care. Adapted with permission from Maitra (2021)

without any misrepresentation or personal cost. There was an increased awareness of eye health, disease prevention and access to affordable care. Additionally, the discourse mediated by the audio-visual capability of WhatsApp, and the facilitation of a two-way communication channel, was well accepted amongst the members of this group. The ease of use of the WhatsApp interface was of considerable appeal and encouraged participants to consider using WhatsApp for seeking more healthcare information in the future.

The factors that drive this engagement can be grouped under two headings: host factors and end-user factors. Amongst the host factors, content matter and the style of engagement, strongly influenced the creation and sustenance of interest in the subject. In the context of high disease burden, information on eye health-related topics remains highly sought after in these communities as no such information is easily available from alternative sources. Topic relevance has been shown to be an important determinant of community engagement with health communication (Strekalova 2016). Information was provided in an easily understood manner and in the native language, often with the use of audio-visual content. Images and videos have been observed to attract higher levels of user engagement (Strekalova and

Krieger 2017). As the trustworthiness of the online content and the source organisation, is of paramount importance in successful user engagement (Song et al. 2016), and marketing may have negative consequences (Kite et al. 2016), the information provided was factual, accurate, and trustworthy. The pedagogic style of heuristic (experiential) teaching and style of message framing, like storytelling through audio-visual methods that portrayed the personal experiences of individuals belonging to a similar community, was helpful in allowing the participants to relate. Storytelling is a powerful and established pedagogic method of getting the message across to adult informal learners with low levels of literacy (Caminotti and Gray 2012) and remains largely unexplored in social media-based health communication strategy. Message framing was specifically targeted at providing positive messages and called for increased user engagement (Strekalova and Damiani 2018).

End-user factors were based on the intrinsic characteristics of the deprived community. WhatsApp was a novel medium in these communities that stimulated curiosity about the new way of communication. Coupled with an information gap on eye care, motivated the community to engage and increase their personal knowledge of eye health. Long-standing dissatisfaction with health professionals, increased the desire to remain better informed and self-manage simpler eye health issues, as noted elsewhere (Li et al. 2014). Participants also engaged in emotional and social support: they bonded with the healthcare providers and their fellow participants, factors that facilitated their sustained involvement (Lu et al. 2017). Most participants were of younger age, and a younger age has been associated with increased social media usage for health communication (Huo et al. 2019).

In contrast to other studies where increased digital literacy has been a favourable factor in engagement, despite this lack of health literacy, participants exhibited a strong interest in digital health communication through WhatsApp. This may reflect the innate curiosity of the population studied or other less-understood sociocultural traits that were not explored in this study.

Importantly, the impact of this interventional study was not restricted to improvements in awareness of eye care practices. By empowering women, it made a modest contribution to challenging centuries of gender bias. This bias is particularly strong in deprived settings with deep-rooted ignorance and sociocultural prejudices, that have ensured that women remain subjugated, and their voices and concerns unheard (Nussbaum and Glover 1995). Empowered with newly acquired knowledge, and the resultant boost in confidence regarding eye care and the use of new technology, participants were considered resourceful within their families and the wider community. Information on eye care was actively shared by the participants with friends and relatives. Participants reported an increase in self-respect and power over their own lives, society and in their own community.

7 Benefits of WhatsApp in Health Communication for Eye Care Over Other Traditional Mediums

From a healthcare provider's perspective, WhatsApp may be used primarily for information dissemination purposes to improve health education, and health literacy and for health advocacy on eye care. By sharing high-quality, easy-to-understand, unbiased and relevant scientific information on eye health issues, healthcare providers can expect deprived communities to be better informed about the problems associated with visual impairment and blindness. The multimedia capability of WhatsApp is more suitable than printed matter for sharing audio-visual information on eye care for informal learners with literacy issues. It is expected that WhatsApp would enable the information to reach further transgressing geographical barriers, be repeated frequently thereby reinforcing key messages and facilitating sharing amongst the community members. Consequently, it may be envisaged that this increased awareness of eye care would enable the deprived community to better manage their eye health issues. The two-way communication capability of WhatsApp may be particularly valuable during situations of health crises and eye health-related emergencies, where time is of the essence for successful clinical outcomes. Additionally, through information gathering and surveillance capabilities, WhatsApp may benefit healthcare providers in understanding the specific health needs of this hidden and hard-to-reach community, and develop responsive services, more suited to the needs of the disadvantaged population. Finally, with improved uptake and utilisation of eye care services, members of the deprived community are expected to achieve better eye health outcomes, as envisaged by NCPB and WHO.

WhatsApp is perceived to be a low-cost platform requiring a modest investment in existing infrastructure and this makes it attractive to healthcare organisations. WhatsApp may also be used to initiate and sustain professional communication amongst healthcare providers. By addressing the information gap on good eye care WhatsApp has the potential to bridge a link between healthcare providers and deprived community members. This approach would be expected to mitigate the mistrust and distrust between the stakeholders, facilitate better relationships through mutual respect and understanding, and foster favourable attitudes, ultimately steering them to be strategically aligned in the process of addressing the burden of eye care. Such generous acts of breaking down sociocultural barriers in health communication may also have the added benefit of increasing organisational visibility and promoting its philanthropic image.

For the members of the deprived community, WhatsApp appears to be an appealing tool for information seeking on eye care. Unlike other mediums, WhatsApp can be accessed at a time and place of convenience, which suits well the needs of individual community members who have significant constraints on personal time, space and privacy. The audio-visual nature of information, accessed through an easy interface requiring minimal ICT skills, and availability at no extra cost, is expected to be easily assimilated by informal learners, thereby significantly impacting their information gap on eye care. The two-way communication facility of WhatsApp encouraging the deprived community members to address queries and concerns in real-time is expected to reassure them, and consequently, this benefit can help change attitudes favourably towards healthcare providers whilst facilitating bonding. Overall, positive experience gained from affordable eye care, is also expected to be shared widely, possibly through WhatsApp, and this is expected to have a wider impact on society. This enhanced engagement is expected to benefit these individual members and the society as a whole in reducing the disease burden of visual impairment and blindness. Beyond eye health, reinforced with this newly acquired knowledge and social network creation, WhatsApp would also have the wider benefit of empowering individuals, particularly women from these disadvantaged backgrounds, and would encourage individuals to challenge prevailing sociocultural norms.

For policymakers, WhatsApp can provide a unique medium that requires a minimal skill set for effective use, utilises a low-cost infrastructure and has a wider penetration into the disadvantaged 'hard to reach' population where the eye disease burden remains disproportionately high. WhatsApp-based health communication is highly acceptable to both healthcare providers and deprived community members, which makes it an ideal medium for information dissemination. On a wider note, these research findings may be extrapolated to other areas of public health importance.

8 Barriers of WhatsApp in Health Communication for Eye Care

Despite the perceived benefits of WhatsApp-based health communication on eye care, considerable barriers remain. From the healthcare provider's perspective, although a lack of financial and logistical support, stands out as the most important barrier to challenging implementation, other factors may also have an important role to play. Structural organisational difficulties include limited ICT infrastructure, an underdeveloped communication strategy and a lack of prioritisation. Social media is not embedded in most healthcare providers' organisational strategy and many express reservations about its use in healthcare settings. A lack of clear guidance from higher authorities, coupled with the absence of a lack of social media-based communication strategy, remains an important structural barrier in the implementation of WhatsApp. Cultural factors like a rigid organisational hierarchy, may not encourage staff keen on newer ideas, to voice out concerns in support of including WhatsApp in the existing communication strategy. Irrational concerns and widespread apprehension on issues surrounding misinformation, privacy and legality, a lack of necessary skills, perceived ineffectiveness of social media, damage to institutional reputation, and negative publicity, continue to be strong cultural barriers to widespread WhatsApp use amongst healthcare providers.

For the members of the deprived community, the main barriers to implementation relate to extreme poverty and low levels of smartphone ownership, illiteracy (including non-existent digital literacy), and geographical remoteness. Non-familiarity with this form of communication may also result in a mismatch and reduced acceptance amongst members of the deprived community. Additionally, sociocultural factors including a lack of personal time and privacy, deep-rooted mistrust towards healthcare providers, sociocultural taboos, and fear of being rendered a social outcaste and of reprisal from powerful community members, may alone or in combination, contribute to acting as effective barriers in WhatsApp's wider implementation.

Amongst the members of the deprived community, there is a genuine worry that familiarity with the new technology and the new form of communication will result in increased confidence levels, social empowerment, and possibly, social mobility. There is an unease about this newly acquired social empowerment, which may enable them to challenge deeply rooted sociocultural practices, prejudices and taboos, perpetuated by ignorance and individuals with vested interests. There is palpable apprehension of the consequences of the inevitable friction that would result from unsettling the community hierarchy and long-established practices, and about retribution in the form of domestic abuse and psychosocial isolation; an ill-fated and unwanted consequence of a strategy intended at improving lives.

9 Conclusion and Future Research

In conclusion, through the use of a qualitative case study, this chapter described the innovative use of social technology in addressing the unmet information needs on eye care in deprived communities. Specifically, it demonstrated that WhatsApp can be effectively used as a suitable vehicle for information dissemination on eye care in mediating behavioural change in deprived settings. These findings may be considered in developing policies that develop and disseminate eye care information. On a broader note, the benefits of WhatsApp-based health communication can be realised in other public health settings and can be an invaluable tool to address topical issues in healthcare disaster management. Despite the benefits, considerable barriers remain, which relate to structural and cultural constraints within the healthcare provider organisation, and may preclude its implementation. Perceived ineffectiveness of social media, lack of financial and logistical support, limited ICT infrastructure, underdeveloped communication strategy, lack of necessary ICT skills, rigid organisational hierarchy and lack of prioritisation, lack of guidance from higher authorities, irrational concerns over misinformation, privacy and legality, and damage to institutional reputation, and its consequent negative publicity, are some of the barriers identified. From the perspectives of the deprived community, extreme poverty, low levels of smartphone ownership, illiteracy (including non-existent digital literacy), geographical remoteness, lack of personal time and privacy, deep-rooted mistrust towards healthcare providers, sociocultural taboos, and fear of reprisal contribute as effective barriers in WhatsApp's wider implementation.

Future research should improve the methodology by exploring the perspectives of male participants, assessing information literacy competencies of the study population, systematically investigating the influence of sociocultural determinants underpinning outcomes, including principles of health economics and undertaking a formal cost–benefit analysis to understand the financial benefits of this portal in resource restraint settings. Future research should also explore post-interventional changes in the organisational culture and attitude towards a new form of communication. As other mediums of health communication exist, a comparative study should evaluate these various modalities from the perspectives of efficiency, acceptability, reach and their ability to attain the desired outcomes in eye care. Moreover, WhatsApp may be studied as an information-gathering tool to inform healthcare providers about the specific health needs of society and develop responsive and resource-efficient services. Finally, the effectiveness of WhatsApp-based health communication strategy needs to be studied on other public health conditions of concern.

References

Aleo CL, Hark L, Leiby B et al (2014) Ophthalmic patients' utilization of technology and social media: an assessment to improve quality of care. Telemed J E Health 20:887–892. https://doi.org/10.1089/tmj.2013.0365

Augestad LB (2017) Self-concept and self-esteem among children and young adults with visual impairment: a systematic review. Cogent Psychol 4:1319652. https://doi.org/10.1080/23311908.2017.1319652

Bae SS, Baxter S (2018) YouTube videos in the English language as a patient education resource for cataract surgery. Int Ophthalmol 38:1941–1945. https://doi.org/10.1007/s10792-017-0681-5

Bambara JK, Wadley V, Owsley C et al (2009) Family functioning and low vision: a systematic review. J vis Impair Blind 103:137–149

Bathelt J, de Haan M, Dale NJ (2019) Adaptive behaviour and quality of life in school-age children with congenital visual disorders and different levels of visual impairment. Res Dev Disabil 85:154–162. https://doi.org/10.1016/j.ridd.2018.12.003

Bathelt J, de Haan M, Salt A, Dale NJ (2018) Executive abilities in children with congenital visual impairment in mid-childhood. Child Neuropsychol 24:184–202. https://doi.org/10.1080/09297049.2016.1240158

Bowen M, Edgar DF, Hancock B et al (2016) The Prevalence of Visual Impairment in People with Dementia (the PrOVIDe study): a cross-sectional study of people aged 60–89 years with dementia and qualitative exploration of individual, carer and professional perspectives. NIHR Journals Library, Southampton (UK)

Brunes A, Heir T (2018) Sexual assaults in individuals with visual impairment: a cross-sectional study of a Norwegian sample. BMJ Open 8:e021602. https://doi.org/10.1136/bmjopen-2018-021602

Caminotti E, Gray J (2012) The effectiveness of storytelling on adult learning. J Work Learn 24:430–438. https://doi.org/10.1108/13665621211250333

Cevik M, Ong DSY, Mackenzie G (2019) How scientists and physicians use Twitter during a medical congress. Clin Microbiol Infect 25:1561.e7-1561.e12. https://doi.org/10.1016/j.cmi.2019.04.030
Chanfreau J, Cebulla A (2009) Educational attainment of blind and partially sighted pupils. Royal National Institute of Blind People, London
Côté IM, Darling ES (2018) Scientists on Twitter: preaching to the choir or singing from the rooftops? FACETS. https://doi.org/10.1139/facets-2018-0002
Cumberland PM, Rahi JS, UK Biobank Eye and Vision Consortium (2016) Visual function, social position, and health and life chances: the UK biobank study. JAMA Ophthalmol 134:959–966. https://doi.org/10.1001/jamaophthalmol.2016.1778
Cupples ME, Hart PM, Johnston A, Jackson AJ (2012) Improving healthcare access for people with visual impairment and blindness. BMJ 344:e542. https://doi.org/10.1136/bmj.e542
Dada T, Aggarwal A, Bali SJ et al (2013) Caregiver burden assessment in primary congenital glaucoma. Eur J Ophthalmol 23:324–328. https://doi.org/10.5301/ejo.5000238
Deiner MS, McLeod SD, Chodosh J et al (2018) Clinical age-specific seasonal conjunctivitis patterns and their online detection in Twitter, blog, forum, and comment social media posts. Invest Ophthalmol vis Sci 59:910–920. https://doi.org/10.1167/iovs.17-22818
Ehrlich MS, Richard MJ, Woodward JA (2012) Blepharospasm cause and treatment: YouTube as a reflection of current popular thought. Ophthalmic Plast Reconstr Surg 28:475–476. https://doi.org/10.1097/IOP.0b013e318263c46d
Finkelstein J, Knight A, Marinopoulos S et al (2012) Enabling patient-centered care through health information technology. Evid Rep Technol Assess (Full Rep) 1–1531
Frick KD, Joy SM, Wilson DA et al (2015) The global burden of potential productivity loss from uncorrected presbyopia. Ophthalmology 122:1706–1710. https://doi.org/10.1016/j.ophtha.2015.04.014
Fung IC-H, Tse ZTH, Fu K-W (2015) The use of social media in public health surveillance. Western Pac Surveill Response J 6:3–6. https://doi.org/10.5365/WPSAR.2015.6.1.019
GBD 2017 Disease and Injury Incidence and Prevalence Collaborators (2018) Global, regional, and national incidence, prevalence, and years lived with disability for 354 diseases and injuries for 195 countries and territories, 1990–2017: a systematic analysis for the Global Burden of Disease Study 2017. Lancet 392:1789–1858. https://doi.org/10.1016/S0140-6736(18)32279-7
Gesualdo F, Stilo G, D'Ambrosio A et al (2015) Can Twitter be a source of information on allergy? Correlation of pollen counts with tweets reporting symptoms of allergic rhinoconjunctivitis and names of antihistamine drugs. PLoS One 10:e0133706. https://doi.org/10.1371/journal.pone.0133706
Gupta R, Sankhe S, Dobbs R et al (2014) From poverty to empowerment: India's imperative for jobs, growth, and effective basic services. McKinsey Global Institute
Hodge S, Eccles F (2013) Loneliness, social isolation and sight loss. Lancaster University
Hong T, Mitchell P, Burlutsky G et al (2014) Visual impairment and the incidence of falls and fractures among older people: longitudinal findings from the Blue Mountains eye study. Invest Ophthalmol vis Sci 55:7589–7593. https://doi.org/10.1167/iovs.14-14262
Huo J, Desai R, Hong Y-R et al (2019) Use of social media in health communication: findings from the health information National Trends Survey 2013, 2014, and 2017. Cancer Control 26:1073274819841442. https://doi.org/10.1177/1073274819841442
Jaggernath J, Øverland L, Ramson P et al (2014) Poverty and eye health. Health 06:1849. https://doi.org/10.4236/health.2014.614217
Kite J, Foley BC, Grunseit AC, Freeman B (2016) Please like me: Facebook and public health communication. PLoS One 11:e0162765. https://doi.org/10.1371/journal.pone.0162765
Li N, Orrange S, Kravitz RL, Bell RA (2014) Reasons for and predictors of patients' online health information seeking following a medical appointment. Fam Pract 31:550–556. https://doi.org/10.1093/fampra/cmu034
Lim WM (2016) Social media in medical and health care: opportunities and challenges. Mark Intell Plan 34:964–976. https://doi.org/10.1108/MIP-06-2015-0120

Lu Y, Wu Y, Liu J et al (2017) Understanding health care social media use from different stakeholder perspectives: a content analysis of an online health community. J Med Internet Res 19:e109. https://doi.org/10.2196/jmir.7087
Lupón M, Armayones M, Cardona G (2018) Quality of life among parents of children with visual impairment: a literature review. Res Dev Disabil 83:120–131. https://doi.org/10.1016/j.ridd.2018.08.013
Maitra C (2021) WhatsApp in health communication: the case of eye health in deprived settings in India. Doctoral Thesis, Manchester Metropolitan University
McGregor F, Somner JEA, Bourne RR et al (2014) Social media use by patients with glaucoma: what can we learn? Ophthalmic Physiol Opt 34:46–52. https://doi.org/10.1111/opo.12093
McMullan D (2011) Newspapers and Social Media: From Monologue to Dialogue. INMA
Misra V, Vashist P, Malhotra S, Gupta SK (2015) Models for primary eye care services in India. Indian J Commun Med 40:79. https://doi.org/10.4103/0970-0218.153868
Moorhead SA, Hazlett DE, Harrison L et al (2013) A new dimension of health care: systematic review of the uses, benefits, and limitations of social media for health communication. J Med Internet Res 15:e85. https://doi.org/10.2196/jmir.1933
Mosca R, Kritzinger A, van der Linde J (2015) Language and communication development in preschool children with visual impairment: a systematic review. S Afr J Commun Disord 62:e1–e10. https://doi.org/10.4102/sajcd.v62i1.119
Naidoo KS, Fricke TR, Frick KD et al (2019) Potential lost productivity resulting from the global burden of Myopia: systematic review, meta-analysis, and modeling. Ophthalmology 126:338–346. https://doi.org/10.1016/j.ophtha.2018.10.029
Nair G (2015) Eye care deficit makes India the blind capital of the world|EYEWAY. http://www.eyeway.org/?q=eye-care-deficit-makes-india-blind-capital-world. Accessed 23 Sep 2017
National Library of Medicine (2019) Health Communication—MeSH—NCBI. https://www.ncbi.nlm.nih.gov/mesh/?term=health+communication+. Accessed 14 Mar 2020
Neena J, Rachel J, Praveen V et al (2008) Rapid assessment of avoidable blindness in India. PLoS One 3:e2867. https://doi.org/10.1371/journal.pone.0002867
Nussbaum M, Glover J (eds) (1995) Women, culture, and development: a study of human capabilities. Oxford University Press
O-Adewuyi E (2016) Behavior change communication using social media: a review. Int J Commun Health 9:109–116
Owsley C, McGwin G, Scilley K et al (2007) The visual status of older persons residing in nursing homes. Arch Ophthalmol 125:925–930. https://doi.org/10.1001/archopht.125.7.925
Şahin A, Şahin M, Türkcü FM (2019) YouTube as a source of information in retinopathy of prematurity. Ir J Med Sci 188:613–617. https://doi.org/10.1007/s11845-018-1902-2
Sanguansak T, Morley KE, Morley MG et al (2017) Two-way social media messaging in postoperative cataract surgical patients: prospective interventional study. J Med Internet Res 19:e413. https://doi.org/10.2196/jmir.8330
Song H, Omori K, Kim J et al (2016) Trusting social media as a source of health information: online surveys comparing the United States, Korea, and Hong Kong. J Med Internet Res 18:e25. https://doi.org/10.2196/jmir.4193
Statista (2019a) Social media usage worldwide. Statista Research Department
Statista (2019b) Social media usage in India. Statista Research Department
Statista (2019c) WhatsApp. Statista Research Department
Stevens GA, White RA, Flaxman SR et al (2013) Global prevalence of vision impairment and blindness: magnitude and temporal trends, 1990–2010. Ophthalmology 120:2377–2384. https://doi.org/10.1016/j.ophtha.2013.05.025
Strekalova YA (2016) Emergent health risks and audience information engagement on social media. Am J Infect Control 44:363–365. https://doi.org/10.1016/j.ajic.2015.09.024
Strekalova YA, Damiani RE (2018) Message design and audience engagement with tobacco prevention posts on social media. J Cancer Educ 33:668–672. https://doi.org/10.1007/s13187-016-1135-x

Strekalova YA, Krieger JL (2017) A picture really is worth a thousand words: public engagement with the national cancer institute on social media. J Cancer Educ 32:155–157. https://doi.org/10.1007/s13187-015-0901-5
Susrut Eye Foundation (2019) Susrut Eye Foundation and Research Centre. Susrut Eye Foundation & Research Centre. http://www.susrut.org/. Accessed 20 Oct 2019
Swenor BK, Muñoz B, West SK (2014) A longitudinal study of the association between visual impairment and mobility performance in older adults: the salisbury eye evaluation study. Am J Epidemiol 179:313–322. https://doi.org/10.1093/aje/kwt257
Ulldemolins AR, Lansingh VC, Valencia LG et al (2012) Social inequalities in blindness and visual impairment: a review of social determinants. Indian J Ophthalmol 60:368–375. https://doi.org/10.4103/0301-4738.100529
van der Aa HPA, Comijs HC, Penninx BWJH et al (2015) Major depressive and anxiety disorders in visually impaired older adults. Invest Ophthalmol vis Sci 56:849–854. https://doi.org/10.1167/iovs.14-15848
Vemparala R, Gupta P (2017) National Programme for Control of Blindness (NPCB) in the 12th five year plan: an overview. Offi Sci J Delhi Ophthal Soc 27:290–292. https://doi.org/10.7869/djo.271
VISION2020 India (2018) VISION 2020: The right to sight—India. A snapshot of our journey 2004–2018, Delhi, India
Whitehead L, Seaton P (2016) The effectiveness of self-management mobile phone and tablet apps in long-term condition management: a systematic review. J Med Internet Res 18:. https://doi.org/10.2196/jmir.4883
WHO (2019) World Report on Vision. World Health Organization, Geneva, Switzerland
WHO (2013) Universal eye health: a global action plan 2014–2019. WHO Publication, Spain
WHO (2023) Vision impairment and blindness. https://www.who.int/news-room/fact-sheets/detail/blindness-and-visual-impairment. Accessed 22 Oct 2023
Wrzesińska M, Kapias J, Nowakowska-Domagała K, Kocur J (2017) Visual impairment and traits of autism in children. Psychiatr Pol 51:349–358. https://doi.org/10.12740/PP/OnlineFirst/61352

Towards a Multidimensional Deployment Framework for the Use of Drones in Disaster Response

Uttam Pudasaini and Pradeep Kumar Ray

1 Introduction

1.1 Background

The number of disaster events has increased significantly in recent decades around the world. From 2001 to 2022, there have been an average of 350 to 500 disaster incidents reported per year, compared to an average of 90 to 100 per year between 1970 and 2000 (UNDRR 2022). These disaster incidents have caused significant loss of life, injuries, displacement of people, and destruction of property, making their prevention and management necessary. The proper management of these disasters requires different strategies at all four phases of the disaster cycle including disaster response, recovery, mitigation, and preparedness (Coppola 2006). Of the different phases of disaster management, the disaster response is a very sensitive stage that involves numerous activities, such as saving lives and providing basic human needs (food, shelter, clothing, health, and safety) to the affected population, restoring damaged infrastructure (including electricity, communications, etc.) and assessing the damages (Cumbane and Gidófalvi 2019). Rapid and effective response within no more than 72 h is crucial to assist the affected population (Erdelj et al. 2017). For this

U. Pudasaini (✉)
Nepal Flying Labs, Kathmandu, Nepal
e-mail: uttam@flyinglabs.org
URL: https://flyinglabs.org/nepal/; https://www.naxa.com.np/

NAXA, Kathmandu, Nepal

P. K. Ray
University of Michigan—Shanghai Jiao Tong University Joint Institute, Shanghai, China
e-mail: p.ray@unsw.edu.au

UNSW Business, Sydney, Australia

P. K. Ray et al. (eds.), *Technology Innovation for Sustainable Development of Healthcare and Disaster Management*, Disaster Risk Reduction,
https://doi.org/10.1007/978-981-97-2049-1_11

kind of timely and effective mobilization to take place, individuals, and organizations in charge of responding to emergencies must have detailed accurate data regarding the situation in the field (Kedia et al. 2020). With the advancement in geospatial technologies and their ability to offer and process up-to-date imagery, remote sensing[1] technology via both satellite and airborne platforms has become common in the field of natural hazards and disaster management (Tarolli 2014). Among different geospatial tools, drones offer a unique opportunity to support disaster response and to make low-cost observations on a local scale, complementing satellite and aircraft-based data collection (Kucharczyk and Hugenholtz 2021). Drones by definition are pilotless aircraft that can be operated remotely, semi-autonomously, autonomously, or have a combination of all these capabilities (Van Blyenburgh 1999). According to a study that compared satellite imagery taken by traditional aircraft with those taken by drones, drones have proven to be a more effective tool for on-demand imageries during a disaster situation, such as identification of flooded areas (Giordan et al., 2018) and landslide locations (Fiorucci et al. 2018).

Drones are becoming more popular and are being used for many civil applications because of their ease of deployment, operational flexibility, low maintenance cost, high mobility, hovering capability, and ability to provide high-quality spatial and temporal resolution datasets in both 2D and 3D environments (Agüera-Vega et al. 2017; Hildmann and Kovacs 2019). Regarding the applications of drones in disaster response, some of the popular applications of drones during disaster response are the assessment of damaged buildings and infrastructures (Hildmann and Kovacs 2019); identifying affected areas in a short time (Antoine et al. 2020; Ezequiel et al. 2014); establishing vital wireless communication links between the survivors and the nearest available cellular infrastructure (Erdelj et al. 2017) and even transportation of medical and emergency items to the areas affected (Yakushiji et al. 2020). The main reasons for this increasing use of drone technology in emergencies are its affordability, potential to be quickly deployed, its ability to capture high-resolution information of affected areas for emergency responders who need that level of detail to prioritize their response activities, and most importantly its ability to operate in hazardous environment reducing the possible risks to human responders (Greenwood and Joseph 2020a, 2020b; Kucharczyk and Hugenholtz 2021). In addition to the availability of cheap drone platforms and multiple types of sensors, Antoine et al. (2020) emphasize that recent advancements in image processing technology and drone photogrammetry[2] software are also the major reasons that have made the production of centimetric-level high-resolution outputs possible using drone technology.

[1] Remote sensing is defined as the art and science of obtaining information, without physical contact, from the object under considerations.

[2] Photogrammetry is the process by which images and digital representations of energy patterns derived from noncontact sensor systems are captured, measured, and interpreted to obtain reliable information about physical objects and the environment.

1.2 Research Objective

Drone-based remote sensing-related applications in general require more research to demonstrate and verify their utility (Kucharczyk and Hugenholtz 2021). In the context of disaster management, Tanner (2018) notes that despite drones' versatility, their use for disaster management still faces several challenges. Daud et al. (2022) highlighted that though the number of research regarding the application of drones in disaster management is increasing, there is still not sufficient research evidence that focuses on understanding the underlying challenges and applications of using drones during immediate disaster scenarios. In Kucharczyk and Hugenholtz (2021)'s review of drones in disaster management-relevant articles, the author identified the global trend of using drones before and after disasters, bias, and research opportunities in the field. The research found that most of the study areas were in high-income countries, indicating that further research is needed to understand their use, impacts, and existing challenges in the context of developing countries. Also, Greenwood et al. (2020) note that there is a paucity of literature addressing the experiences of people working on the front lines who use drones in disaster response exercises.

All the instances mentioned in the above section showed that there is a need for research to develop a systematic understanding of the existing challenges of using drones in a real disaster scenario in the context of developing countries. This chapter presents our research into the challenges of using drones for disaster management in the context of a Nepal, a developing country from South Asia.

The major aim of this research study is to document and advance awareness regarding major challenges encountered in the deployment of drones during the disaster response phase in a developing country context. This study uses Nepal as a case study to investigate the existing challenges experienced at both individual and organizational levels. Nepal is among the first countries that witness drone applications during a mega-disaster incident. It was during the 2015 Gorkha earthquake, that several drones were deployed for crisis mapping and humanitarian activities to capture images of the destruction (Meier 2015). Though there were a number of issues such as overcrowding of drones, drone flights by inexperienced pilots, and more experimentation than actual use cases, the usage during the earthquake triggered increased adoption of drones in the humanitarian and disaster management sector globally (Pageni et al. 2020).

In particular, this study aims to identify the existing challenges by directly interacting with the relevant professionals involved in disaster response activities and also individuals working on the design of drone-related policies in the country. This includes drone pilots, project managers from both non-government and private organizations, government stakeholders concerned with providing drone permissions, emergency responders from security agencies, and senior management team members from drones as a service startup company.

The chapter is organized as follows. The next section (#2) provides a summary of the drone applications for disaster management in Nepal. This is followed by Section #3 on our research methodology including data collection and analysis based on the

HOTGIIF (Human Resources (H), Operational (O), Technological (T), Geography (G), Institutional (I), Instrumental (I), and Financial (F)) multidimensional framework for a holistic analysis of the challenges to evolve a robust deployment framework. Section #4 presents our findings based on the qualitative research involving several key stakeholders in Nepal. Finally, Section #5 corroborates our findings with those reported by other researchers all over the world so that our findings can be useful for developing a holistic drone deployment framework for disaster-prone communities that are affected by disasters quite frequently.

2 Drones for Disaster Response in Nepal

The first recorded use of drones in Nepal was related to its applications in environment and climate change-related research. Popular early usage included monitoring animals and spotting poachers by the World Wildlife Fund (WWF), Nepal (Poppleton 2012), investigating the impact of climate change in glaciers by ICIMOD Nepal (Kraaijenbrink et al. 2016), and estimating the amount of biomass in forest areas by Kathmandu University (Baral 2015).

Regarding the applications in the disaster context, drones were widely used in Nepal in the aftermath of the April 2015 earthquake (Pageni et al. 2020). An unprecedented number of small-sized drones were flown over earthquake-affected areas by media agencies and humanitarian responders particularly to capture information on the extent of damage caused by the earthquake, aid search and rescue operations, and supply relief items (Winchester 2015). Crowe (2015) highlights two major reasons behind drones becoming popular in the post-earthquake scenario: the low resolution of satellite imageries and the rough terrain of Nepal that makes it difficult to gather ground information using other methods. Pageni et al. (2020) argue that although a large number of drone-based activities were introduced, most of them were just experimental. But the experimental usage also gained huge international attention and humanitarian organizations began exploring the applicability of drones in post-disaster situations (Meier 2015; Wang 2020). This led to the rapid proliferation of drones in Nepal and consequently prompted policymakers to act swiftly and enforce strict drone flight operation guidelines that even banned the use of drones over sensitive areas like heritage sites, government offices, etc. (Hern 2015; MoHA 2015).

Despite strict operational guidelines from the government, the post-earthquake time in Nepal witnessed a rise in the use of drones at both individual and organizational levels (NFL 2020). Pageni et al. (2020) highlight that actors from both private and non-government sectors started using drones for various purposes like journalism, surveying, and mapping, improving access to healthcare in remote areas, agriculture, risk assessments and disaster preparedness and damage assessments, biodiversity conservation, ecological monitoring, etc. Due to the increase in positive and useful application of the technology, the Ministry of Home Affairs (MoHA) and the Civil Aviation Authority of Nepal (CAAN) revised and updated the existing drone procedural guidelines with several improvements including decentralization of the drone flight permission process and providing authority to district offices all over the country to review and issue permissions for drone flights for drones that weights less than 2 Kg (CAAN 2021; MoHA 2019). Pageni et al. (2020), highlighted that there were some positive aspects as the government agencies now started using drones as a part of their disaster response activities. Some of the popular examples of drone use by government agencies include the use of drones to capture images and study the details after the March 2019 Southern Nepal Tornado by the National Department of Hydrometeorology (Khatri 2019); search and rescue missions for a missing person in the Mahakali river by the national security forces (RSS 2021); and to identify damaged houses, map out inundation areas, and record the changes in topography caused by massive erosion and deposition during the massive 2021 Melamchi flood by the National Disaster Risk Reduction And Management Authority (NDRRMA) with the support from the World Bank (Takamatsu et al. 2022).

Pudasaini (2022) points out a number of challenges including lack of skilled manpower, lack of a digital system for preparing and submitting flight permissions, lack of drone pilot licensing policies, drone insurance, hardware issues, and cost of sensors as reasons that are limiting the growth and adoption of drones for disaster response in Nepal. This study dives deeper into systematically identifying and categorizing the challenges, interacting with the disaster response professionals and grouping them into different thematic areas.

3 Research Methodology

3.1 Theoretical Framework and Research Design

The systematic categorization of challenges of using drones in the disaster response phase requires the use of a theoretical framework or a model. As per the review of ICT4D literature, the challenges of using technologies in a developing country context are often analyzed using popular ICT4D frameworks such as the ITPOSMO checklist as proposed in the design reality gap (Heeks 2006) and the STOPE framework (Choi et al. 2016). However, both frameworks don't seem to fully help the systematic categorization of challenges related to drone use in a disaster context. Some of the highlighted challenges include bad weather situations that could complicate drone flights; ethical issues regarding data privacy and issues with the actual value of the technology after translating high-volume drone-captured data into different sorts of analysis outputs. In this regard, this study demands a new theoretical framework to systematically understand and analyze these sorts of unique challenges as well as others.

Based on the reviews of academic and practitioner literature (Li 2023), a first set of thematic categories "HOTGIIF" were inductively extracted for further study. The details of each theme of this framework are highlighted in the table (Table 1):

Table 1 Thematic categories representing different challenges concerning drone use in a disaster context ("HOTGIIF" framework)

SN	Challenge category	Meaning
1	Human resources (H)	Resources such as skilled human resources available to be quickly deployed in any disaster scenarios.
2	Operational (O)	Any sort of issues that hamper smooth operations of a disaster response intervention with drones.
3	Technological (T)	The overall technical capacity of the drone hardware and software along with the status of available technological infrastructures.
4	Geography (G)	Geographical details such as topography, elevation, and other geographical conditions of a given area.
5	Institutional (I)	Institutional issues such as existing legal permissions to fly, the status of organizational setup to look after drone activities in the country, and institutional mindsets.
6	Instrumental (I)	Status of drone-captured datasets being translated to action.
7	Financial (F)	All financial resources related to setting up necessary infrastructure (hardware, software, human resource, logistics, etc.) for a drone in a disaster response mission.

3.2 *Data Collection*

This study involved the collection of both primary and secondary datasets. Regarding the secondary datasets, a review of existing drone-related procedural guidelines, practitioner reports by different humanitarian and disaster management-focused organizations, media articles, and other publications relating to the use of drones during disaster situations in Nepal was reviewed to broaden the understanding of the given topic. The primary data collection further involved the following steps.

3.2.1 Interview Design

A qualitative interview with concerned professionals was selected as a method for primary data collection as it is one of the widely used methods associated with qualitative types of research that involves interacting with people who have lived experiences of a given context (Cavaye 1996). The first set of interview questions was drafted based on the individual parameters of the 'HOTGIIF' framework. Regarding the type of interview, a semi-structured qualitative interview approach was chosen for the interview process. The semi-structured interview was selected because it is flexible and allows certain open-ended nature of the questions to arise as per the flow of conversation thus providing more opportunity for both researchers and respondents in the interview to discuss issues in more detail. As it was a semi-structured interview, the questions were modified based on the flow of the conversation while keeping them within the overall main framework.

3.2.2 Participant Recruitment and Sampling

A total of 13 individuals were identified for the interview. The purposive and snowball sampling methods were selected for the sampling of participants for the interview. The purposive sampling method was adopted for selecting the first eight people based on the first author's work contacts. A snowball sampling approach was then used to select the remaining five individuals based on the recommendations of the first eight participants. Below are the details of all participants consulted for the interviews (Table 2).

Though the selection of individuals was based on networks and referrals, there were specific criteria used for the finalization to select only those who could provide more relevant information to help find the answers to the research questions. The first criterion was their involvement in previous real disaster response activities relating the use of drones. The assumption was that the individuals who were directly involved in the design and implementation of drone-based interventions in real disaster response activities are more connected to the projects and can provide details based on lived experiences as opposed to individuals who have worked in a simulated environment or unfamiliar with the details of drone operations in a real disaster context. The next criteria was the diversity of the participants. A diverse group of participants involving people from multiple sectors were selected to gather perspectives from different user groups (Fig. 1).

The last criterion was the selection of participants available and willing to participate in the interview process and discuss the relevant experiences.

Table 2 List of interview respondents

Code	Respondents	Roles	Institutions
PM1	R1	Project manager	Private
PLOT1	R2	Drone pilot	Private
PLOT2	R3	Drone pilot	Private
PLOT3	R4	Drone pilot	Private
PM2	R5	Project manager	Non-profit
PM3	R6	Project manager	Non-profit
AVI1	R7	Policy makers-aviation	Government
AVI2	R8	Policy makers-aviation	Government
SEC1	R9	Policy maker-security	Government
PM4	R10	Project manager	Non-profit
AVI3	R11	Policy makers-aviation	Government
MFCT1	R12	Drone manufacturer, repair and maintenance	Private
SEC2	R13	Policy makers-security	Government

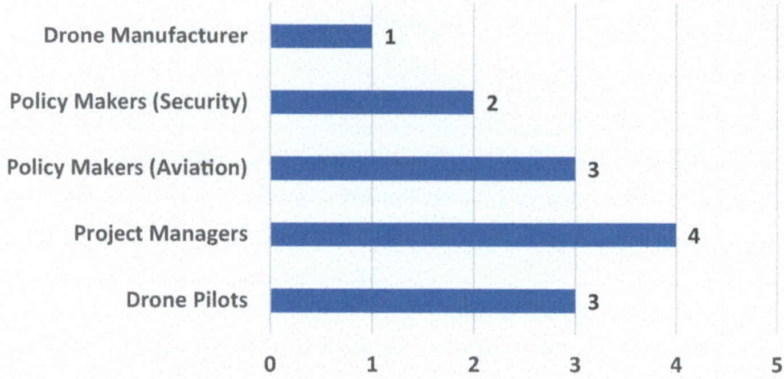

Fig. 1 Type of respondents selected for the interview

3.3 *Data Analysis*

After the collection of the interview datasets, the next step was to conduct a thematic data analysis method (Pudasini 2022). As the major themes were already there in the used "HOTGIIF" framework, the coding process focused more on identifying sub-themes belonging to each of the themes in the framework. This process involved importing all transcribed datasets into Nvivo, a computer-assisted qualitative data analysis software tool that was used for accurate data analysis in a digital environment (Welsh 2002).

First, the individual themes of the "HOTGIIF" framework were used as an initial codebook in NVivo. Further, the analysis of responses from the initial interactions also showed that both human and financial resources can be represented under the same thematic category "resource." Hence they were removed and combined as a single parameter "resources" for further analysis. After this initial revision, the coding was continued for all remaining eight interviews using the revised "ROTIDI" (R: Resource Challenges; O: Operational Chalenges T: Technological Challenges I: Infrastructure Challenges D: Data Privacy and Security Concerns I: Information Challenge) framework. There was a set of sub-themes generated for each of the themes of the framework. These sub-themes were revised iteratively as coding was continued in other remaining interview transcriptions. A final set of the ROTIDI framework along with the identified sub-themes is included in the findings chapter.

4 Findings

The analysis of the response from all interviews was coded in NVivo and specific sub-themes were identified for each category. This section now elaborates on the major findings across each theme of the ROTIDI framework as follows.

4.1 *Theme 1-Resource Challenges*

Almost all the respondents reported human resources and financial resources as two major resource challenges. The section below elaborates on the observations in detail for each of the issues.

Theme 1.1 Human Resources

The results from the interviews showed that both government and non-government sectors face problems finding skilled and qualified human resources. Project managers interviewed reported that Nepal has a very limited number of qualified trained drone pilots. Drone pilots with a specialized domain knowledge of disaster context were even less.

Only one of the interviewed project managers (PM4) had training in disaster management while none of the pilots did. Most of the project managers mentioned a common struggle in finding human resources with good drone data interpretation and analysis skills. Similarly, the government institutions also lacked human resources with knowledge of drone policies and regulations (AVI1, AVI2, SEC2), and analysis of outputs generated from drone data processing to turn into disaster response actions (SEC1). One of the government officers (AVI1) even claimed that he was the only government staff in Nepal who had a postgraduate-level academic certification course on drones.

To overcome the human resource challenges, the drone pilots suggested further actions like the provision of a drone pilot certification course with a specific disaster management component and organizing more technical training with practical sessions for both drone piloting and data processing. Most of the respondents from government organizations stressed on the need for more capacity-building courses focusing on drone piloting, regulatory policies, and data analytics.

Theme 1.2: Financial Resources

The major challenges under this category are primarily related to limited investment and expensive setup requirements. Most of the project managers in the interview mentioned that it was expensive for their organizations to set up all necessary hardware and software infrastructure for complete drone-based solutions. The cost grew once they started drone missions. Setting up high-tech processing devices is another area that is important but cost-intensive.

Besides hardware financing issues, most project managers, with a few exceptions, admitted using pirated versions of software to save costs. This is an interconnected challenge due to the monetary policy of Nepal that makes it difficult to pay for software licenses outside of Nepal (PM1).

Besides issues with hardware and software pricing, two drone pilots (PLOT1, and PLOT3) explicitly called for an increase in their salaries to match the new skills and amount of dedicated work they do both in the office and field environment. On the other hand, a project manager from the private sector mentioned that companies faced financial challenges because clients did not pay for consulting work on time. Drones were also often deployed on a pro bono basis to develop a proof of concept for future opportunities (PM2). A respondent from a humanitarian organization mentioned that though there were limited drone-focused programs, his organization made sure that finances were available for all necessary operational activities.

Few pilots in the study from the private sector complained that lack of insurance policies for drone hardware is a factor that made them feel financially unsecured in going to risky environments during a disaster.

4.2 Theme 2: Operational Challenges

The operational challenges reported by the respondents relate to weather and environmental issues, lack of coordination among actors, and issues with monitoring drone missions. Most of the drone pilots and project managers reported that the available consumer drones in the market could not operate well in cloudy and foggy weather. Moreover, disasters often damage infrastructure like roads hence creating logistical challenges particularly for the drone pilots to reach the disaster-affected regions on time. In such situations, even though drone missions are carried out, most of the pilots highlighted lack of coordination as a major issue while working in the field as there are multiple flights for rescue operations being conducted in the same area and also other drone pilots trying to do a similar work.

Besides the coordination problem, most of the pilots (PM1, PM2, PM4) highlighted the lack of detailed Standard Operating Procedures (SOP) on how to operate drones in disaster response as a major challenge. To overcome this challenge, one of the pilots (PLOT3) responded that their team often reached out to local governments for coordination support. Local governments helped them in establishing proper coordination among air traffic control at the federal level and security forces on the ground. The challenge on the government side is more about getting the information on who is flying where. In this regard, a government officer stressed the need for a digital system that can help government agencies easily coordinate among drone pilots and rescue helicopters and help both teams to smoothly run their operations.

4.3 Theme 3: Technological Challenges

The technological challenges reported by the respondents are technology infrastructure challenges, hardware challenges, and software-related issues. Most of the drone pilots reported that it became difficult to charge the batteries and conduct enough drone flights as the disaster-affected communities often suffered from damaged telecommunication and electricity infrastructures.

Another pilot (PLOT1) however mentioned that it was not possible for all drone teams to manage an extra pair of batteries as they were quite expensive and, in many cases, not readily available in the market due to hardware import issues. In addition to this, advanced sensors that can support SAR operations during a disaster were also not easily available in Nepal (SEC1). As a result, the outputs captured by drones often got limited to good-quality pictures and videos of the affected areas. Difficulty in repairing hardware and importing spare parts in case of issues during operations is a bigger and more frequent problem.

It is however the combination of hardware and software issues that frequently leads to delays in generating outputs. Most project managers said that they struggled with completing the reports of drone missions also because they did not have high-end processing tools (computers and software).

4.4 Theme 4: Institutional Challenges

The institutional challenges reported by the respondents relate to Nepal's sociocultural context, lack of a dedicated institutional setup and legal provisions for drone operations, and lack of certification institutions.

The general institutional challenge in Nepal was the skepticism of concerned organizations toward the significance of drones in disaster management. At the government level (AVI1, SEC1), it was difficult to convince the leadership to design and invest in long-term programs to promote the use of drones. At the non-governmental level (PM2, PM5), the senior management was often more interested in the traditional ways of disaster response. Respondents mentioned that the major issue was the lack of awareness and understanding of the possible benefits of drone technology in saving lives and properties.

Project managers highlighted that the major institutional challenge in Nepal was the time-consuming manual drone flight permission-securing process. In addition to not having any digital systems to apply for and receive flight permission, there were no special provisions to ease the process of using drones in a disaster context.

Most of the pilots mentioned that the ad hoc flight approvals coordinating with local agencies have been an easy solution for them. In contrast, the respondent from the aviation sector was concerned that the ad hoc permissions from local governments made it difficult for the central-level authorities to provide information about flying objects to bigger aircraft operators (AVI3). Most respondents from government agencies reported that Nepal neither had a single government body handling drone operations nor had the necessary laws and regulations.

Other issues caused by institutional challenges as raised by most of the respondents include existing issues with importing drones and necessary hardware into the country (MFCT1) and challenges setting up dedicated agencies to pilot certification courses in general and more specifically to disaster context (PM2, PLOT3).

4.5 Theme 5: Data Privacy and Safety Concerns

The challenges under this theme primarily relate to aviation safety challenges, human safety concerns, and ethical concerns related to data privacy. Most of the pilots reported that drone flights in a disaster-affected area are challenging as the airspace over such locations is usually busy with rescue helicopters flying very close to the ground. Respondents reported that most uncoordinated drone missions are of high risk to both the aircraft and the human population living on the ground below especially when there are semi-skilled and amateur pilots flying drones.

Besides the risk of drone crashes hitting them, most of the pilots mentioned the community being doubtful of the purpose of drone missions. Community members have even reported it to security agencies.

Though none of the pilots mentioned the community members raising concerns around ethics and data privacy, all of them agreed that capturing pictures and videos of residents in temporary shelters during a disaster context had privacy issues. Hence, taking consent from the community wherever possible was important for them. They reported that they normally conducted community engagement exercises wherever possible with representatives from local governments or community groups before the drone flights.

4.6 Theme 6: Instrumental Challenges

More than half of the respondents found instrumental challenges concerning the translation of huge amounts of drone data into disaster response actions. A project manager highlighted that the problem with data starts right at the time of collection (PM4). A major problem was data redundancy and duplication as multiple teams conducted drone flights in a disaster-affected area in parallel. Most of the pilots were found to often be storing the captured datasets in their own devices in the absence of any data sharing guidelines and platforms.

Upon further investigation, most of the government officials mentioned the lack of expertise and knowledge as being a barrier to the analytical interpretation of datasets. There were very limited examples of the use of drone data in better planning disaster response operations and other activities afterwards during disaster management phase. A drone project manager (PM1) highlighted that their team had high-end processing machines and a team of data analysts and could process the images to generate outputs that assisted in activities like rapid damage assessment and identification of areas that required the most attention. In a similar vein, another respondent pilot reported engaging a team of experts in case of sensitive disaster events and provided recommendations quickly.

When asked about how to deal with this challenge effectively, all the project managers and a few respondents from the government highlighted the need for proper drone data-to-action guidelines along with the provision of a certification course to build the necessary skills to utilize data and come up with quick recommendations.

The findings from this study in terms of major sub-themes under each theme of the "ROTIDI" framework as discussed in the above section can be summarized as follows (Fig. 2).

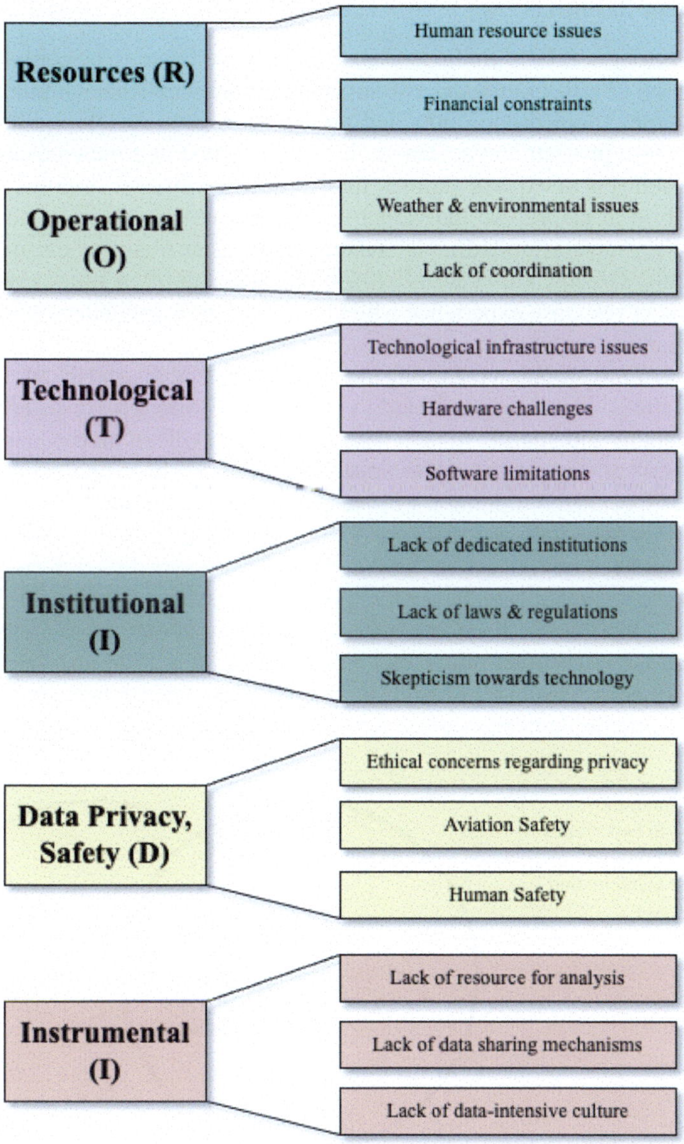

Fig. 2 Major themes and sub-themes of existing challenges as per the ROTIDI framework

5 Discussions

The section below systematically discusses in detail the findings of the study as mapped onto the "ROTIDI" framework.

a. **Resources Challenges**

i. **Human Resources**

The human resource-related findings from this study are found to be consistent with Jeyabalan et al. (2020) findings as participants reported challenges findings expert human resources (both drone pilots and project managers) to take on drones for disaster response projects. Jeyabalan et al. (2020) argued that the challenge exists because the skillsets of experts required for drone-based disaster response missions are quite hybrid in nature. An individual with drone piloting skills cannot operate properly if they have a limited understanding of disaster management in general. Similar to this, Greenwood et al. (2020), found that the often times drone users lack the disaster response experience while the emergency responders working in a disaster scenario lack the drone use experience (Fig. 3).

Upon further investigation regarding the reasons, the study found that the majority of the pilots and project managers couldn't find certified drones for disaster response training programs and courses to help them build their skills in their country. Some of them specifically highlighted the need for a certified licensing program with provisions of specialized training for drone use in a disaster context.

Existing literature such as Iqbal et al. (2022) found that in most cases, standard certification and licenses are a must-have document to operate drones during disaster context and the countries even have strict policies regarding pilot skills. In the case of Nepal, the adoption of a licensing system similar to that of automobiles, limiting drone operations to certified pilots, had already been brought to and discussed on several occasions by Nepalese government officials Wang (2020). However, it has not been brought into implementation yet. A participant in this study mentioned that the

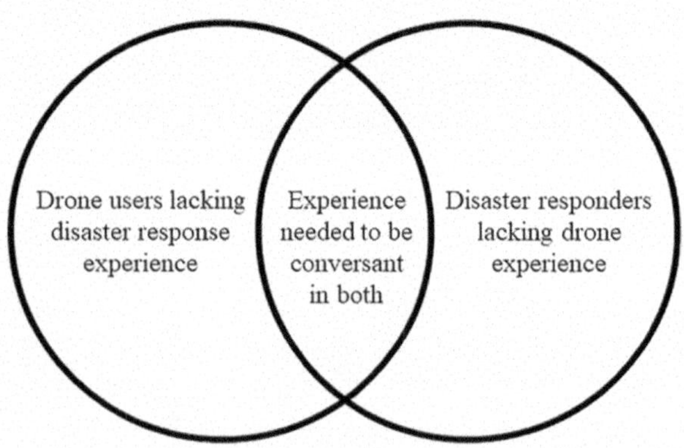

Fig. 3 Venn diagram depicting gaps in experience between different user groups. *Source* (Greenwood et al. 2020)

GoN has already included provisions regarding the need for certification with some preliminary criteria and guidelines in their recent release of the updated drone use guidelines (CAAN, 2021). However, no criteria were specified for use in a disaster scenario.

ii. **Financial Resources**

The findings from this study are consistent with Greenwood and Joseph (2020b) findings as respondents reported challenges in managing adequate financing for the necessary hardware and software tools required for efficient data capture and processing with drones in a disaster context. Greenwood and Joseph (2020b) argued that the emergency response organizations from humanitarian agencies find it challenging to justify the return on investment in case of drone use in disaster as the likelihood of disaster is uncertain resulting in drones staying idle at the office most of the time without work. On the contrary, this study found some different results in the case of non-government sectors. The respondents from the private sector in the study reported that drones don't need to stay idle in their case because they can use them for any other sort of consulting work and are not restricted to disaster situations only like for humanitarian agencies. For such a situation, it is suggested to conduct a cost–benefit analysis prior to integrating drone technology in the humanitarian action (Rejeb et al. 2021). Besides investment issues, the project managers in the study reported issues around insurance and hesitation from drone pilots to not travel to the field without insurance.

In addition to a lack of investment and insurance policies, the study also found that organizations in Nepal find drone data processing software very expensive. Though there are commercial software solutions capable of quickly doing data processing in a disaster context available in the market, none of the respondent agencies were found to be using them. Upon further investigation, it was found that investment in a drone-based data collection and processing system requires the set-up of multiple elements which increases the cost.

b. **Operational Challenges**

The study perceived a lack of coordination among different actors as a major operational challenge for both government officials and drone users. A review of multiple studies of drone use in Nepal revealed that this situation has persisted since the onset of drone application in Nepal. During the 2015 Gorkha earthquake, several organizations were interested in using drones over the affected area which created coordination challenges (Greenwood 2015). Wang (2020) in another study regarding drone use for post-disaster mapping after the earthquake reflected on multiple drone-based mapping operations resulting in duplicated datasets for the same area because of a lack of coordination. This accords with the observations made in this study as the participants reported their experiences of multiple drone operators capturing datasets in the field and them waiting for several hours for drone operations in a disaster-affected area due to multiple actors. Vizvári et al. (2019) point out that multi-organization coordination in a disaster situation is challenging and suggested the development of a standard set of rules to facilitate comprehensive collaboration between all stakeholders on the ground. In this regard, the recommendations made during this study are also consistent with Vizvári et al. (2019) findings as participants in this study also highlighted the need for a detailed SOP consisting of step-by-step guidelines on removing any sort of data duplication.

Another operational challenge identified by this study was the frustration shared by drone pilots regarding the inability to take off flights during different disaster situations due to the inability of hardware to perform well and do the data capture during bad weather. This result to some extent contradicts other findings that claim drones to be a better-performing technology in disaster contexts due to their ability to fly below the cloud and capture ground information in detail (Greenwood 2015). Though this is the rationale behind the use of drones, most of the respondents belonging to drone use groups reported that they can only afford easily available general consumer-graded drones which unfortunately don't have a mechanical setup to cope with extreme weather conditions. In such scenarios, the drone gets grounded which ultimately reduces its value in response missions (Greenwood and Joseph 2020b).

c. **Technological Challenges**

The findings from this study are consistent with DeBusk (2010) findings as participants reported that the use of drones is often hindered by the immaturity of the technology itself. Most of the respondents belonging to drone user groups in this study reported that one of the major drawbacks of commonly available small-sized drones is their limited flight time due to the battery capacity. This gets more difficult in a disaster situation as disaster often damages power infrastructures (Mohsan et al. 2022). Sun et al. (2020) in another study found that such a situation makes it difficult for drone team to charge their batteries and this ultimately limits the coverage of data collection in the affected areas. Even if the pilot team manages somehow to deploy the drones in a disaster-affected area, Sun et al. (2020) warn that operating small drones during challenging weather might cause some damage to the hardware thus requiring repair and proper maintenance. This also accords with the observations made in the study as respondents reported the case of damage to drones during project operations in areas with challenging weather conditions such as high-altitude settlements and even in urban areas during foggy weather in the immediate aftermath of disaster events. In case of damage to the drones, most of the respondents belonging to drone use groups highlighted that Nepal still lacks proper repair and maintenance centers for drone repairs.

Besides hardware instruments and infrastructures, the study also found that drone use during a disaster situation in Nepal is also limited by software constraints, particularly due to the time-long data processing time with existing hardware setup at most of the institutions. This result reflects those of observations made by WeRobotics (2017) during their drones for disaster response coordination training where the data capture took a few minutes while its processing took several hours. This time-consuming data processing process means the emergency response team in a disaster situation will be waiting for necessary results and insights for a longer time which would result in other consequences such as people in need not getting help on time, the media starting to criticizing slow disaster response, politicians criticizing the responders, etc. (WeRobotics 2017). To tackle this challenge, the review of different literature found that there have been different works such as the development of artificial (AI) based solutions to automate the data analysis process for the informed disaster management (Sun et al. 2020). However, all these reports and studies also warned that the AI-based accurate drone imagery analysis for different geographical settings is still a topic of research and is in infancy.

d. **Institutional Challenges**

Similar to the findings of this study, other researchers such as Aiyetan and Das (2022) and Kucharczyk and Hugenholtz (2021) also found that a major challenge to the use of drones in developing countries is the lack of proper government policies, regulations, and guidelines. Most of the respondents belonging to the drone use group in the study reported that there is no standard process to follow regarding drone use in a disaster context and they often follow the same time-consuming manual process for flight permission requests. However, some of them were found to be able to navigate through the process utilizing their personal relationships at government offices and historical evidence to get flight permission missions. Similar to these findings, other studies have also found that in other developing countries, people utilize personal connections with officials at concerned government agencies to secure permissions (Greenwood et al. 2020). The study found that even the most recent version of procedural guidelines still doesn't have any specific clauses to ease drone operations in disaster situations. Wang (2020) reported that several developing countries still have many gaps in their drone use guidelines and argue that as drone technology has emerged rapidly over the last decades, any guidelines or regulations defined by the national and international bodies are not mature and irregularities are observable. Greenwood and Joseph (2020b) on the other hand mentioned that the case depends on countries as some countries have essentially no regulations at all, while others have exceedingly strict drone regulations.

The potential of drones in disaster relief operations is stunted by a multitude of institutional challenges hampering the wide-scale adoption of the technology (Shamsoshoara et al. 2020). Greenwood et al. (2020) mentioned that the absence of a standardized accreditation mechanism for drone pilots operating in post-disaster environments is a critical gap that needs to be addressed. Hence, it is recommended that the GoN dedicate a priority unit to regulating drones instead of distributing the authority to the CAAN and the MoHA's existing departments (Pageni et al. 2020).

e. **Data Privacy and Safety Issues**

Several researchers highlighted privacy as a key ethical issue with the use of drones involving data collection of disaster-affected population (Greenwood et al. 2020; Jeyabalan et al. 2020; Wang 2020). This however was not raised as a significant concern by most of the respondents in this study. Some drone pilots reported that communities occasionally have concerns regarding the purpose of drone intervention. Gevaert et al. (2018) supported this and argued that in some communities, residents are not capable of identifying the ethical concerns that emerge from the use of drones but might raise generic concerns regarding the purpose of the mission, particularly in situations when drones fly over private spaces.

Regarding lessons learned on how to deal with the community concerns for a drone mission, the result showed that community consultation helped the drone teams get approval to conduct the mission from the community. This is consistent with the literature that the general public is more likely to accept UAVs when they perceive them as a technology that is being utilized by actors who are working to support their community (Greenwood et al. 2020). This is actually the best practice as it is important for all drone-related stakeholders (operators, project managers, and community members) to be involved in identifying the impact the technology has on individuals and the potential consequences of the technology (Peckham and Sinha 2019).

Regarding safety concerns, the review of previous studies regarding security-related concerns such as Righetti et al. (2018), and Shakhatreh et al. (2019) indicated and categorized safety concerns as humans, data, and aviation safety, However, this study only found aviation safety-related concerns from the respondents. Most of the government officials from the study were found to be more cautious about it than all other stakeholders. Though there have been no officially reported cases of collisions among drones or with crewed aircraft, government officials were found to be more concerned about the possible risks associated with collisions. Willens et al. (2017) argue this sort of concern to be logical and further added that the safety concerns in the case of small drones are more serious as there lack of clear flight regulations and standards for operations.

f. **Instrumental Challenges**

The respondents in the study highlighted that they face problems ranging from proper data storage to efficient data processing and scientific analysis. Other researches such Greenwood et al. (2020) and Velev et al. (2019) also found similar results that mentioned the impact of drones on disaster response outcomes has not been properly quantified yet.

Regarding the possible reasons, there exist very few methodologies that guide the use of drone data in a systematic way for activities like pre-disaster resilience mapping as well as post-disaster damage assessment (Greenwood and Joseph 2020b). As a result, datasets often get collected but are not extensively used. The participants in this study also reported that there exists no methodological guide to utilize the big data from drones in disaster response decision-making. In such cases, the drone users themselves come up with methodologies and systems for using drone data which are not always effective and bring disparity and variation in results. This issue can be further understood if the problem is analyzed over the information value chain (Fig. 4).

Summary–Discussions

The value of any technology either drone or anything else is determined by whether the use of technology has created any real-life impact. In the case of drone use during a disaster situation, a careful analysis of findings and discussions can summarize the

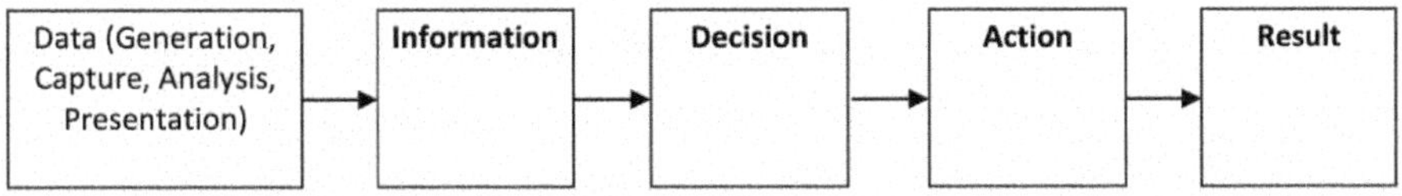

Fig. 4 Basic information value chain

actual value of using a drone during disaster response operations. Based on the findings and discussions stated above, the value of drone technology can be assessed in a set of four actions: the ability to swiftly deploy the technology; the ability to capture required datasets in the field, and the ability to analyze those datasets on time and the ability to use the analytic results in disaster response activities. In this regard, the presence of different challenges limits the successful functioning of the activities highlighted above. The table below maps out the major challenges found in the study under each activity (Table 3).

Regarding further steps in mitigating the identified challenges, there are several small interventions required, some already being carried out by different stakeholders. A careful analysis of those activities and existing challenges showed that the challenges are interconnected to one another. For example, the availability of skilled human resources determines the effectiveness of the data collection and analysis of the processed outputs during disaster response operations. Likewise, financial status and available budget determine the technology capacity of drone operators which ultimately decides how quickly drone datasets can be processed generating information for emergency response teams. This relationship among different challenges as highlighted by the ROTIDI framework has been presented in the form of a diagram below (Fig. 5).

In this regard, no specific challenge can be addressed by interventions targeted to that specific issue only. Due to the holistic nature and interrelationship among the

Table 3 Challenges encountered under different phases of drone use from data capture to action

Ability to deploy the drones	Ability to capture datasets	Ability to analyze datasets on time	Will drones make a difference?
• Institutional challenges (timely flight permissions) • Human resource challenges (availability of expert human resources locally)	• Operational challenges (weather conditions, coordination issues) • Technological challenges (hardware limitations) • Financial challenges (Purchasing advance sensors for detailed datasets)	• Human resource challenges (lack of expert human resources) • Technological challenges (advance software systems for automatic analysis) • Financial (hiring expert human resources)	• Human resource challenges (qualified human resource to analyze outputs and integrated with response actions) • Institutional challenge (drone skepticism, data-driven decision-making culture)

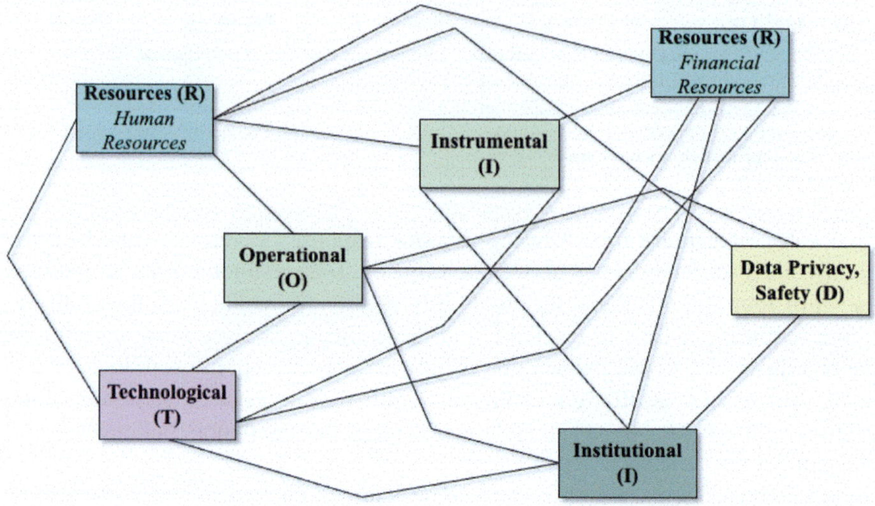

Fig. 5 Interrelationship among different thematic challenges

challenges, there should be an integrated approach to deal with multiple challenges either together or one after another.

6 Conclusions and Future Work

The research showed that drones are increasingly popular with relative advantages such as affordability, deployment flexibility, and centimeter-level spatial resolution data capture, however, their use in a disaster situation is limited due to several challenges. In the case of a country like Nepal, which is one of the earliest countries to introduce drones for disaster response activities, the study found only a limited use such as capturing images and videos to prepare maps and rapid assessment of damages as the common use of drones during disaster events. Besides, drone applications faced many challenges. Given the growing, worldwide popularity of drones in various types of applications (involving substantial investments), it is important to develop robust guidelines for the successful deployment of drones for various applications including disaster management.

This chapter has discussed a systematic study on the challenges facing the large-scale deployment of drones for disaster management in developing countries. The research led to a theoretical framework 'ROTIDI' to evaluate the different sorts of challenges of implementing drones for disaster response projects in developing countries context taking the case of a disaster-prone, developing country, such as Nepal. We carried out qualitative research involving a large cross-section of stakeholders of drone applications for disasters in Nepal. We discussed our findings and corroborated them with such studies in other parts of the world. However, more research is necessary to evaluate the effectiveness of this ROTIDI framework for Drone-based Disaster Management in other developing countries.

This study has generalized the situations during all disaster scenarios and evaluated the challenges of deploying drones during such situations. However, more research is needed to investigate specific types of disasters and that may need a modification/enhancement of the ROTIDI framework proposed by us. Also, it is important to note that most of the drones in disaster response projects discussed by the respondents in this study were short-term pilot projects mostly carried out by agencies to test whether the technology will meet their requirements. Further research is needed to determine if all the challenges mentioned earlier persist or not once the projects are designed on a long-term basis Besides, more research is needed to validate the ROTIDI framework for evolving drone applications other than disaster management (e.g., healthcare, agriculture, urban planning, etc.)

References

Agüera-Vega F, Carvajal-Ramírez F, Martínez-Carricondo P (2017) Accuracy of digital surface models and orthophotos derived from unmanned aerial vehicle photogrammetry. J Surv Eng 143(2)

Aiyetan A, Das D (2022) Use of drones for construction in developing countries: barriers and strategic interventions. Int J Constr Manag 1–10

Antoine R et al (2020) Geoscientists in the sky: unmanned aerial vehicles responding to geohazards. Surv Geophys 41(6)

Baral S (2015) Uav for biomass estimation: geospatial lab. Kathmandu University. https://geospatiallab.wordpress.com/2015/07/15/uav-for-biomass-estimation/. Accessed July 11

Van Blyenburgh P (1999) Uavs: an overview. Air & Space Europe 1(5–6):43–47. https://doi.org/10.1016/s1290-0958(00)88869-3. Available at: https://doi.org/10.1016/s1290-0958(00)88869-3

CAAN Civil Aviation Authority of Nepal (2021) Unmanned aircraft system requirements (uasr). Civil aviation Authority of Nepal, Babarmahal, Kathmandu. https://caanepal.gov.np/storage/app/media/drone/UAS%20Requirement%20final.pdf

Cavaye AL (1996) Case study research: a multi-faceted research approach for is. Inf Syst J 6(3):227–242

Choi H, Park MJ, Rho JJ, Zo H (2016) Rethinking the assessment of e-government implementation in developing countries from the perspective of the design–reality gap: applications in the indonesian e-procurement system. Telecommun Policy 40(7)

Coppola DP (2006) Introduction to international disaster management. Elsevier Science

Crowe S (2015) Nepal earthquake: drones helping recovery efforts. https://www.roboticsbusinessreview.com/rbr/nepal_earthquake_drones_helping_recovery_efforts/. Accessed July 3

Cumbane SP, Gidófalvi G (2019) Review of big data and processing frameworks for disaster response applications. ISPRS Int J Geo Inf 8(9):387

Daud SMSM et al (2022) Applications of drone in disaster management: a scoping review. Sci Justice 62(1):30–42

DeBusk W (2010) Unmanned aerial vehicle systems for disaster relief: Tornado alley. [Online]. https://ntrs.nasa.gov/api/citations/20090036330/downloads/20090036330.pdf

Erdelj M, Natalizio E, Chowdhury KR, Akyildiz IF (2017) Help from the sky: leveraging uavs for disaster management. IEEE Pervasive Comput 16(1):24–32

Ezequiel CAF et al (2014) Uav aerial imaging applications for post-disaster assessment, environmental management and infrastructure development. In: 2014 international conference on unmanned aircraft systems (ICUAS). IEEE, pp 274–283

Fiorucci F et al (2018) Criteria for the optimal selection of remote sensing optical images to map event landslides. Nat Hazard 18(1):405–417

Gevaert CM, Sliuzas R, Persello C, Vosselman G (2018) Evaluating the societal impact of using drones to support urban upgrading projects. ISPRS Int J Geo Inf 7(3):91

Giordan D et al (2018) Low cost, multiscale and multi-sensor application for flooded area mapping. Nat Hazard 18(5):1493–1516

Greenwood F, Joseph D (2020a) Aid from the air: a review of drone use in the rcrc global network. Red Cross Red Crescent 3. https://americanredcross.github.io/rcrc-drones/Aid_from_the_Air.pdf. Accessed 22 July 2022

Greenwood F, Joseph D (2020b) Aid from the air: a review of drone use in the rcrc global network: the American red cross. [Online]. https://americanredcross.github.io/rcrc-drones/Aid_from_the_Air.pdf. Accessed 16 June 16.

Greenwood F, Nelson EL, Greenough PG (2020) Flying into the hurricane: a case study of uav use in damage assessment during the 2017 hurricanes in texas and florida. PLOS One 15(2)

Greenwood F (2015) How drones can help nepal recover from the earthquake. https://slate.com/technology/2015/04/how-drones-can-help-nepal-recover-from-the-earthquake.html

Heeks R (2006) Implementing and managing egovernment: an international text. Sage

Hern A (2015) Nepal moves to limit drone flights following earthquake: the Guardian. https://www.theguardian.com/technology/2015/may/06/nepal-moves-to-limit-drone-flights-following-earthquake. Accessed 1 July

Hildmann H, Kovacs E (2019) Using unmanned aerial vehicles (uavs) as mobile sensing platforms (msps) for disaster response, civil security and public safety. Drones 3(3):59

Iqbal et al (2022) https://www.sciencedirect.com/science/article/abs/pii/B9780323911665000070

Jeyabalan V, Nouvet E, Meier P, Donelle L (2020) Context-specific challenges, opportunities, and ethics of drones for healthcare delivery in the eyes of program managers and field staff: a multi-site qualitative study. Drones 4(3):44

Jorz V (2019) Open aerial map, drones and archaeology: the implications of using drones to contribute and share aerial data on an open data repository. University of Waterloo

Jupp V (2006) The sage dictionary of social research methods. Sage

Kedia T et al (2020) Technologies enabling situational awareness during disaster response: a systematic review. Disaster Med Public Health Prep 1–19

Khatri PP (2019) Had tornado hit bara- parsa, southern nepal? Meteorologists are not sure: nepal24hours. https://nepal24hours.com/had-tornado-hit-bara-parsa-southern-nepal-meteorologists-are-not-sure/. Accessed 5 July

Kraaijenbrink P et al (2016) Seasonal surface velocities of a himalayan glacier derived by automated correlation of unmanned aerial vehicle imagery. Ann Glaciol 57(71):103–113

Kucharczyk M, Hugenholtz CH (2021) Remote sensing of natural hazard-related disasters with small drones: Global trends, biases, and research opportunities. Remote Sens Environ 264

Li J, Seale H, Ray P, Wang Q, Yang P, Li S, Zhang Y, MacIntyre CR (2023) e-Health preparedness assessment in the context of an influenza pandemic: a qualitative study in China. Br Med J (BMJ) Open 3:e002293. https://doi.org/10.1136/bmjopen-2012-002293

Meier P (2015) Humanitarian uav missions in nepal: early observations. https://irevolutions.org/2015/05/03/humanitarian-uav-missions-nepal/. Accessed 16 June

MoHA (2015) Unmanned aerial vehicle (uav) udaan sambhandi karyabidhi, 2072: ministry of home affairs. https://nepalindata.com/media/resources/items/15bUnmanned_Aerial_Vehicle_UAV.pdf

MoHA (2019) Drone related procedure, 2019: ministry of home affairs. https://moha.gov.np/en/post/drone-related-procedure-2019. Accessed 15 July

Mohsan SAH et al (2022) Towards the unmanned aerial vehicles (uavs): a comprehensive review. Drones 6(6):147

NFL (2020) Launching a digital solution to coordinate drone flights for disaster response: Nepal flying labs. https://blog.werobotics.org/2020/12/07/launching-a-digital-solution-to-expedite-drone-flight-permissions-for-disaster-response/. Accessed 17 May

Pageni B, Uttam Pudasaini DP, Pradhan S (2020) Evolution of unmanned aerial vehicles in nepal. J GeoInformatics Nepal. https://www.researchgate.net/profile/Biplav-Pageni/publication/354477449_Evolution_of_Unmanned_Aerial_Vehicles_in_Nepal/links/613a51cdee29a068c715277e/Evolution-of-Unmanned-Aerial-Vehicles-in-Nepal.pdf

Peckham R, Sinha R (2019) Anarchitectures of health: futures for the biomedical drone. Glob Public Health 14(8):1204–1219

Poppleton L (2012) *Wwf nepal launches conservation drones*: world wildlife fund (WWF). WWF Nepal Launches Conservation Drones. Accessed 3 July

Pudasaini U, MSc in Digital Development Dissertation (2022) The University of Manchester, UK

Rejeb A, Rejeb K, Simske S, Treiblmaier H (2021) Humanitarian drones: a review and research agenda. Internet Things 16

Righetti F, Vallati C, Anastasi G (2018) Iot applications in smart cities: a perspective into social and ethical issues. In: 2018 IEEE international conference on smart computing (SMARTCOMP). IEEE, pp 387–392

RSS (2021) Authorities to use drone camera to search for missing dhami. https://nepalnews.com/s/nation/authorities-to-use-drone-camera-to-search-for-missing-dhami. Accessed 5 July

Shakhatreh H et al (2019) Unmanned aerial vehicles (uavs): a survey on civil applications and key research challenges. IEEE Access 7:48572–48634

Shamsoshoara A et al (2020) An autonomous spectrum management scheme for unmanned aerial vehicle networks in disaster relief operations. IEEE Access 8:58064–58079

Sun W, Bocchini P, Davison BD (2020) Applications of artificial intelligence for disaster management. Nat Hazards 103(3):2631–2689

Takamatsu M, Karelia HD, Lnu TO, Dahal RK (2022) Melamchi flood disaster in nepal: damage and risk quantification with drone survey, satellite-based land displacement analysis, and 2d flood modeling. [Online]. https://policycommons.net/artifacts/2478052/melamchi-flood-disaster-in-nepal/3500175/

Tanner V (2018) Using drones in disaster areas: perspectives of disaster responders in north carolina, virginia and maryland. East Carolina University

Tarolli P (2014) High-resolution topography for understanding earth surface processes: opportunities and challenges. Geomorphol 216:295–312

UNDRR (2022) Global assessment report on disaster risk reduction. Geneva 2, Switzerland: united nations office for disaster risk reduction [Online]. https://reliefweb.int/report/world/global-assessment-report-disaster-risk-reduction-2022-our-world-risk-transforming. Accessed 15 June

Velev D, Zlateva P, Steshina L, Petukhov I (2019) Challenges of using drones and virtual/augmented reality for disaster risk management. Int Arch Photogramm, Remote Sens Spat Inf Sci 42:437−440

Vizvári B et al (2019) Top-down approach to design the relief system in a metropolitan city using uav technology, part i: the first 48 h. Nat Hazards 99(1):571–597

Wang N (2020) We live on hope…: ethical considerations of humanitarian use of drones in post-disaster Nepal. IEEE Technol Soc Mag 39(3):76−85

Welsh E (2002) Dealing with data: Using nvivo in the qualitative data analysis process. Forum Qual SozForschung/Forum: Qual Soc Res

WeRobotics (2017). Building expertise in humanitarian drone coordination in malawi. https://blog.werobotics.org/2017/12/14/humanitarian-drone-coordination-malawi/. Accessed July 20

Willens DD, Proudlove C, Miller T (2017) Drone away: bombs, photographs, pizzas, hearts, and headaches. The Brief 46(4)

Winchester L (2015) Tragic drone footage reveals devastating aftermath of nepal earthquake. https://www.express.co.uk/news/world/573439/Nepal-earthquake-devastating-drone-footage. Accessed 3 July

Yakushiji K et al (2020) Short-range transportation using unmanned aerial vehicles (uavs) during disasters in japan. Drones 4(4):68

Information Technologies Used

A Generic Model to Capture Essential Information on Limited Capacity Mobile Devices During a Disaster

Teuku Aulia Geumpana and Pradeep Kumar Ray

1 Introduction

The rapid advancement of technology and the proliferation of interconnected devices have led to a surge in data generation and the demand for efficient and timely processing. In response to this evolving landscape, the integration of cloud computing and edge computing paradigms has gained significant attention. One noteworthy solution that has emerged in this context is the Priority Over Cloudlet Terminal (POCkET) Framework (Geumpana and Rabhi 2020), complemented by the Generic Information Model (GIMo). This framework offers a comprehensive approach to resource allocation and task scheduling in a cloudlet-edge environment, while GIMo provides a standardized representation of information exchanged between cloudlets and edge devices.

Traditional cloud computing models face challenges in meeting the latency requirements and resource constraints of latency-sensitive and resource-intensive applications (Shukla 2021; Zhang et al. 2010). The POCkET Framework addresses these challenges by incorporating cloudlets, which are lightweight computing infrastructures located at the edge of the network. By strategically placing cloudlets closer to edge devices, the framework aims to minimize latency, reduce network congestion, and enable efficient data processing (Geumpana and Rabhi 2020).

T. A. Geumpana (✉)
School of Information and Physical Sciences, University of Newcastle, Newcastle, Australia
e-mail: Teuku.Geumpana@newcastle.edu.au

P. K. Ray
University of Michigan—Shanghai Jiao Tong University Joint Institute, Shanghai, China

UNSW Business, Sydney, Australia

P. K. Ray
e-mail: p.ray@unsw.edu.au

P. K. Ray et al. (eds.), *Technology Innovation for Sustainable Development of Healthcare and Disaster Management*, Disaster Risk Reduction,
https://doi.org/10.1007/978-981-97-2049-1_12

At the core of the POCkET Framework lies the Generic Information Model (GIMo), which plays a pivotal role in facilitating effective communication and coordination between cloudlets and edge devices. GIMo provides a standardized representation of information, ensuring compatibility and interoperability across heterogeneous systems. It establishes a common language that enables seamless integration and cooperation, ultimately enabling efficient resource allocation and task scheduling (Geumpana and Rabhi 2020).

The motivation behind the development of the POCkET Framework is rooted in the need to address the latency requirements in a low bandwidth environment during the disaster emergency response. In other contexts, real-time applications, such as augmented reality, autonomous vehicles, and telemedicine, also require low-latency processing capabilities that traditional cloud computing models struggle to provide (Attaran 2021; Singh and Gill 2023; Mulimani and Rachh 2021). By bringing computing resources closer to edge devices, the POCkET Framework significantly reduces the latency involved in data transmission and processing, enhancing the user experience and enabling the deployment of time-critical applications.

Resource utilization in distributed computing environments is another challenge that the POCkET Framework aims to tackle. With the exponential growth of connected devices and the sheer volume of data generated, efficient resource management becomes paramount especially in the hostile environment during the emergency response (Haworth and Bruce 2015). The POCkET Framework optimizes resource utilization by intelligently allocating tasks to the most suitable computing entities, striking a balance between cloud servers and edge devices. By offloading certain tasks to cloudlets, the framework reduces the burden on the central cloud infrastructure, enabling better scalability and improved performance (Geumpana and Rabhi 2020).

The heterogeneity of edge devices and cloudlets is also a significant consideration within the POCkET Framework. Variations in processing capabilities, energy constraints, and network connectivity are taken into account (Bilal et al. 2018), and GIMo plays a critical role in capturing and representing these diverse characteristics (Geumpana and Rabhi 2020). Through this standardized representation, the framework ensures seamless interoperability across different edge devices and cloudlet configurations, facilitating a plug-and-play environment for distributed computing.

The potential impact of the POCkET Framework and GIMo extends across various industries and domains. In disaster response scenarios, for instance, the integration of cloudlets and edge devices through this framework can enhance information dissemination, resource allocation, and decision-making, improving the effectiveness and efficiency of emergency management (Geumpana and Rabhi 2020). Additionally, this can also be extended to various applications in healthcare, transportation, smart cities, and industrial environments which can benefit from the reduced latency, improved resource utilization, and standardized information exchange provided by the POCkET Framework and GIMo.

2 Fundamentals of POCkET Framework in Cloud Computing and Edge Computing Perspectives

Cloud computing is a paradigm that involves delivering computing resources, including storage, processing power, and software applications, over the Internet (Bhardwaj et al. 2010). In the context of the "Priority Over Cloudlet Terminal (POCkET) Framework: Generic Information Model (GIMo)," cloud computing represents the centralized infrastructure that hosts cloud servers responsible for processing and storing data (Geumpana and Rabhi 2020). The POCkET Framework recognizes the significance of cloud computing while aiming to address its limitations, such as high latency and resource constraints for real-time and latency-sensitive applications.

Similarly, edge computing is known as a distributed computing paradigm that enables computation and data storage closer to the data source, typically at the edge of the network (Yu et al. 2017). In the context of the POCkET Framework and GIMo, edge computing represents the integration of cloudlets within the network infrastructure (Geumpana and Rabhi 2020). Cloudlets are localized computing resources strategically placed closer to edge devices to minimize latency and optimize resource allocation. By leveraging edge computing, the POCkET Framework aims to enhance the efficiency and performance of distributed computing, especially in scenarios where real-time data processing, low-latency communication, and reduced reliance on distant cloud servers are critical.

In the Internet world, the integration of cloud computing and edge computing, often referred to as cloud-edge convergence, combines the advantages of both paradigms to create a more efficient and effective computing environment. Cloud-edge convergence seeks to balance the benefits of centralized cloud computing with the low-latency and localized processing capabilities of edge computing (Ai et al. 2018). This convergence enables tasks to be processed at the most suitable location based on their requirements, optimizing resource utilization, and reducing network congestion.

The POCkET Framework embodies the cloud-edge convergence concept by integrating cloudlets within the network infrastructure. These cloudlets act as localized computing resources that enhance the performance of edge devices by offloading computation and data processing tasks. The framework leverages the benefits of both cloud computing and edge computing to provide a scalable, low-latency, and efficient computing environment.

Resource allocation and task scheduling are fundamental aspects of both cloud computing and edge computing (Alizadeh et al. 2020). In cloud computing, resource allocation involves assigning virtual machines, storage, and network bandwidth to applications based on their resource requirements (Hameed et al. 2016). Task scheduling, on the other hand, determines the order and allocation of computing tasks to available resources for efficient execution (Buyya et al. 2009).

In the context of the POCkET Framework and GIMo, resource allocation and task scheduling play crucial roles in optimizing the utilization of cloudlets and edge devices. The framework allocates resources based on factors related to data dimensions such as proximity, frequencies, recencies, and others (Geumpana and Rabhi 2020). Another fundamental aspect related to the POCkET Framework and GIMo is the standardized representation of information. In cloud computing and edge computing, the representation of information exchanged between different components and systems is crucial for interoperability and compatibility (Roman et al. 2018). Standardized information models ensure seamless communication, data exchange, and coordination between cloud servers, cloudlets, and edge devices.

Within the POCkET Framework, GIMo plays a pivotal role in facilitating effective communication and coordination between cloudlets and edge devices. It enables seamless integration and cooperation by establishing a standardized format for exchanging information. By adhering to the GIMo specifications, different components within the POCkET Framework can understand and interpret the exchanged data correctly, regardless of their individual characteristics.

The standardized representation of information through GIMo ensures that cloudlets and edge devices can communicate efficiently, share data, and coordinate tasks effectively. It enables uniformity in how information is structured, communicated, and processed, which is crucial in distributed computing environments where diverse systems and technologies coexist.

3 The POCkET Framework

The Priority Over Cloudlet Terminal (POCkET) Framework is an innovative approach that addresses the challenges of resource allocation and task scheduling in distributed computing environments, particularly in the context of cloud-edge convergence. The framework aims to optimize resource utilization, reduce latency, and enhance the overall performance and scalability of distributed systems (Geumpana and Rabhi 2020). In conjunction with the POCkET Framework, the Generic Information Model (GIMo) provides a standardized representation of information exchanged between cloudlets and edge devices, ensuring interoperability and compatibility.

At its core, the POCkET Framework leverages the concept of cloudlets, which are lightweight computing infrastructures located at the edge of the network. By strategically placing cloudlets closer to the edge devices, the framework aims to minimize latency and reduce network congestion. Cloudlets act as localized computing resources that can offload computationally intensive tasks from resource-constrained mobile devices, reducing their energy consumption and improving their overall performance (Babar et al. 2021). This integration of cloudlets within the network infrastructure brings computing resources closer to the edge devices, facilitating faster data processing and enabling real-time applications.

Resource allocation is a critical aspect of the POCkET Framework. The framework intelligently allocates computing resources based on various factors, such as proximity, computing capabilities, and energy constraints. By considering these factors, the framework ensures efficient utilization of resources and minimizes the impact of resource limitations in resource-constrained environments. This allocation strategy helps optimize the performance and scalability of distributed systems, allowing for better overall system efficiency.

Task scheduling is another key component of the POCkET Framework. The framework aims to schedule tasks in a way that minimizes latency and enhances overall system performance. It determines the order and allocation of computing tasks to available resources, taking into account factors such as task requirements, resource availability, and network conditions. By intelligently scheduling tasks, the framework ensures that computation is offloaded to cloudlets when necessary, reducing the burden on edge devices and improving the response time for latency-sensitive applications. Figure 1 shows the high-level interaction between users and different layers of the POCkET Framework.

GIMo, the Generic Information Model, as one of the main components in the POCkET Framework plays a crucial role in facilitating effective communication and coordination between cloudlets and edge devices within the POCkET Framework. GIMo provides a standardized representation of information, ensuring interoperability and compatibility across heterogeneous systems. It establishes a common language that enables seamless integration and cooperation between different components, enhancing the efficiency and effectiveness of distributed computing. GIMo captures essential information such as network topology, available resources, and

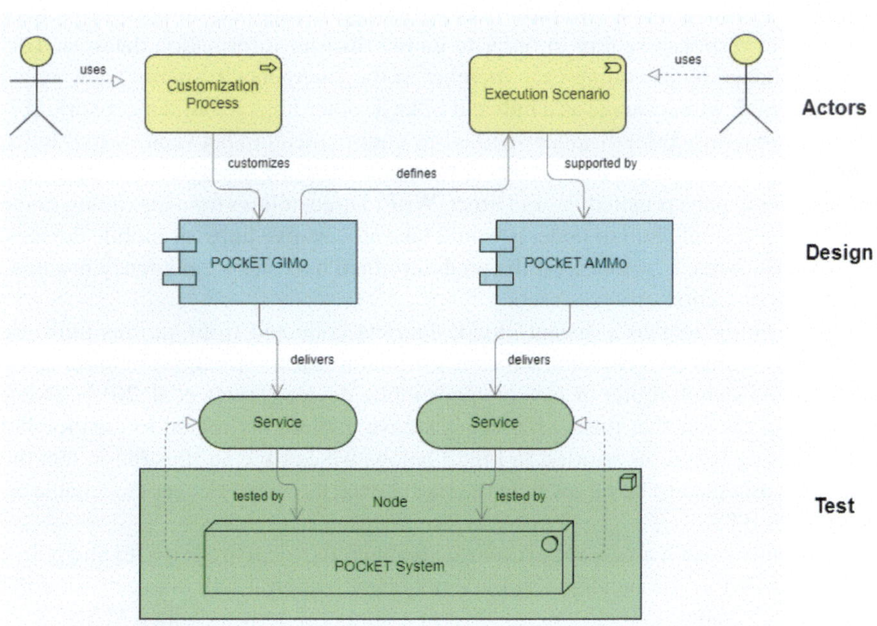

Fig. 1 POCkET framework

task requirements, enabling efficient resource allocation and task scheduling within the POCkET Framework.

The integration of the POCkET Framework and GIMo offers numerous benefits in various domains and industries. For example, in disaster response scenarios, the framework can enhance information dissemination, resource allocation, and decision-making, improving the effectiveness and efficiency of emergency management. In health care, the POCkET Framework can enable real-time applications and improve patient monitoring and diagnostics by leveraging cloudlets for computationally intensive tasks. Similarly, in transportation systems, the framework can optimize resource allocation and reduce latency, leading to improved traffic management and safety.

4 Generic Information Model (GIMo) Assumptions

Making assumptions is a crucial first step in the creation of GIMo. This aids in the understanding of the circumstances in a disaster response setting by the designer; certain environmental assumptions must be made to guarantee that the system is designed to meet the response needs.

The suggested framework first presupposes that the emergency response team creates an ad hoc network (Braunstein et al. 2006) utilizing peer-to-peer technology in a disaster response setting in order to gather disaster information datasets. This is accomplished by providing each member of the emergency response team with a mobile device, which serves as a hub and links to other hubs to build a network. It is envisaged that only brief distances will allow communication between various hubs. A peer-to-peer technique that creates a temporal network through the relationship between many hubs is called Wi-Fi Direct. Wi-Fi Direct allows for a maximum range of 200 m for a single hub or node. It would take at least five hubs, or mobile devices, for a dataset to reach a cloudlet if the predetermined maximum emergency response distance of one kilometre was used.

Before being sent to a distant cloud, datasets collected from various hubs are temporarily stored on a cloudlet, which is a miniature cloud storage system set up by a specialized switch router or just another mobile device (Lewis et al. 2014). Every information dataset that the source devices have gathered is routed to the cloudlet for processing based on priority. In Fig. 2 below, we can see an illustration that the bushfire is assumed to be far away in the rural hence the network is not reachable by any nearest BTS.

To connect to each emergency response member, the local fire brigades and police who responded to the incidents as the emergency response will connect using the Wi-Fi Direct feature in their phones to form a temporal network called MANET.

Nonetheless, certain datasets, such as those pertaining to the current state of hazards at a specific place, will undergo updates after a certain amount of time. The status of a bush fire at a specific site X, for instance, is being sent by emergency response officer Bob. 30 min later, Alice, another emergency officer, is sending

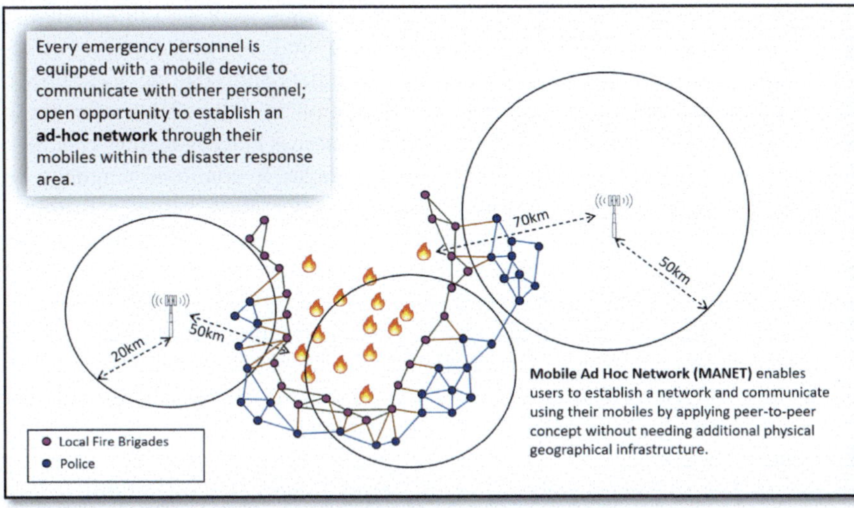

Fig. 2 Assumptions for disaster response environment

information on the amount of danger associated with the neighboring industrial plant at a location Y that might be affected by the wildfire hazard from place X. The state of the bush fire at site X is being updated by Bob. After 60 min, Alice updates the status of the industrial plant's risk level at the Y location. Despite having identical information, the two new datasets must be prioritized differently.

A dataset will be traveling over the ad hoc network in the direction of the closest cloudlet as part of the process of reaching one. Initially, a gadget looks around it to find the closest cloudlet. The hub forwards the dataset to a neighboring hub if it cannot locate the cloudlet. The dataset will arrive at the desired cloudlet as a result of this technique.

In conclusion, we take into account that a dataset may take a while to reach a cloudlet, there is a limit on network capacity, and the disaster area is significant. Owing to the restricted infrastructure on site, network disruptions can occur during the setup of the temporal network. For connecting devices on-site to the Internet, Cloudlet, one of the infrastructure facilities in a temporal network environment, is the only data storage option. To prevent the risk, it ought to be maintained within a safe range.

Assuming these conditions are met, the POCkET architecture must enable the subsequent primary execution scenario for mobile ad hoc network-based disaster response communication. In brief, the duties in executing this scenario are to ensure that the source device level is where datasets are gathered. Then, by means of other devices or directly, datasets are sent to the cloudlet, in which the process of prioritizing datasets occurs at this level before they are sent to the remote server.

In the context of the "Priority Over Cloudlet Terminal (POCkET) Framework: Generic Information Model (GIMo)," GIMo plays a vital role in facilitating effective communication and coordination between cloudlets and edge devices. GIMo provides a standardized representation of information exchanged between these entities, ensuring interoperability and compatibility across heterogeneous systems.

The Generic Information Model (GIMo) serves as a common language or framework that enables seamless integration and cooperation between different components within the POCkET Framework. It captures essential information about network topology, available resources, task requirements, and other relevant parameters, facilitating efficient resource allocation and task scheduling.

GIMo establishes a standardized format for representing information, which allows cloudlets and edge devices to communicate effectively and exchange information seamlessly. By providing a common understanding of the data exchanged, GIMo eliminates potential interoperability issues that may arise from differences in data formats or protocols.

One of the key advantages of GIMo is its ability to ensure compatibility across heterogeneous systems. In distributed computing environments, where different cloudlets and edge devices may have varying architectures, capabilities, and communication protocols, GIMo acts as a unifying model. It defines a standardized information representation that enables different components to understand and interpret the exchanged data correctly, regardless of their individual characteristics.

By using GIMo, the POCkET Framework can achieve efficient coordination and cooperation between cloudlets and edge devices. For example, when a task needs to be offloaded from an edge device to a cloudlet, GIMo provides a standardized representation of the task's requirements, enabling the cloudlet to evaluate its suitability and allocate resources accordingly. Similarly, when a cloudlet needs to communicate with an edge device to exchange data or coordinate tasks, GIMo ensures that both parties understand the exchanged information, regardless of their underlying technologies or protocols.

GIMo also enables the POCkET Framework to handle dynamic changes in the network and resource availability. By representing network topology and resource availability in a standardized manner, GIMo allows the framework to adapt and make informed decisions based on real-time information. This flexibility ensures that the POCkET Framework can optimize resource allocation and task scheduling even in dynamic and evolving environments.

Furthermore, GIMo supports interoperability with other existing standards and frameworks. It can be integrated with industry standards, such as those related to Internet of Things (IoT) or cloud computing, to facilitate seamless data exchange and collaboration between different systems and domains. This interoperability ensures that the POCkET Framework can leverage existing technologies and infrastructures, making it adaptable and compatible with various applications and industry-specific requirements.

5 Implementation of GIMo and Integration to POCkET

The process of developing GIMo component into POCkET will go through the following steps:

1. Identifying disaster information characteristics for the emergency response phase from an existing metamodel in Othman et al. (2014).
2. Defining appropriate dataset attributes.
3. Identifying and modeling the dataset priority elements.

5.1 Disaster Information Characteristic Identification

Based on the metamodel used in Othman et al. (2014), we identified different concepts that stimulate the generation of information that is regularly gathered during the emergency response operation. The following are among the top concepts that appeared frequently during the incident, aid, and rescue response phase: disaster, victim, emergency plan, response organization, coordination, communication, situational analysis, food aid, medical aid, and rescue.

5.2 Dataset Definition

A POCkET dataset is a structured collection of data that is organized and presented in a specific format, containing essential disaster response information. It represents a unified and cohesive set of data points or observations grouped together based on a shared theme or purpose. These datasets consist of individual data elements that contain specific pieces of information, such as text, images, and videos. The datasets are derived from diverse sources, including situational reports, past surveys, or data collected from sensors and devices.

To define the POCkET dataset based on the identified concepts, we establish dimensions that capture the acquired information:

1. Unique Dataset Identifier (DsID): To differentiate one dataset from another, each dataset is given a unique identification (DsID). Similar DsIDs are allocated to datasets when the same type of data is updated or resent.
2. Dataset Source Location Information (DsLOC): It contains metadata that shows where the source device was located in relation to the specific dataset it gathered. It helps calculate the distance from the source of the data and gives details about its origin.
3. Dataset Creation Time (DsCT): The timestamp of the dataset's original generation at the source device level is recorded in this feature.
4. The date and time that the dataset was received correctly by the cloudlet is indicated by the Dataset Arrival Time (DsAT).

5. Dataset Category (DC): Based on its nature or intent, the collected information is categorized into distinct groups or categories using the dataset category attribute.
6. Dataset Type (DT): The dataset type attribute identifies the format types of the information within the dataset. Information can be represented in various formats, such as text, sound, images, and videos.
7. Dataset Size (DS): The dataset size attribute quantifies the size or volume of the collected information within the dataset.
8. Dataset Frequency (DF): Dataset Frequency represents the time window in which information of the same kind becomes available at the cloudlet.
9. Dataset Recency (DR): This metric shows how recent or fresh the data was when it arrived at the cloudlet.
10. Distance to Cloudlet (D2C): The Distance to Cloudlet attribute identifies the remoteness of the source device's information from the cloudlet, considering the geographic distance.
11. Hubs to Hop (H2H): This measure of the communication architecture indicates how many intermediate hubs a dataset needs to pass through in order to reach its closest cloudlet.
12. Dataset Priority Value (DPV): The Dataset Priority Value is a ranking value that is obtained by adding up the outcomes of the priority functions for each attribute. It aids in assessing a dataset's significance or urgency inside the POCkET Framework.

5.3 Dataset Priority Model

The Dataset Priority Model in the POCkET Framework refers to a mechanism or approach used to assign priority values to datasets based on their significance or urgency within the context of disaster response.

The Dataset Priority Model considers various factors and attributes associated with a dataset to determine its priority value. These attributes may include dataset recency, dataset size, dataset category, distance to the cloudlet, dataset frequency, and other relevant parameters. Each attribute is assigned a weight or importance based on its relevance to the specific disaster response scenario.

The model utilizes these attribute weights and combines them using a predefined priority function to calculate the dataset's overall priority value. The priority function may involve aggregating the attribute values or applying specific algorithms to derive a composite priority score.

By assigning priority values to datasets, the POCkET Framework can facilitate resource allocation, decision-making, and task scheduling based on the relative importance of each dataset. This prioritization ensures that critical or time-sensitive datasets receive appropriate attention and resources, leading to more efficient and effective disaster response operations.

It's important to note that the specifics of the Dataset Priority Model may vary based on the requirements and context of the POCkET Framework implementation.

The weights assigned to attributes and the priority function used may differ depending on the specific needs and priorities of the disaster response scenario.

Certain qualities have priority levels, ranging from 1 to 10, where 1 represents the lowest priority and 10 represents the most, as will be explained subsequently.

The default priority value ranges for each dataset type are listed in the first dimension. The default priority range denotes the anticipated level of importance assigned to every attribute in the information category. Furthermore, the expected update frequency and transmission to the cloudlet for each category characteristic are represented by the default frequency range.

Furthermore, the dataset type dimension encompasses various formats, including text, image, voice, and video, each with its own defined characteristics and default priority ranges. This dimension provides an understanding of the available dataset types and their corresponding priority levels. Additionally, the default size range serves as estimated sizes specific to each data type.

The third dimension, dataset size, refers to a range of sizes representing the required disk storage capacity. Default values for dataset size ranges were determined using information from Reynisson (2015). The paper's context depicts a disaster area scenario characterized by limited network capabilities and vulnerable infrastructures. For every dataset size range, default priority range assignments have been made to guarantee effective data transfer. The basic premise is that smaller datasets should be prioritized over bigger datasets since they are less likely to cause delays to other datasets because of bandwidth constraints.

As previously indicated, the Dataset Frequency (DF) characteristic uses the time window to ascertain when the cloudlet becomes accessible for the same sort of information. Only once a dataset (called j) reaches the cloudlet is this characteristic computed. Assuming the relationship DsIDj = DsIDi holds true, the previously received dataset with the same dataset ID (let's call it i) is retrieved in order to determine the time window. The difference in arrival times between the two datasets, represented as DsATj–DsATi, is then used to define the temporal frame. Therefore, DsATj–DsATi = DFj.

Every dataset gets updated by default every 20 min, resulting in a maximum default period of three hours. It is thought under these default conditions that datasets that need to be updated frequently are potentially significant and should be monitored immediately.

The freshness of the data received at the cloudlet is indicated by the next dimension, dataset recency. It is determined by subtracting the Dataset Creation Time (DsCT) from the Dataset Arrival Time (DsAT). At design time, default settings for dataset recency are specified. Datasets with more recent arrival times are given higher priority values than datasets with less recent arrival times in the default scenario.

By assessing the difference between the source location of the dataset and the cloudlet, the distance attribute is computed at the cloudlet. As previously noted in the assumptions, the distance to cloudlet characteristic is important because decision makers would value information that is closer to the hazard more. Longer distances to the cloudlet can cause delays, even if some datasets might include important insights for decision-making. Priorities need to be set in order to solve this problem and achieve a balance.

The final attribute associated with a dataset is the number of hops to reach the cloudlet. Like the rationale behind the distance to the cloudlet attribute, the number of hops is utilized to gauge the distance between the information source and the cloudlet. When the dataset is formed, this attribute is set to 0, and it gets larger each time it goes via a hub and reaches the cloudlet.

5.4 Priority Level Modeling for Datasets

A multi-dimensional aggregation function is used to determine a dataset's priority level when it is received by the cloudlet. These dimensions include dataset category, dataset size, dataset frequency, dataset type, dataset recency, hubs to hop, and distance to cloudlet. Each attribute is assigned a weight value that influences its contribution to the overall priority calculation. We can see the interaction between each dimension in the following model as shown in Fig. 3 on the next page.

Moreover, POCkET GIMo is an adaptable information model that lets users change information values according to various scenarios. However, it is important for developers to tailor POCkET GIMo to specific disaster contexts. To accomplish this, developers should follow three key steps outlined in Fig. 4:

1. Examine the DMM idea categories.
2. Look at the relevant dataset attributes' default values.
3. Establish a priority function for these attributes in the dataset.

Developers can efficiently adapt POCkET GIMo to a particular scenario by following these steps.

6 Evaluation Based on Case Studies and Applications of POCkET and GIMo

In the context of disaster response, the Generic Information Model (GIMo) and the Priority Over Cloudlet Terminal (POCkET) Framework can greatly improve information distribution, resource allocation, and the ability to make decisions. Here, we evaluate GIMo by exploring case studies and applications that demonstrate the effectiveness of POCkET and GIMo in disaster scenarios:

Case Study 1: Real-time Data Analysis in Disaster Response:

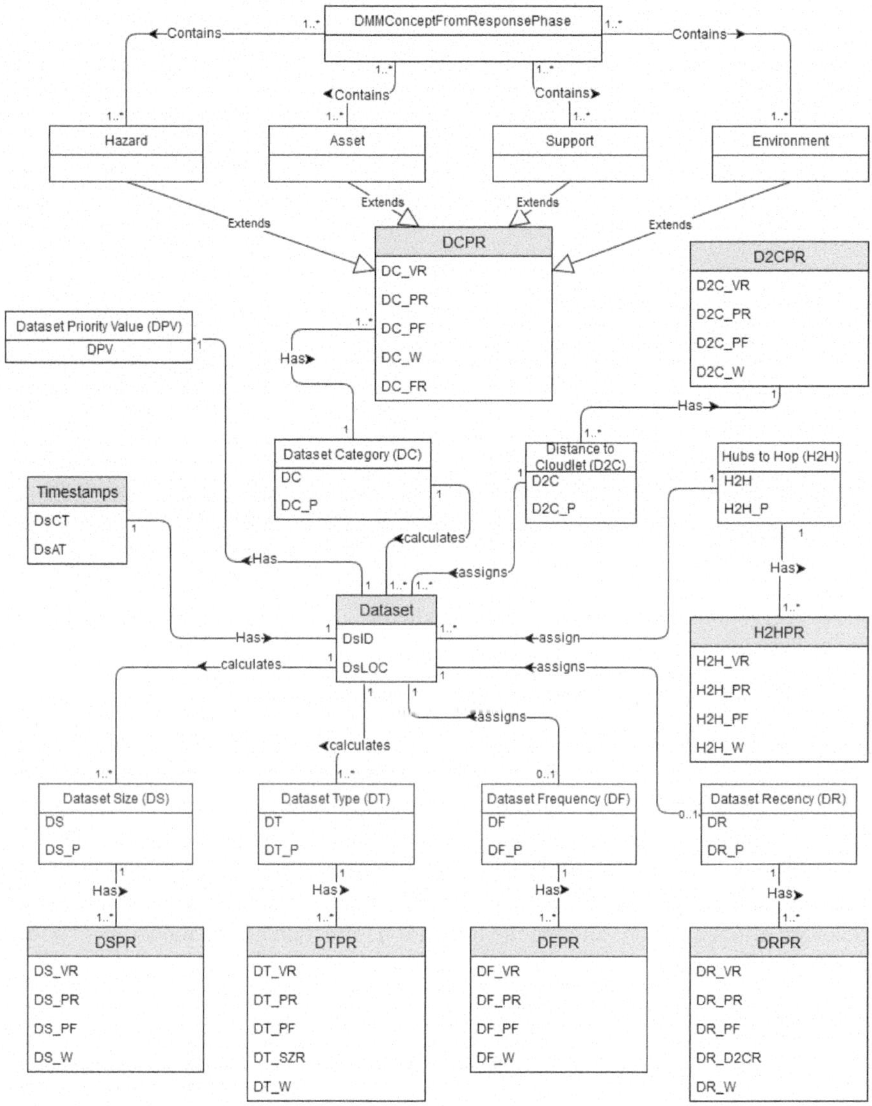

Fig. 3 POCkET GIMo (Geumpana and Rabhi 2020)

In a disaster situation, timely analysis of data plays a crucial role in decision-making and resource allocation. By leveraging the POCkET Framework and GIMo, real-time data collected from various sources such as sensors, social media, and satellite imagery can be efficiently processed and analyzed. Cloudlets situated in proximity to the affected area can offload the data analysis tasks from resource-constrained edge devices, reducing latency and enabling faster decision-making. GIMo facilitates

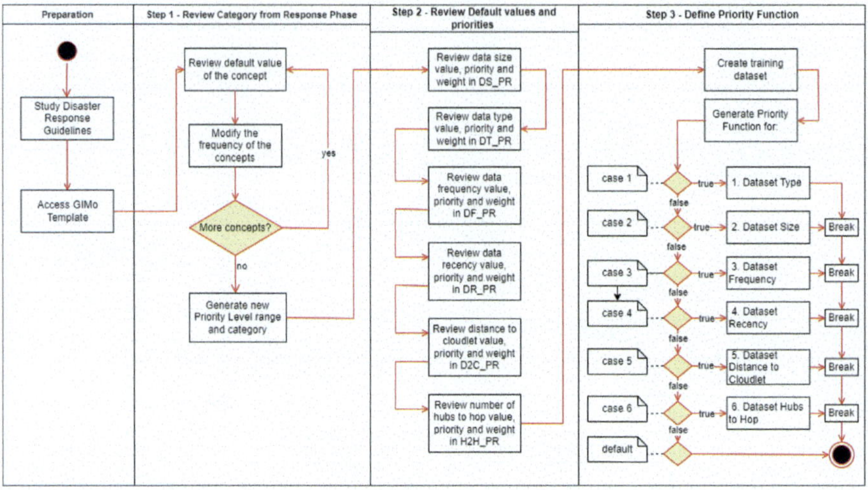

Fig. 4 GIMo customization

seamless communication between cloudlets and edge devices, ensuring standardized data exchange and coordination.

Case Study 2: Resource Allocation and Task Offloading:

During disaster response, there is often a strain on available resources. The POCkET Framework, integrated with GIMo, enables optimized resource allocation and task offloading. For example, in a scenario where multiple edge devices are collecting data and require computationally intensive analysis, cloudlets equipped with the POCkET Framework can dynamically allocate resources based on task requirements, available resources, and proximity. This ensures that the most critical tasks are prioritized and processed efficiently, maximizing the utilization of resources and reducing response time.

Case Study 3: Collaboration and Coordination:

Effective collaboration and coordination among responders are essential in disaster scenarios. The POCkET Framework, along with GIMo, facilitates seamless communication and coordination between different entities involved in the response efforts. Cloudlets can serve as central coordination points, collecting and disseminating information to relevant edge devices in a standardized format defined by GIMo. This enables responders to exchange critical information, share situational awareness, and coordinate their actions in real time, enhancing the overall effectiveness of the response.

Case Study 4: Mobile Healthcare Support:

Disaster situations often result in a surge in demand for healthcare services. The POCkET Framework, integrated with GIMo, can support mobile healthcare applications in such scenarios. By offloading computationally intensive healthcare tasks, such as real-time patient monitoring and diagnostics, to cloudlets, the framework reduces the burden on resource-constrained edge devices. GIMo ensures a standardized representation of patient data and healthcare requirements, facilitating interoperability and seamless communication between edge devices and cloudlets. This enables healthcare providers to deliver timely and accurate care, even in challenging and resource-constrained environments.

Case Study 5: Efficient Resource Management:

Optimal utilization of resources is critical in disaster response. The POCkET Framework, integrated with GIMo, allows for efficient resource management. By continuously monitoring resource availability, network conditions, and task requirements, cloudlets can dynamically allocate resources to edge devices based on their proximity and capabilities. GIMo provides a standardized representation of resource information, enabling efficient resource allocation and utilization. This ensures that resources, such as processing power, storage, and communication bandwidth, are effectively distributed, leading to improved overall system performance and response efficiency.

These case studies highlight the effectiveness of implementing the POCkET Framework and GIMo in disaster response scenarios. By leveraging cloudlets, optimizing resource allocation, enabling real-time data analysis, and facilitating seamless communication and coordination, POCkET and GIMo enhance the overall effectiveness, efficiency, and resilience of disaster response operations. The integration of POCkET and GIMo provides a standardized and interoperable framework that enables seamless collaboration among different entities involved in the response efforts.

7 Challenges and Future Directions

There are unique difficulties in implementing the Generic Information Model (GIMo) and the Priority Over Cloudlet Terminal (POCkET) Framework in emergency situations. Additionally, exploring the future directions of GIMo within the POCkET Framework can shed light on potential advancements and improvements.

The implementation of the POCkET Framework faces several challenges. One challenge is the heterogeneity of devices, networks, and communication technologies in disaster response scenarios. Ensuring seamless interoperability and compatibility between the framework's components is crucial. GIMo needs to address the heterogeneity of devices, data formats, and communication protocols to facilitate effective information exchange and coordination.

Another challenge lies in the dynamic nature of disaster environments. Conditions such as network topology, resource availability, and task requirements constantly change. GIMo must be capable of adapting to these dynamic environments, providing real-time updates on resource states and tasks. It requires mechanisms to handle frequent network updates and changes, supporting dynamic resource allocation and task scheduling accordingly.

Limited resources pose a significant challenge in disaster scenarios, with constraints on computing power, energy, and bandwidth. GIMo needs to consider these limitations and optimize resource allocation and utilization for efficient operation within the framework. Employing strategies like load balancing, task prioritization, and resource sharing can effectively address these resource constraints.

Additionally, security and privacy are paramount in disaster response operations. Sensitive information, such as personal data, medical records, and critical infrastructure details, is involved. To safeguard data during information exchange and ensure secure communication within the POCkET Framework, GIMo must incorporate robust security and privacy mechanisms. Encryption, authentication, access control, and data anonymization techniques are essential to address these concerns effectively.

Incorporating machine learning and artificial intelligence techniques into GIMo can enhance its capabilities in disaster response. By leveraging AI algorithms, GIMo can intelligently analyze data, predict resource requirements, and optimize resource allocation and task scheduling. This integration enables proactive decision-making and enhances the overall efficiency of the POCkET Framework.

The future of GIMo lies in leveraging edge intelligence capabilities. By enabling edge devices to perform data analysis and decision-making locally, GIMo reduces latency and dependency on cloudlets. GIMo can incorporate edge intelligence algorithms, allowing edge devices to make informed decisions based on localized data and resources, improving responsiveness and adaptability in disaster scenarios.

Enhancing the resilience and fault tolerance of the POCkET Framework is crucial in disaster response. GIMo can incorporate mechanisms to handle node failures, network disruptions, and resource inconsistencies. By implementing fault-tolerant strategies and redundancy mechanisms, GIMo ensures continuous operation and reliable communication within the POCkET Framework, even in challenging and unpredictable conditions.

Standardization and interoperability are increasingly important as the field of disaster response evolves. GIMo should align with existing standards and frameworks to ensure seamless integration with other systems and technologies. Collaborating with industry initiatives and organizations can help establish GIMo as a widely accepted standard for information exchange and coordination in disaster response.

Rigorous testing and validation of the POCkET Framework, along with GIMo, in realistic disaster scenarios are crucial for their practical implementation. Establishing dedicated testbeds and simulation environments provides insights into performance, scalability, and effectiveness. This validation process helps identify potential limitations, refine algorithms, and guide future improvements of GIMo in disaster contexts.

8 Conclusion

In conclusion, the integration of the Priority Over Cloudlet Terminal (POCkET) Framework and the Generic Information Model (GIMo) holds significant promise in the realm of disaster response. By leveraging cloudlets, optimizing resource allocation, facilitating real-time data analysis, and enabling seamless communication and coordination, POCkET and GIMo enhance the efficiency, effectiveness, and resilience of disaster response operations.

Throughout this book chapter, we have explored the fundamentals of cloud computing and edge computing, the POCkET Framework, and the Generic Information Model (GIMo) in the context of disaster response. We have discussed the importance of information dissemination, resource allocation, and decision-making in disaster scenarios, highlighting the need for efficient frameworks and models to address the challenges that arise in such contexts.

Furthermore, we have examined case studies and applications that demonstrate the practical implementation of POCkET and GIMo in disaster response. These case studies have showcased the ability of the framework to support real-time data analysis, optimize resource allocation, enable collaboration and coordination, facilitate mobile healthcare support, and enhance resource management.

However, we also acknowledge that implementing POCkET and GIMo in disaster contexts is not without its challenges. The heterogeneity of devices, dynamic environments, limited resources, and security and privacy concerns pose significant obstacles that need to be addressed. Future directions such as the integration of machine learning and AI, edge intelligence, resilience and fault tolerance, standardization and interoperability, and thorough testing and validation offer exciting opportunities for advancing the capabilities of POCkET and GIMo in disaster response.

In conclusion, the POCkET Framework and GIMo provide a standardized and interoperable framework that enables seamless collaboration among different entities involved in disaster response. By harnessing the power of cloudlets, optimizing resource allocation, and facilitating effective communication and coordination, POCkET and GIMo contribute to the overall efficiency, resilience, and success of disaster response efforts.

As we move forward, it is crucial to continue research and development in this field, fostering collaborations between academia, industry, and government agencies. By collectively working toward refining and advancing the POCkET Framework and GIMo, we can empower disaster response teams with powerful tools to mitigate the impact of disasters, save lives, and support affected communities more effectively.

References

Ai Y, Peng M, Zhang K (2018) Edge computing technologies for internet of things: a primer. Digit Commun Netw 4(2):77–86
Alizadeh MR et al (2020) Task scheduling approaches in fog computing: a systematic review. Int J Commun Syst 33(16):e4583
Attaran M (2021) The impact of 5G on the evolution of intelligent automation and industry digitization. J Ambient Intell HumIzed Comput 1–17
Babar M et al (2021) Cloudlet computing: recent advances, taxonomy, and challenges. IEEE Access 9:29609–29622
Bhardwaj S, Jain L, Jain S (2010) Cloud computing: a study of infrastructure as a service (IAAS). Int J Eng Inf Technol 2(1):60–63
Bilal K et al (2018) Potentials, trends, and prospects in edge technologies: fog, cloudlet, mobile edge, and micro data centers. Comput Netw 130:94–120
Braunstein B et al (2006) Challenges in using distributed wireless mesh networks in emergency response. In: Proceedings of the 3rd international ISCRAM conference
Buyya R, Ranjan R, Calheiros RN (2009) Modeling and simulation of scalable cloud computing environments and the CloudSim toolkit: challenges and opportunities. In: 2009 international conference on high performance computing & simulation. IEEE
Geumpana TA, Rabhi F (2020) Generic disaster information model for message delivery triage in disaster response. In: Mobile technologies for delivering healthcare in remote, rural or developing regions. Institution of Engineering and Technology, pp 191–206
Hameed A et al (2016) A survey and taxonomy on energy efficient resource allocation techniques for cloud computing systems. Computing 98:751–774
Haworth B, Bruce E (2015) A review of volunteered geographic information for disaster management. Geogr Compass 9(5):237–250
Lewis G et al (2014) Tactical cloudlets: moving cloud computing to the edge. In: 2014 IEEE military communications conference, IEEE
Mulimani MS, Rachh RR (2021) Edge computing in healthcare systems. In: Deep learning and edge computing solutions for high performance computing, pp 63–100
Othman SH, Beydoun G, Sugumaran V (2014) Development and validation of a disaster management metamodel (DMM). Inf Process Manage 50(2):235–271
Reynisson JÁ (2015) Performance of mobile GIS in conjunction withinternet bandwidth in rural areas
Roman R, Lopez J, Mambo M (2018) Mobile edge computing, fog et al.: a survey and analysis of security threats and challenges. Futur Gener Comput Syst 78:680–698
Shukla S et al (2021) Improving latency in internet-of-things and cloud computing for real-time data transmission: a systematic literature review (SLR). Clust Comput 1–24
Singh R, Gill SS (2023) Edge AI: a survey. Internet Things Cyber-Phys Syst
Yu W et al (2017) A survey on the edge computing for the internet of things. IEEE Access 6:6900–6919
Zhang Q, Cheng L, Boutaba R (2010) Cloud computing: state-of-the-art and research challenges. J Internet Serv Appl 1:7–18

Intelligent Eye-Tracking for the Early Diagnosis of Autism: A Mental Health Disaster with Families

Guangtao Zhai, Huiyu Duan, Xiaoyu Ren, and Pradeep Kumar Ray

1 Introduction

Autism spectrum disorders (ASD) represent a spectrum of neurodevelopmental conditions characterized by challenges with social interaction, communication, and tendencies toward repetitive and restricted behaviors (Robertson and Baron-Cohen 2017). The term "spectrum" in autism spectrum disorder emphasizes the vast diversity of challenges and strengths faced by each person with autism. The disease affects individuals of all ethnic, racial, and socioeconomic groups, although its presentation and severity vary from person to person.

The impact of autism spectrum disorder on an individual is profound and multifaceted. For individuals, it may require navigating a world that is not always adapted to their unique ways of processing and responding to social cues and sensory information. Challenges often include difficulties in understanding social norms, building relationships, and effectively communicating needs and emotions. In addition, people with autism spectrum disorder may have extraordinary abilities in specific areas such as memory, art, music, and mathematics, which further adds to the complexity of the condition. For families, having a member with autism can present a range of emotional, logistical, and financial challenges. Diagnosis of autism spectrum disorder often requires finding appropriate treatments, educational programs, and support services, which can be time-consuming and expensive. Emotional impacts

G. Zhai (✉) · H. Duan · X. Ren
Shanghai Jiao Tong University, Shanghai 200240, China
e-mail: zhaiguangtao@sjtu.edu.cn

P. K. Ray
University of Michigan—Shanghai Jiao Tong University Joint Institute, Shanghai, China

UNSW Business, Sydney, Australia

P. K. Ray
e-mail: p.ray@unsw.edu.au

P. K. Ray et al. (eds.), *Technology Innovation for Sustainable Development of Healthcare and Disaster Management*, Disaster Risk Reduction,
https://doi.org/10.1007/978-981-97-2049-1_13

include coping with the diagnosis, adjusting expectations for a child's development, and dealing with society's misunderstandings about autism.

Family members often become advocates for their loved ones, navigating a range of care options and educational settings while also educating others about the condition. The need for a strong support network is critical as families may experience stress, anxiety, and emotional strain. However, families also often report experiencing deep joy, pride, and a deepened sense of empathy and understanding while living with ASD.

The social impact of autism spectrum disorder is also significant, as communities and institutions must adapt to be more inclusive and supportive of people with autism. This includes creating education programs tailored to different learning needs, developing workplace accommodations, and raising public awareness to promote greater understanding and acceptance in society.

Baron-Cohen and Wheelwright (2004) and Simmons et al. (2009) have proven that early diagnosis of autism spectrum disorder (ASD) is a critical factor that can significantly impact a patient's developmental trajectory and quality of life. The earlier autism spectrum disorders are identified and addressed, the greater the chance that children will develop basic social, communication, and cognitive skills. Early intervention has been shown to significantly improve children's development, including their ability to communicate, express emotions, develop relationships, and learn. A major challenge in the early diagnosis of autism spectrum disorders is the variability of presentation and subtlety of early symptoms. Traditional diagnostic methods often rely on behavioral observations and parent interviews, which can be subjective and may not capture the nuances of early developmental stages. As a result, many children are not diagnosed until preschool age or older, potentially missing a critical window for early intervention (Association et al. 2013).

Intelligent eye-tracking technology represents a major advance in the early detection of ASD. The technology works on the idea that certain visual and attention patterns in people with autism spectrum disorder are different from those in people with neurotypical development (Simmons et al. 2009; Wang et al. 2014; Osterling and Dawson 1994). Eye-tracking involves measuring and recording an individual's eye movements and gaze patterns while receiving visual stimuli. These patterns can provide valuable insights into an individual's cognitive and social processing, areas that are frequently affected in autism spectrum disorders.

Recent research has shown that children with autism spectrum disorder show distinct differences in eye-tracking measures, such as reduced eye contact, atypical facial scanning, and differences in attention to social stimuli (Dawson et al. 2005; Sasson et al. 2011; Corden et al. 2008; Klin and Jones 2008; Jiang and Zhao 2017). These variants can be detected as early as infancy, long before traditional behavioral symptoms of autism spectrum disorder become apparent. The non-invasive nature of eye-tracking technology, combined with its objective and quantifiable data, makes it a particularly promising tool for early ASD screening and diagnosis. By leveraging intelligent eye-tracking, clinicians can identify ASD earlier than traditional methods. Early identification through this advanced technology can provide early intervention, tailored support, and ultimately better outcomes for children with autism spectrum disorder. Additionally, it provides a more objective and standardized approach to ASD diagnosis, reducing the variability and subjectivity associated with traditional diagnostic methods.

In summary, early diagnosis of autism spectrum disorder is critical to maximizing the developmental potential of individuals with autism. Against this background, intelligent eye-tracking technology has become a breakthrough tool that promises to provide earlier and more precise diagnosis and open the way for timely and effective intervention strategies. Intelligent eye-tracking technology brings significant benefits to the early diagnosis of autism and can positively change the coping mechanisms and support structures of families of children with autism.

2 Atypical Visual Attention of Individuals with Autism

Understanding the visual attentional responses to naturalistic stimuli in people with autism is critical to understanding the sensory processing differences inherent in the disorder (Itti 2015; Tseng et al. 2013). Visual attention in individuals with autism has been the subject of extensive research, revealing unique patterns that differ significantly from neurotypical individuals. This difference is particularly evident in their responses to natural, everyday stimuli, including factors ranging from social to non-social contexts.

Research shows that people with autism spectrum disorder often display atypical visual attention patterns. These patterns may include altered attention to specific details rather than the whole, a preference for non-social stimuli over social stimuli, and differences in tracking moving objects or navigating complex visual environments. For example, neurotypical people may naturally focus on faces in social situations, whereas people with autism spectrum disorder may pay more attention to inanimate objects or specific details of the environment. We will discuss these features in detail as follows.

2.1 Atypical Visual Attention on Natural Stimuli

People with autism may display unique attentional behaviors when exposed to natural stimuli such as landscapes, people, and everyday scenes. These can include.

(1) Excessive attention to detail: The tendency to focus intensely on specific details rather than the overall background or scene (Ren et al. 2022).
(2) Decreased attention to social cues: There may be less attention to social elements, such as facial expressions or body language, which are often the focus in neurotypical individuals (Duan et al. 2018; Duan et al. 2019a, b; Fang et al. 2020; Gutiérrez et al. 2021).

(3) Sensory sensitivity: Increased or decreased sensitivity to certain visual aspects of natural stimuli, such as brightness, color intensity, and motion, may affect how a person with autism interacts with the environment (Wang et al. 2014).

Central bias in autism spectrum disorder is characterized by the tendency of individuals to focus primarily on central aspects of a visual scene or stimulus, often ignoring peripheral details. This pattern contrasts with typical visual exploration strategies, in which attention is more evenly distributed across the visual field.

The characteristics of center bias include.

(1) Central focus of visual scenes: People with autism spectrum disorder may show a preference for viewing central elements of an image or scene and may miss contextual information presented peripherally (Wang et al. 2015; Duan et al. 2019a, b; Ren et al. 2022).
(2) Impact on social perception: In social situations, this bias may result in limited attention to central features of the face (such as the nose or mouth) while ignoring more expressive features (such as the eyes). This may affect the interpretation of social cues and emotional expressions (Duan et al. 2019a, b).
(3) Processing of non-social stimuli: Central bias also affects the way people with autism spectrum disorder process non-social stimuli. They may focus intensely on core aspects of the task or object, which may affect their overall understanding of and interaction with it (Ren et al. 2022).

The effects of central bias in autism spectrum disorders are multifactorial:

(1) Social Interactions: Reduced attention to peripheral facial features or social situations may impact the ability to fully participate in and interpret social interactions.
(2) Information Processing: There can be challenges in processing and integrating contextual information, which is often critical to understanding complex scenes.
(3) Educational and therapeutic approaches: Recognizing central bias can inform tailored educational strategies and therapeutic interventions, emphasizing the need to encourage broader visual exploration and processing.

Understanding these differences in visual attention is critical to developing effective support and intervention strategies for individuals with autism spectrum disorder. It highlights the need for a tailored approach to educational settings, therapeutic interventions, and the design of learning materials and environments. For example, for people with autism spectrum disorder, learning resources that minimize overwhelming sensory input and focus on clear, detailed visual information may be more effective. Moreover, understanding central biases in visual attention in individuals with autism spectrum disorder is critical to developing effective communication strategies, educational approaches, and therapeutic interventions. It highlights the need for a nuanced approach to supporting people with autism spectrum disorder, taking into account the unique ways in which they perceive and interact with the world around them.

2.2 *Atypical Visual Attention on Face Stimuli*

According to Duan et al. (2019a, b), one of the most interesting and telling aspects of the field of autism spectrum disorder is how individuals with the disorder display atypical visual attention, particularly when viewing human faces. The human face is a primary source of social and emotional information, and typical development involves a natural tendency to focus on faces from an early age. However, this tendency is often different in people with autism spectrum disorder, affecting their social interactions and emotional understanding.

Research using eye-tracking technology has revealed the unique patterns in how people with autism spectrum disorder view faces. The main differences include.

(1) Reduced eye contact: A hallmark feature of autism spectrum disorder is reduced eye contact when looking at faces. This is often one of the earliest indicators of autism spectrum disorder in young children.
(2) Focus on non-social features: People with autism spectrum disorder may not focus on the eyes or mouth, which convey important social information, but instead focus more on parts of the face that convey less information, such as the forehead or chin.
(3) Limited attention to expressive areas: Often little attention is paid to areas of the face where emotions are expressed, such as the eyes and mouth. This can create challenges in interpreting facial expressions and understanding social cues.

Atypical visual attention to faces in people with autism spectrum disorder has profound consequences:

(1) Social Communication Challenges: Difficulty reading facial expressions may contribute to social communication challenges for people with autism.
(2) Emotion Recognition and Empathy: The ability to recognize and respond to the emotions of others may be affected, which may impact the development of empathy and emotional intelligence.
(3) Intervention Strategies: These findings have implications for developing intervention strategies tailored to the unique ways in which people with autism spectrum disorders process visual information.

Eye-tracking studies are critical to understanding these atypical patterns. By precisely measuring where individuals look when looking at faces, researchers have gained insights into cognitive processes and social attention in people with autism. These studies can help develop targeted therapies and educational strategies to enhance social understanding and communication skills.

Atypical visual attention to faces in people with autism spectrum disorder emphasizes the need for a deeper understanding of how they perceive and interpret social information. Recognizing these patterns is critical to developing more effective communication strategies and interventions that improve social interactions and emotional understanding in people with autism.

2.3 Atypical Visual Attention on Joint-Attention Stimuli

According to Dawson et al. (2005) and Fang et al. (2020), atypical visual attention in responses to jointly attended stimuli in autistic individuals is an important aspect of the disorder affecting social communication and interaction. Joint attention refers to the joint attention of two people on an object or event, which is crucial for social learning and interaction.

Understanding atypical joint attention in ASD is important. Joint attention is a basic social communication skill in which two people use gestures and gaze to share an experience or focus on the same object or event. It is essential for language development, social interaction, and learning. However, people with autism spectrum disorder often show atypical responses to joint attention stimuli, which may affect their social and communication development.

Characteristics of atypical joint attention responses include.

(1) Decreased Initiation of Joint Attention (IJA): People with autism spectrum disorder are less likely to initiate joint attention. This includes behaviors such as showing, pointing, and carrying items to share with others.
(2) Responsive Joint Attention (RJA) Challenges: They may also have difficulty responding to others' requests for joint attention, such as following someone else's gaze or pointing gestures.
(3) Eye contact and facial expressions: Difficulty making and maintaining eye contact and interpreting facial expressions may affect the ability to engage in joint attention interactions.

Atypical joint attention behaviors in autism spectrum disorder (ASD) have important developmental implications, which include.

(1) Language and communication delays: Joint attention is closely related to language development. Challenges in this area may result in delays in speech and language acquisition.
(2) Difficulties with social interaction: Shared attention is the cornerstone of more complex social interactions. Difficulties in this area may create challenges in establishing and maintaining social relationships.
(3) Learning and cognition: The role of joint attention in learning from social contexts. Impaired joint attention affects the ability to learn from and interact with others.

Eye-tracking technology provides valuable insights into the joint attention behaviors of people with autism. By tracking gaze patterns, researchers can objectively study how people with autism spectrum disorder process joint attention cues in different contexts. These studies reveal differences in the way people with ASD attend to and process social information, providing potential avenues for intervention and support.

Understanding atypical visual attention in joint attention environments in individuals with autism spectrum disorder is critical to developing effective educational and therapeutic strategies. Developing interventions that target these specific challenges can significantly enhance social communication and interaction skills, ultimately improving the quality of life for individuals with ASD.

2.4 *Atypical Visual Attention on Video Stimuli*

According to Pierce et al. (2011), Tseng et al. (2013), Shi et al. (2015), and Bovery et al. (2019), atypical visual attention in people with autism while watching videos is an important area of research because it can provide insights into how they process and interpret dynamic visual information. Unlike static images, videos contain a wealth of moving visual details and auditory cues, making them complex stimuli for understanding the perceptual and cognitive aspects of ASD.

People with autism spectrum disorder often exhibit unique patterns of visual attention when watching videos that differ significantly from those of neurotypical individuals. These patterns could provide important insights into the sensory processing and cognitive challenges faced by people with autism. Characteristics of atypical visual attention on videos include.

(1) Different focus: People with autism spectrum disorder may focus on different aspects of the video than neurotypical viewers. For example, they may focus more on the background or inanimate objects than on the people or main action in the video.
(2) Difficulty with rapid scene changes: Rapid scene changes common in videos can be challenging for people with autism spectrum disorder. They may not be able to handle quick transitions, or may focus on a specific scene and miss the broader narrative.
(3) Overwhelmed by multisensory input: Videos often combine visual and auditory stimulation. This multisensory input can be overwhelming for some people with autism spectrum disorder, causing difficulty concentrating or understanding content.

Impact on learning and social interaction include.

(1) Educational content: Atypical visual attention patterns affect how people with autism learn from videos, a common educational tool.
(2) Social cues and interactions: Because videos often depict social interactions, atypical attention can impact the understanding and interpretation of social cues and behaviors.
(3) Media consumption: These patterns also influence how people with ASD consume and interact with media, thereby influencing their leisure activities and preferences.

Eye-tracking technology is particularly useful in studying how people with autism view videos. These studies track where and when viewers look at different parts of the screen, providing objective data on their visual attention patterns. The results of these studies help us understand the unique ways in which people with autism process dynamic visual information. The impacts and interventions include.

(1) Tailoring educational materials: Understanding these attention patterns can guide the development of video-based educational materials that better meet the learning needs of people with autism.
(2) Supporting media literacy: Insights from these studies can also help design interventions to support media literacy and social understanding in people with autism spectrum disorder.

Atypical visual attention when viewing videos in individuals with autism spectrum disorder highlights the need for a deeper understanding of their perceptual and cognitive processes. This understanding is critical to creating more accessible and effective educational tools, therapeutic interventions, and media content that address the unique needs of individuals with autism spectrum disorder.

2.5 *Atypical Visual Attention in Open World*

Combining Virtual Reality (VR) with eye-tracking technology to study the visual pattern of autism in the real omnidirectional world is the latest research direction. To support the research, Ren et al. (2022) established a dataset of eye movements for autistic children on panoramic stimuli, which is the first large-scale omnidirectional ASD eye-movement dataset. The dataset includes 300 omnidirectional images, covering a wide range of pixel-level, object-level, and semantic-level information. They recruited fifteen high-functioning children with autism who met DSM-V diagnostic criteria for autism, and sixteen typically developing (TD) controls. They used HTC VIVE Pro Eye, which displayed omnidirectional stimuli and collected eye-tracking data. The software system was built through Unity3D, which was used to control the experimental process and record all data.

As for the methods, Xu et al. (2014) proposed a three-layer saliency calculation model, considering three levels of features: pixel-level features (color, brightness, orientation), object-level features (size, complexity, convexity, solidity, eccentricity), semantic-level features, and background features.

Duan et al. (2022a, b, c) improved the image features considered by the three-level saliency computation model for the proposed omnidirectional ASD eye-movement dataset: Firstly, as mentioned earlier, individuals with ASD exhibit atypical attention to human-relevant stimuli. Therefore, for semantic-level features, only “human” and “others” were considered. Secondly, Wang et al. (2015) demonstrated that human visual attention is more attracted to the central region of planar images, while individuals with ASD show a stronger central bias in their gaze patterns. Rai et al. (2017) confirmed that for panoramic images, people gaze more frequently at the equatorial region of the image, without showing a preference for specific longitudes. Hence, equator features of panoramic images were considered. Lastly, research (Wolfe et al. 2004; Desgingh et al. 2013) has shown that depth cues play an important role in human understanding of the surrounding environment in three-dimensional scenes and contribute to visual saliency. Therefore, the depth features of panoramic images were considered. In summary, they modified the semantic-level features to consider only humans and others, and introduced equator features and depth features of panoramic images.

Duan et al. (2022a, b, c) also improved the method for extracting object features in the three-level saliency computation model: To extract object-level, semantic-level, and background features, it is necessary to compute the precise contours of objects. Xu et al. (2014) used an interactive manual object segmentation tool for manual annotation and feature extraction. However, manual extraction of object features requires a significant amount of time and resources and may be influenced by individual preferences. Therefore, Duan et al. (2022a, b, c) chose the Cascade Mask R-CNN object detection algorithm, and used the pre-trained model Cascade Mask R-CNN X-101–32 × 4d-FPN from the open-source codebase MMlab's mmdetection module (Chen et al. 2019) to compute masks, followed by Gaussian blurring. To extract depth features, they experimented with five depth estimation algorithms specifically designed for panoramic images: SliceNet (Pintore et al. 2021), BiFuse (Wang et al. 2020), UniFuse (Jiang et al. 2021), Ho-HoNet (Sun et al. 2021), and ACDNet (Zhuang et al. 2022). And they selected ACDNet which is the best-performing algorithm. However, this method still caused distortion at the north and south poles of panoramic images. In actual data collection experiments, participants rarely observed the positions of the north and south poles in panoramic images. Therefore, they used the grayscale value of the deepest color in each depth map to cover the distortions at the poles, resulting in the final panoramic depth feature map.

The model used Support Vector Machine (SVM) to evaluate the impact of six factors on visual attention allocation: equator features of panoramic images, depth features of panoramic images, background features, pixel-level features, object-level features, and semantic-level features. To train and test this model, the dataset was divided into 240 training images and 59 testing images. It is important to note that during the actual feature extraction process, cubic projection was employed to reflect the real scenes observed by the participants in the VR environment and reduce distortion. The overall structure is shown in Fig. 1.

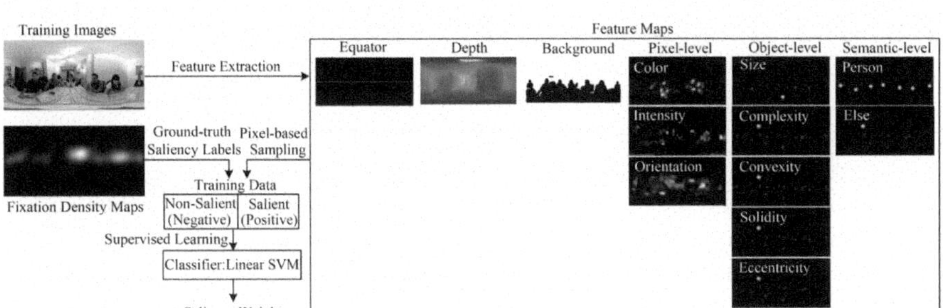

Fig. 1 Overview of the computational saliency model

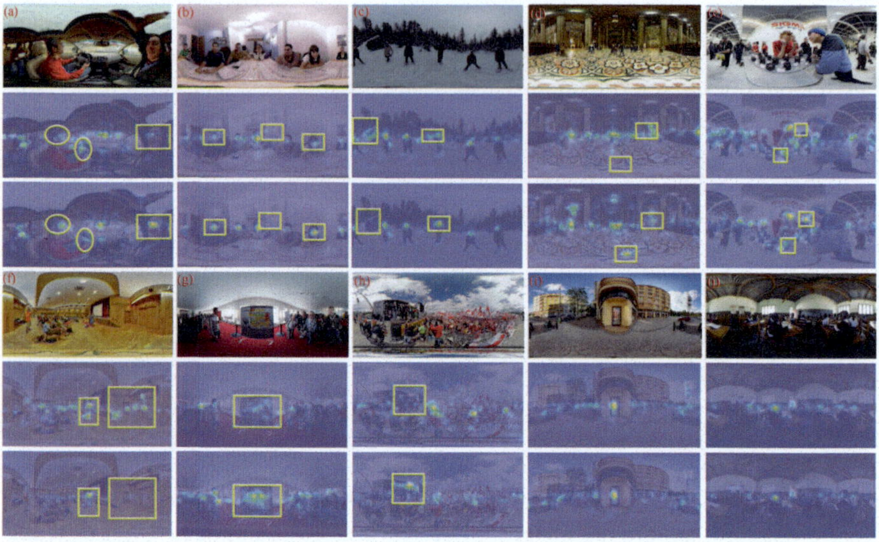

Fig. 2 Visual attention of autistic children (top) and healthy controls (bottom)

The study of atypical visual attention in autistic individuals in the real omnidirectional world further justifies previous findings on natural stimuli, face stimuli, joint-attention stimuli, and video stimuli. It also suggests some new ideas. Characteristics of atypical visual attention in the open world include.

(1) Decreased attention to social cues: Compared to typically developing children, children with autism exhibit lower concentration of visual attention on faces and other social information. Additionally, they are also attracted to other contents in the same stimuli, which is shown in Fig. 2a–d.
(2) Decreased exploratory behavior toward the entire scene: Compared to typically developing children, children with autism often restrict their attention to specific parts or elements, especially in scenes where faces are not prominent and do not express an inclination to interact with the observer, which is shown in Fig. 2d and g.
(3) Decreased initiation of joint attention: Compared to typically developing children, children with autism have lower attention toward the main focus of the scene, disclaiming they exhibit a reduced ability to integrate information, which is shown in Fig. 2g and h.
(4) Different center bias (equator-bias in panoramic images): Compared to typically developing children, children with autism also exhibit the highest visual attention distribution near the equator of panoramic images, gradually decreasing toward the poles. Additionally, both groups showed a tendency to look downward. Nevertheless, they have lower center bias in outdoor scenes with lower levels of high-level semantic information, which is also aligned with previous studies (Fang et al. 2020). Furthermore, the preference for looking downward

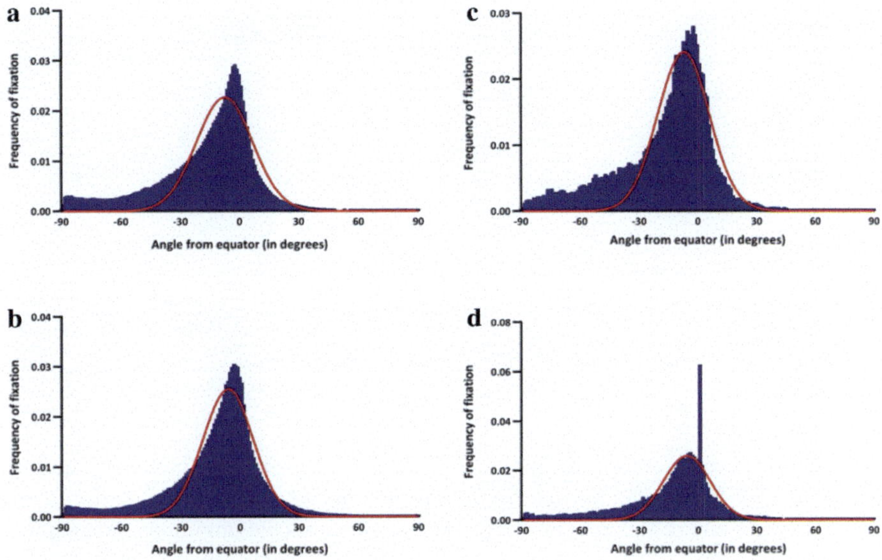

Fig. 3 Frequency of fixations for each latitude and Gaussian curve approximation (**a**: ASD; **b**: TD; **c**: ASD outdoor; **d**: TD outdoor)

was more pronounced in children with autism. All the results can be seen in Fig. 3; Gaussian fitting analysis also revealed the point.

(5) Lower coordination of head and eye movements: Compared to typically developing children, children with autism have more erratic eye movements, indicating lower coordination between head movements and eye movements, as shown in Figs. 4 and 5.

3 Intelligent Eye-Tracking for the Early Diagnosis of ASD

3.1 Eye-Tracking

Eye-tracking technology is a method for measuring where and how a person's eyes move and focus. The technology is widely used in various fields including psychology, neuroscience, marketing, and user experience design. The following is an overview of its key aspects.

The basic principles of eye-tracking include the following aspects. The first is eye-movement recording. The eye-tracking device captures the person's gaze position (fixation point) and tracks the eye movement over time. This is usually done using a camera and infrared light. Next is pupil dilation, and some advanced systems also measure changes in pupil size, which can indicate cognitive and emotional responses. Types of eye-tracking technology include the following, starting with remote eye

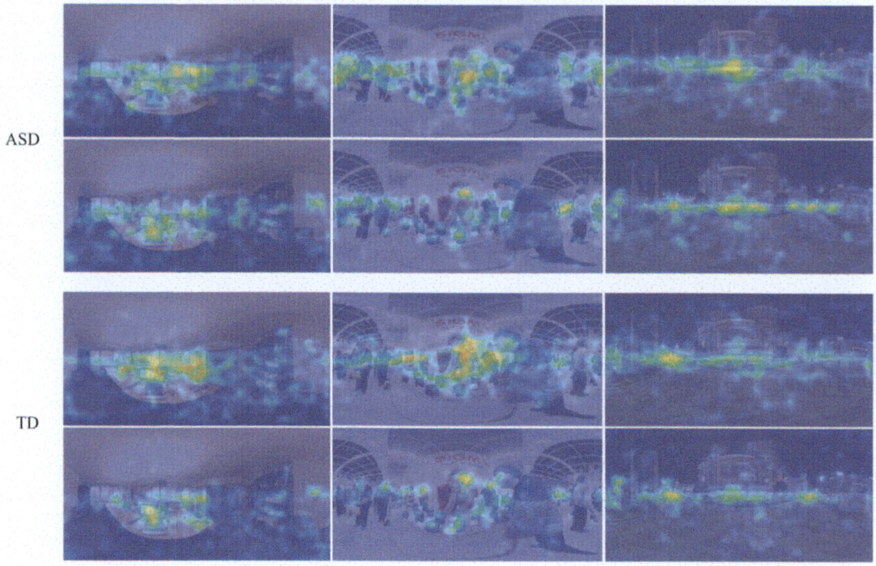

Fig. 4 Heat map of head movement (top) and heat map of eye movement (bottom)

Fig. 5 CC between the fixation density map of head movement and the fixation density map of eye movement

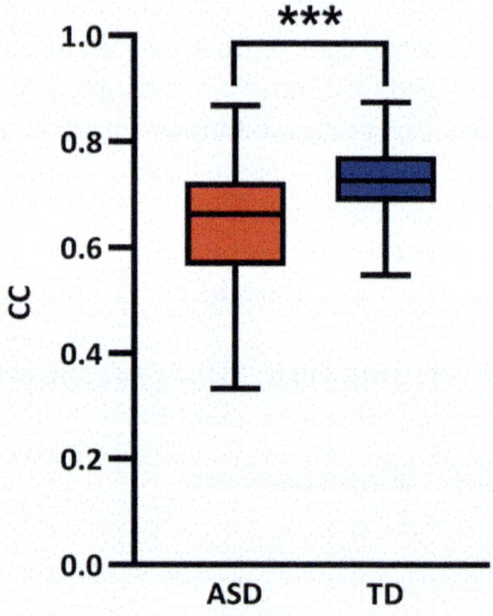

trackers, which are devices that are typically mounted on a computer or screen and use cameras aimed at the eyes to track eye movements. Next are head-mounted eye trackers, which are worn like glasses, are more mobile, and can be used for research in natural environments. Finally, there are eye-tracking glasses, which are specialized glasses with integrated cameras and sensors for maximum mobility.

Eye-tracking measures data such as, first, fixation point, which is the point at which gaze remains stable, indicating that attention is focused on a specific item;

saccades, rapid movements of the eyes between fixations; scan path, a sequence of fixations and glances, shows how someone visually explores an area; pupil dilation, a change in pupil size, may be related to cognitive load or emotional arousal.

Eye-tracking application areas include the following aspects: Used for usability and user experience, to understand how users interact with a website, software, or product; used in psychological research to help study attention, perception, and cognition; used in marketing and advertising to analyze consumer attention to advertisements and products; used in clinical and neuroscience research to help understand conditions such as ASD, Parkinson's disease, and more; for use in education and training to understand reading patterns and learning behaviors.

The advantages of eye-tracking are first of all objective data, providing quantitative and unbiased data on visual attention; to gain insight into unconscious processes and reveal automatic and subconscious visual patterns; versatility for a wide range of research and commercial environments.

Challenges of eye-tracking include accuracy and calibration, which require precise calibration and are sensitive to motion or lighting conditions. Data interpretation, the meaning behind eye movements can be complex and require careful analysis. Ethical and Privacy Issues, Collecting eye-tracking data, especially in public or semi-public, environments, raises privacy concerns.

In summary, eye-tracking technology provides a unique window into where and how we direct visual attention, providing valuable insights in a variety of fields. However, its application requires careful consideration of methodological, interpretive, and ethical aspects.

3.2 Early Diagnosis of ASD

Early diagnosis of ASD is an area of critical focus in developmental medicine and psychology because of its significant implications for intervention and management. Early diagnosis is important, and early diagnosis of autism spectrum disorder can lead to early intervention, which is critical to improving developmental outcomes. Early intervention can significantly enhance communication skills, social skills, and learning abilities. Young children have greater brain plasticity, which means their brains are more capable of adapting and responding to interventions, which can alleviate some of the challenges associated with autism spectrum disorder. Early diagnosis can also help families understand their child's needs, access resources, and plan for the future.

The challenges of early diagnosis are also many. Symptoms of autism spectrum disorder vary from child to child, making early detection challenging. Some children may show obvious symptoms of autism spectrum disorder early on, while others may develop symptoms gradually. Some symptoms of autism spectrum disorder overlap with other developmental disorders, which can lead to misdiagnosis or delayed diagnosis. In some areas, a lack of awareness and diagnostic resources for autism spectrum disorders results in delays in diagnosis.

There are many diagnostic methods for early detection. Routine developmental screening during pediatric exams can help identify children who may have autism spectrum disorder. These screenings usually involve questionnaires and observations. Tools such as the Autism Diagnostic Observation Schedule (ADOS) and the Autism Diagnostic Interview-Revised (ADI-R) are used for a more comprehensive assessment. The role of parents and caregives is to watch for early signs, and they are often the first to notice the early signs of autism spectrum disorder, such as a lack of eye contact, delayed speech, and repetitive behaviors. When parents notice developmental delays or unusual behaviors, it is critical to take prompt action to seek evaluation and diagnosis. Additionally, emerging technologies such as eye-tracking are being explored for their potential in early ASD detection as they can reveal atypical patterns of visual attention and social engagement.

Early diagnosis of ASD plays a key role in enabling effective intervention, supporting families, and improving the overall quality of life of individuals with ASD. Despite the challenges, continued efforts in research, awareness, and resource allocation are critical to improving early diagnostic capabilities.

3.3 Application of Eye-Tracking Technology in Autism Diagnosis

The mechanism of using eye-tracking in assisting autism diagnosis is easy to understand. Eye-tracking technology captures and analyzes eye movement and focus. This is typically done using cameras or sensors that track corneal and pupillary reflections, providing data on gaze direction, gaze duration, and blink patterns.

It is important to then analyze the collected data. Data collected can reveal patterns of visual attention and engagement. In the context of autism spectrum disorder, this may include how individuals attend to social stimuli (such as faces) versus non-social stimuli (such as objects).

Then it can be used for early detection. Multiple studies have shown that eye-tracking can be used to identify atypical visual patterns in young children (sometimes as early as six months old) with ASD. These patterns include reduced eye contact, limited attention to social situations, and a preference for geometric patterns.

The index includes comparative accuracy and specificity and sensitivity. Studies comparing eye-tracking to traditional diagnostic methods have found it to be a useful complementary tool. Eye-tracking provides quantitative data that can support observations in behavioral assessments. Several studies have evaluated the sensitivity (correctly identifying people with ASD) and specificity (correctly identifying people without ASD) of eye-tracking and found it to be promising for differentiating ASD from other developmental disorder tools.

Advantages over traditional methods exist in the following aspects:

(1) Objective measurement: Eye-tracking provides objective, quantifiable data. Traditional methods often rely on behavioral observations and parent reports, which can be subjective and subject to observer bias.
(2) Non-invasive and child-friendly: The technology is non-invasive and more engaging for children, often requiring them to simply view an image or video.
(3) Potential for early diagnosis: Eye-tracking has the potential to identify risk factors for autism spectrum disorder at a much younger age than when many behavioral symptoms become apparent.

4 Discussion

Eye movements of autistic children significantly differentiate from those of typically developing children, and can be captured and analyzed with evolving eye-tracking technology. As a result, eye-tracking can be used as an effective tool to aid in the early diagnosis of autism. Several studies have demonstrated its potential in recognizing specific gaze patterns and abnormalities in individuals with autism. Below are some of the key findings of eye-tracking technology in early autism diagnosis:

(1) Atypical gaze patterns: Compared to typically developing children, children with autism have atypical gaze patterns. They may have a preference for looking at non-social objects or repetitive patterns rather than faces or socially relevant information. They also may have decreased initiation of joint attention, different visual bias toward image features, different center bias, lower coordination of head and eye movements, etc. This deviation from typical gaze patterns can serve as an indicator for early autism diagnosis.
(2) Differences in gaze fixations: Compared to typically developing children, children with autism often have decreased exploratory behavior toward the entire scene.
(3) Impaired gaze following: Compared to typically developing children, children with autism have difficulties in gaze following tasks, which can contribute to challenges in social interaction and communication.
(4) Predictive value: Eye-tracking measures have demonstrated good predictive value for autism diagnosis. Research has shown that eye-tracking data combined with machine learning algorithms can accurately distinguish between individuals with autism and typically developing individuals, even in early infancy.
(5) Early detection potential: Eye-tracking offers opportunities for early detection of ASD. Studies have shown that eye-tracking measures can differentiate infants at high risk for ASD from low-risk infants as early as 6 to 12 months of age, providing an opportunity for early intervention and support.

Overall, eye-tracking has emerged as a valuable tool in the early diagnosis of autism. By capturing and analyzing gaze patterns and deviations, it offers insights into the atypical visual attention characteristics associated with autism spectrum disorder.

Early diagnosis of ASD through eye-tracking has multiple benefits for families. Below are some of them:

(1) Early intervention opportunities: Early diagnosis facilitates early intervention, which is critical to improving life quality for individuals with ASD. With an eye-tracking-based diagnosis, families can detect ASD at an earlier stage, allowing them to access specialized interventions and treatments tailored to their child's needs as early as possible. Early intervention programs, such as behavioral and developmental therapy, speech and language therapy, and occupational therapy, can help them develop essential skills and minimize the impact of the challenges they would face.
(2) Better understanding of ASD: Eye-tracking technology can provide data about the visual attention patterns and differences of individuals with ASD. This information can help families better understand their child's unique visual processing and attention characteristics. By gaining insight into their child's atypical gaze patterns and preferences, families can adapt their interactions, environments, and communication strategies to effectively support their child's specific needs.
(3) Improved access to resources and support: Early diagnosis through eye-tracking can facilitate access to a range of resources and support systems. Families can connect with autism-specific organizations, support groups, and online communities, where they can find information, guidance, and emotional support from others who have similar experiences. They can also access educational resources, workshops, and training programs to enhance their understanding of ASD and learn effective strategies to support their child's development.
(4) Reduced parental stress: Early diagnosis helps reduce parental stress by providing clarity and validation. It allows families to understand the root causes of their child's behaviors and challenges, reducing uncertainty and anxiety. Through early intervention and access to resources and support, families can feel more empowered to cope with the complexities of raising a child with ASD, improving the overall well-being of the child and family.

Overall, early diagnosis of ASD through eye-tracking can provide many benefits to families. It opens the door to early intervention opportunities, increases understanding of ASD, provides access to resources and support, and reduces parental stress. Through early detection of ASD, families can take proactive steps to support their child's development and ensure they receive the necessary interventions and services.

The contents above are the reasons why we composed this chapter. Eye-tracking can be used as an effective tool to aid in the early diagnosis of autism, early invention of autism, and benefits autistic families a lot. In this chapter, we have discussed the atypical visual attention of autistic children and related eye-tracking technology.

Specifically, we mainly focus on five different conditions where eye-movement data are collected, including on natural stimuli, on face stimuli, on joint-attention stimuli, on video stimuli, and on the open world. We have compared the visual attention pattern in these five conditions, and we have observed the existence of significant differences in all these five types, covering object bias, center bias, hand bias, the reduction of joint attention, etc.

While eye-tracking holds great promise for early autism diagnosis, there are still challenges and limitations to their use, as well as potential drawbacks and concerns for families. Below are some important considerations:

(1) Complexity of interpretation: Analyzing eye-tracking data requires expertise and careful interpretation. Distinguishing between typical and atypical gaze patterns can be challenging, especially in young children, who may exhibit variability in their attention and visual preferences. Interpretation of these results should be performed by trained professionals to avoid misdiagnosis or misinterpretation.
(2) Limited diagnostic tools: Eye-tracking alone is not sufficient to definitively diagnose autism spectrum disorders. And the appropriate tool to choose is also important. For example, only relying on the eye movement on a single face in an image is insufficient to differentiate between individuals with autism and those in a healthy state, complex stimuli is crucial. This is why we have further conducted studies in a virtual reality environment on omnidirectional stimuli, which covers more complex social scenes. The use of virtual reality technology can shield the experimental environment from distractions, immerse the subjects in the content of the experiment, and provide higher accuracy of eye-movement data collected. However, resistance to wearing VR helmets can be observed in most autistic children, which makes the full participation rate of the experiments low. Overall, eye-tracking is still an aiding tool by far, which can provide valuable information about visual attention patterns, but a thorough evaluation by a multidisciplinary team, including clinical assessment and behavioral observations, is still needed to make an accurate diagnosis.
(3) Equipment and environmental limitations: Eye-tracking requires specialized equipment and a controlled environment to obtain accurate data. The use of eye-trackers could be challenging for young children or people who have difficulty sitting still or wearing the necessary equipment. Setting a standard test environment can be time-consuming and may not always be feasible in the real world.
(4) Ethical considerations: The use of eye-tracking technology raises ethical considerations, particularly regarding privacy and consent. Families may have concerns about the collection and storage of their child's eye-tracking data and its potential use beyond diagnosis. It is crucial for researchers and practitioners to address these concerns and ensure appropriate safeguards are in place to protect privacy and confidentiality.

In summary, eye-tracking is an effective tool for ASD early diagnosis, but the use of it must be very careful. Many studies are working on it, but there is still a long way to go. For further studies, VR videos could be introduced into this region, which also needs to be further explored (Duan et al. 2024; Duan et al. 2022a, b, c; Duan et al. 2022a, b, c; Duan et al. 2023; Zhu et al. 2021; 2023; Duan et al. 2017a, b; Duan et al. 2017a, b; Yang et al. 2019). Considering the previously mentioned resistance of autistic children to VR helmets, the CAVE environment (Cruz-Neira et al. 1992) could be introduced in subsequent studies. Besides, video-based eye-tracking methods have achieved good performance and may be inexpensive alternatives to professional eye-trackers (Tu et al. 2022a, b; Tu et al. 2022a, b). The influence of other modalities such as text-context (Wang et al. 2023), and audio-context (Sun et al. 2023) on visual attention for individuals with autism also needs further exploration.

Overall, using intelligent eye-tracking techniques for diagnosis of autism is a major trend, and can assist early diagnosis and intervention. However, more exploration studies should be conducted. We hope this chapter can help other researchers conduct experiments and facilitate future studies related to this topic.

References

Association AP et al (2013) Diagnostic and statistical manual of mental disorders (DSM-5®). American Psychiatric Pub

Baron-Cohen S, Wheelwright S (2004) The empathy quotient: an investigation of adults with Asperger syndrome or high functioning autism, and normal sex differences. J Autism Dev Disord 34:163–175

Bovery M, Dawson G, Hashemi J, Sapiro G (2019) A scalable off-the-shelf framework for measuring patterns of attention in young children and its application in autism spectrum disorder. IEEE Trans Affect Comput 12(3):722–731

Chen K, Wang J, Pang J, Cao Y, Xiong Y, Li X, Lin D (2019) MMDetection: open mmlab detection toolbox and benchmark. arXiv:1906.07155

Corden B, Chilvers R, Skuse D (2008) Avoidance of emotionally arousing stimuli predict social–perceptual impairment in asperger's syndrome. Neuropsychologia 46(1):137–147

Cruz-Neira C, Sandin DJ, DeFanti TA, Kenyon RV, Hart JC (1992) The CAVE: audio visual experience automatic virtual environment. Commun ACM 35(6):64–73

Dawson G, Webb SJ, McPartland J (2005) Understanding the nature of face processing impairment in autism: insights from behavioral and electrophysiological studies. Dev Neuropsychol 27(3):403–424

Desingh K, Krishna KM, Rajan D, Jawahar CV (2013) Depth really matters: improving visual salient region detection with depth. In: BMVC. pp 1–11

Duan H, Ren X, Wang L, Shi F, Fan L, Zhai G (2022a) Processing and analysis of eye movement signals for autistic children in omnidirectional space. J Signal Process 38(9):1797–1808

Duan H, Zhai G, Yang X, Li D, Zhu W (2017) IVQAD 2017: an immersive video quality assessment database. In: Proceedings of the IEEE international conference on systems, signals and image processing (IWSSIP). pp 1–5

Duan H, Zhai G, Min X, Zhu Y, Sun W, Yang X (2017) Assessment of visually induced motion sickness in immersive videos. In: Proceedings of the Pacific rim conference on multimedia. pp 662–672

Duan H, Zhai G, Min X, Fang Y, Che Z, Yang X, Zhi C, Yang H, Liu N (2018) Learning to predict where the children with asd look. In: Proceedings of the IEEE international conference on image processing (ICIP). pp 704–708

Duan H, Min X, Fang Y, Fan L, Yang X, Zhai G (2019) Visual attention analysis and prediction on human faces for children with autism spectrum disorder. ACM Trans Multimed Comput, Commun, Appl (TOMM) 15(3s):1–23

Duan H, Zhai G, Min X, Che Z, Fang Y, Yang X, Guti´errez J, Le Callet P (2019) A dataset of eye movements for the children with autism spectrum disorder. In: Proceedings of the ACM multimedia systems conference (MMSys). pp 255–260

Duan H, Min X, Zhu Y, Zhai G, Yang X, Le Callet P (2022) Confusing image quality assessment: toward better augmented reality experience. IEEE Trans Image Process (TIP) 31:7206–7221

Duan H, Shen W, Min X, Tu D, Li J, Zhai G (2022) Saliency in augmented reality. In: Proceedings of the ACM international conference on multimedia. pp 6549–6558

Duan H, Min X, Sun W, Zhu Y, Zhang XP, Zhai G (2023) Attentive deep image quality assessment for omnidirectional stitching. IEEE J Sel Top Signal Process

Duan H, Zhu X, Zhu Y, Min X, Zhai G (2024) A quick review of human perception in immersive media. IEEE Open J Immersive Disp (2024)

Fang Y, Duan H, Shi F, Min X, Zhai G (2020) Identifying children with autism spectrum disorder based on gaze-following. In: Proceedings of the IEEE international conference on image processing (ICIP). IEEE, pp 423–427

Gutiérrez J, Che Z, Zhai G, Le Callet P (2021) Saliency4ASD: challenge, dataset and tools for visual attention modeling for autism spectrum disorder. Signal Process: Image Commun 92:116092

Itti L (2015) New eye-tracking techniques may revolutionize mental health screening. Neuron 88(3):442–444

Jiang H, Sheng Z, Zhu S, Dong Z, Huang R (2021) Unifuse: unidirectional fusion for 360 panorama depth estimation. IEEE Robot Autom Lett 6(2):1519–1526

Jiang M, Zhao Q (2017) Learning visual attention to identify people with autism spectrum disorder. In: Proceedings of the IEEE international conference on computer vision (ICCV), pp 3267–3276

Klin A, Jones W (2008) Altered face scanning and impaired recognition of biological motion in a 15-month-old infant with autism. Dev Sci 11(1):40–46

Osterling J, Dawson G (1994) Early recognition of children with autism: a study of first birthday home videotapes. J Autism Dev Disord 24(3):247–257

Pierce K, Conant D, Hazin R, Stoner R, Desmond J (2011) Preference for geometric patterns early in life as a risk factor for autism. Arch Gen Psychiatry 68(1):101–109

Pintore G, Agus M, Almansa E, Schneider J, Gobbetti E (2021) Slicenet: deep dense depth estimation from a single indoor panorama using a slice-based representation. In: Proceedings of the IEEE/CVF conference on computer vision and pattern recognition. pp 11536–11545

Rai Y, Gutiérrez J, Le Callet P (2017) A dataset of head and eye movements for 360 degree images. In: Proceedings of the ACM on multimedia systems conference. pp 205–210

Ren X, Duan H, Min X, Zhu Y, Shen W, Wang L, Shi F, Fan L, Yang X, Zhai G (2022) Where are the children with Autism looking in reality? In: Proceedings of the CAAI international conference (CICAI). pp 588–600

Robertson CE, Baron-Cohen S (2017) Sensory perception in autism. Nat Rev Neurosci 18(11):671

Sasson NJ, Elison JT, Turner-Brown LM, Dichter GS, Bodfish JW (2011) Brief report: circumscribed attention in young children with autism. J Autism Dev Disord 41(2):242–247

Shi L, Zhou Y, Ou J, Gong J, Wang S, Cui X, Luo X (2015) Different visual preference patterns in response to simple and complex dynamic social stimuli in preschool-aged children with autism spectrum disorders. PloS one 10(3)

Simmons DR, Robertson AE, McKay LS, Toal E, McAleer P, Pollick FE (2009) Vision in autism spectrum disorders. Vision Res 49(22):2705–2739

Sun C, Sun M, Chen H (2021) Hohonet: 360 indoor holistic understanding with latent horizontal features. In: Proceedings of the IEEE/CVF conference on computer vision and pattern recognition. pp 2573–2582 (2021)

Sun Y, Min X, Duan H, Zhai G (2023) The influence of text-guidance on visual attention. In: Proceedings of the IEEE international symposium on circuits and systems (ISCAS). pp 1–5

Tseng PH, Cameron IG, Pari G, Reynolds JN, Munoz DP, Itti L (2013) High-throughput classification of clinical populations from natural viewing eye movements. J Neurol 260(1):275–284

Tu D, Min X, Duan H, Guo G, Zhai G, Shen W (2022) End-to-end human-gaze-target detection with transformers. In: Proceedings of the IEEE conference on computer vision and pattern recognition (CVPR). pp 2192–2200

Tu D, Min X, Duan H, Guo G, Zhai G, Shen W (2022) Iwin: human-object interaction detection via transformer with irregular windows. In: Proceedings of the European conference on computer vision (ECCV). pp 87–103

Wang S, Xu J, Jiang M, Zhao Q, Hurlemann R, Adolphs R (2014) Autism spectrum disorder, but not amygdala lesions, impairs social attention in visual search. Neuropsychologia 63:259–274

Wang S, Jiang M, Duchesne XM, Laugeson EA, Kennedy DP, Adolphs R, Zhao Q (2015) Atypical visual saliency in autism spectrum disorder quantified through model-based eye tracking. Neuron 88(3):604–616

Wang FE, Yeh YH, Sun M, Chiu WC, Tsai YH (2020) Bifuse: monocular 360 depth estimation via bi-projection fusion. In: Proceedings of the IEEE/CVF conference on computer vision and pattern recognition. pp 462–471 (2020)

Wang J, Duan H, Liu J, Chen S, Min X, Zhai G (2023) AIGCIQA2023: a large-scale image quality assessment database for AI generated images: from the perspectives of quality, authenticity and correspondence. arXiv:2307.00211

Wolfe JM, Horowitz TS (2004) What attributes guide the deployment of visual attention and how do they do it? Nat Rev Neurosci 5(6):495–501

Xu J, Jiang M, Wang S, Kankanhalli MS, Zhao Q (2014) Predicting human gaze beyond pixels. J vis 14(1):28–28

Yang J, Zhai G, Duan H (2019) Predicting the visual saliency of the people with VIMS. In: Proceedings of the IEEE visual communications and image processing (VCIP). pp 1–4

Zhu Y, Zhai G, Yang Y, Duan H, Min X, Yang X (2021) Viewing behavior supported visual saliency predictor for 360 degree videos. IEEE Trans Circuits Syst Video Technol. pp 4188–4201

Zhu X, Duan H, Cao Y, Zhu Y, Zhu Y, Liu J, Chen L, Min X, Zhai G (2023) Perceptual quality assessment of omnidirectional audio-visual signals. arXiv:2307.10813

Zhuang C, Lu Z, Wang Y, Xiao J, Wang Y (2022) ACDNet: adaptively combined dilated convolution for monocular panorama depth estimation. In: Proceedings of the AAAI conference on artificial intelligence. pp 3653–3661

Blockchain Technology for Preventing Disasters in Surgery: A Case Study on Plastic Surgery

Chongdan Pan and Pradeep Kumar Ray

1 Introduction

In medical terminology, sentinel events define major healthcare disasters that could be prevented (e.g., death, serious body harm, and permanent damage to health). The most common sentinel events are wrong-site surgery, foreign body retention, and falls. This chapter focuses on one type (wrong surgery) (Patra and Jesus 2023). Numerous reports consider wrong-point surgery, wrong-side surgery, and wrong-case surgery independently; this can be useful in performing a root cause analysis and developing and constituting results. The moot question is, "Why do these effects take place?" And maybe just as importantly, "Why do they keep passing despite the acknowledgment that they should never happen?" The simple answer is that doctors, nurses, and other healthcare workers are humans who could make errors. Hence, it is important to design and implement processes to avoid human errors. There are moral, legal, and fiscal imperatives to do so. This chapter discusses the application of blockchain technology to avoid healthcare disasters of wrong surgery (https://blog.thesullivangroup.com/wrong-site-surgery-statistics).

Particularly, we focus on the plastic surgery industry due to its rapidly growing market value and the urgent need for transparency and trust. In 2020, over 24 million surgical procedures were performed by more than 43,750 surgeons (International

C. Pan (✉)
Centre For Entrepreneurship (CFE), University of Michigan-Shanghai Jiao Tong University Joint Institute, 800 Dong Chuan Road, Minhang, Shanghai 200240, China
e-mail: panchongdan@gmail.com

P. K. Ray
University of Michigan—Shanghai Jiao Tong University Joint Institute, Shanghai, China

UNSW Business, Sydney, Australia

P. K. Ray
e-mail: p.ray@unsw.edu.au

P. K. Ray et al. (eds.), *Technology Innovation for Sustainable Development of Healthcare and Disaster Management*, Disaster Risk Reduction,
https://doi.org/10.1007/978-981-97-2049-1_14

Society of Aesthetic Plastic Surgeries 2020). However, the market is highly susceptible to wrong surgeries, especially in emerging markets like China. In 2019, the size of the medical aesthetics market in China amounted to 177 billion yuan, with a yearly growth of 22%. However, there were only less than 24% licensed doctors and less than 12% regulated institutions (BeauCare Clinics First Cosmetic Hospital China Data Research Center 2018; Statista 2012). As a result, numerous unqualified institutions and individuals are performing risky underground plastic surgeries, leading to potential infections, disfigurement, or even fatalities caused by unqualified surgeons or medicine. To exacerbate the issue, patients often overlook these risks due to the lack of information, relying instead on customer reviews that can easily be falsified through false marketing.

For the plastic surgery industry, it is crucial to maintain trustworthy relationships and ensure secure personal data management for stakeholders. Fortunately, blockchain was designed specifically to address these pain points. Introduced to the world as the fundamental technology of Bitcoin by Satoshi Nakamoto in 2008, blockchain aims to record each transaction in a decentralized and immutable manner (Nakamoto 2008). Over the years, Bitcoin has withstood multiple network attacks and reached a total market value of over 500 billion USD, demonstrating the reliability and robustness of blockchain technology. At present, blockchain has evolved to support smart contract functionality, allowing for the automation of multi-step processes that can be recorded on the blockchain. This makes blockchain applicable in various domains beyond just currency transactions (Christidis and Devetsikiotis 2016). For instance, a smart contract can securely record all details of a surgery on the blockchain in an immutable and traceable manner. Such transparency builds trust among stakeholders, provides authentic information, and deters malicious activities.

In our case study on plastic surgery, we leverage blockchain to develop a decentralized application prototype. This prototype includes features such as fair surgeon peer review, medicine authentication, and resistance to cyberattacks. These features play a pivotal role in avoiding healthcare disasters, particularly within the plastic surgery industry. Compared to traditional healthcare applications, our blockchain solution demonstrates superiority in functionality, reliability, and security.

2 Our Solution Based on Blockchain

2.1 Introduction

To solve the problems in the plastic surgery industry and medicine accessibility, we designed a system built upon blockchain, to make the process more transparent, reliable, and smooth for patients to get service from plastic surgeons. The blockchain can not only be used to manage the patient's health data or claims but also their interactions with doctors, insurance companies, as well as medicine manufacturers. In this way, all actions and reviews are recorded on the blockchain so that patients can know how reliable the surgeon is by tracking the historical records. What's more, the blockchain can record important information about prescription medicine to make sure they're qualified and safe.

For now, a prototype has been developed by Solidity, a programming language specialized in designing smart contracts running on the Ethereum Virtual Machine. After generating an identification key, patients can start by proposing a smart contract to a doctor's public address. The confidential information of a smart contract will be encrypted, but the status of each smart contract and the public key of users will be public. Depending on what status a smart contract is in, people's behavior can be traced. If a patient misses an appointment, then the smart contract cannot proceed to the next stage. Since the users' public key and the status of each smart card are open to the public, the system can track how punctual the patient is. What's more, the smart contract ensures a seamless connection between data upload and payment. For example, the patient's medical history record will only be valid and stored on the blockchain when the user makes the necessary payment.

2.2 *Stakeholders*

2.2.1 Doctors

Doctors are using the system to promote themselves and attract customers so that they can get greater revenue. To get certificated and attractable for patients, the doctor needs to make important information public on the system, including digital identity and license. Since the license number is traceable, we can check whether the doctor is using a verified license or a fake one. The digital identity is like the public address or public key of the doctor. Knowing the doctor's public key, patients can either propose a smart contract to request an appointment or trace the doctor's previous activities on the blockchain. In addition, the patient can use the doctor's public key to encrypt the sensitive medical records and send them to the doctor. The transmission of the medical records will be recorded publicly on the blockchain, but its sensitive content is secure and only visible to the doctor, thanks to the property of asymmetric encryption (Simmons 1979). The digital identity will be recorded on the blockchain with all actions that the doctor made, such as surgeries, prescriptions, examinations, and the review that the doctor got. Although the sensitive information in these actions is encrypted, people can still evaluate a doctor based on historical records and reviews.

2.2.2 Patients

Patients are other primary users of the system. The blockchain enables them to manage their data safely, as well as find reliable doctors based on the historical information stored on the blockchain. Before entering the system, the user needs to generate a public–private key pair. After the doctor's office visit, the doctor will use the public key to encrypt the patient's new medical history records and store them on the blockchain. In this way, only the patients and the doctor who uploaded the records can see the sensitive content. For doctors, they won't have any incentive to leak the information because people can easily check who may view the information and such action can injure the reputation. Hence, the patients have full control of the medical history records since nobody can see them later without the patient's private key. The patient's public key can be used to identify the patient as well since it represents the owners of the medical history records. Therefore, doctors can check the miss and reschedule rate of a patient as well as search for the patient's historical actions on the blockchain.

2.2.3 Third Parties

Third parties can be other stakeholders or people who have an interest in the system, such as medicine manufacturers and insurance companies. A lot of plastic surgery malpractice is due to unqualified medicine; hence, manufacturers will be asked to upload the serial number of medicines used in each prescription to the blockchain so that they can be verified and traceable. Storing the medicine information on the blockchain can also help the medicine manufacturers manage the supply chain efficiently. Insurance companies can inject the insurance terms into the smart contract set between patients and doctors. With the help of a smart contract, the insurance can be more clearly identified and executed as automatically as possible.

2.3 *Workflow*

The workflow between patients and doctors based on our system can be divided into the following four stages (Fig. 1).

2.3.1 Information Searching

In this initial stage, patients and doctors need to create their private–public key pair so that the patient can set up a secure account to store health data and the doctor can become searchable. Then, the patient can search for doctors based on the initial public information stored on the blockchain.

The search process is very similar to the existing application, and then the patient can propose a smart contract with an appointment time to the doctor's public address. The creation of a smart contract will cause some transaction fees since it needs to store new data on the blockchain. During the search process, previous patients' reviews can serve as a reference. Patients can make choices either based on the previous reviews or the smart contract records associated with a specific doctor. Different

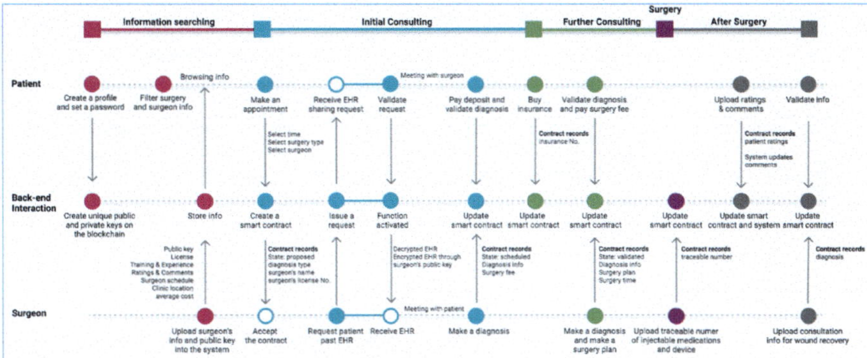

Fig. 1 Example of our solution's workflow

from the traditional system, it's impossible for the doctor to fake the review because they're linked to a smart contract containing all immutable information regarding the treatment. If the doctor wants to make a fake review, he or she must go through the whole process from proposing a smart contract to the end of visiting, which requires a lot of time and money (Figs. 2, 3, and 4).

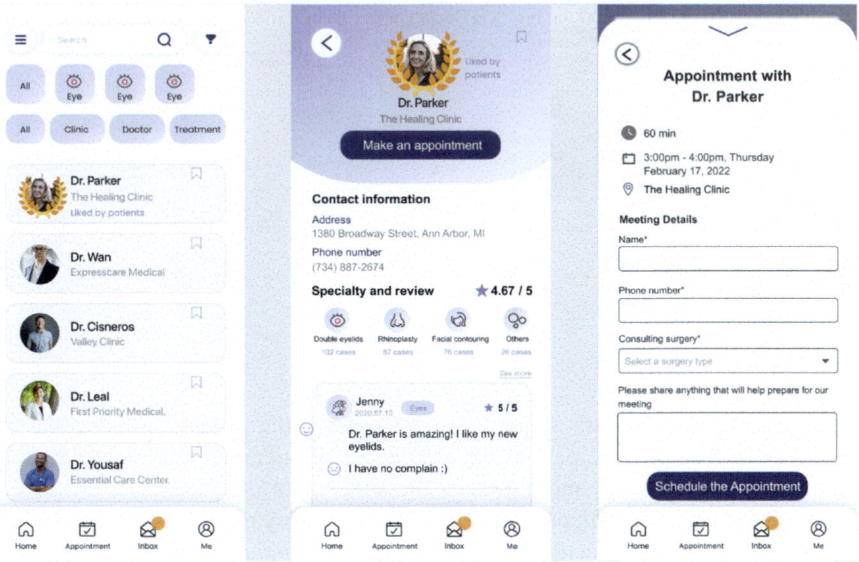

Fig. 2 Information search

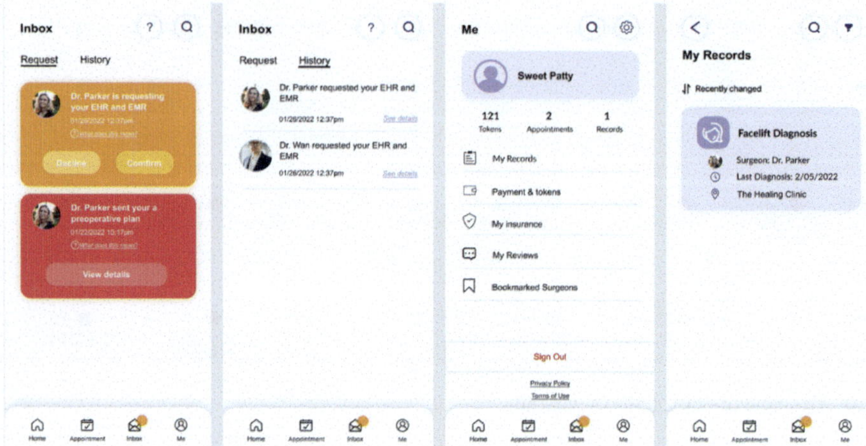

Fig. 3 Initial consulting

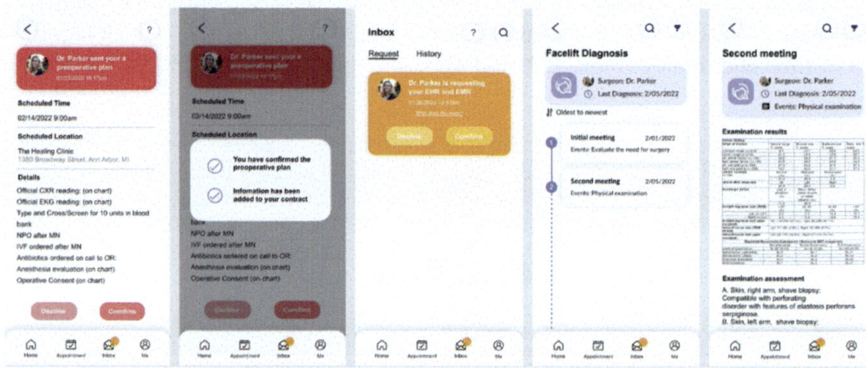

Fig. 4 Further consulting

2.3.2 Initial Consulting

After receiving the patient's request for the surgery, the doctor can schedule the surgery as well as ask the patient for a deposit. The deposit makes sure that the patient will keep the time and appear at the surgery. In this stage, the doctor can also request to view the patient's previous medical history record. Since the medical history records are stored on the blockchain, only the person with the owner's private key can determine whether to share them. If the patient authorizes access, the records will be decrypted by the patient's key and encrypted by the doctor's key again so that only the doctor can view them. Since all this process is done through an atomic smart contract, the content of the smart contract can't be revealed to other people.

If both doctor and patient agree with the schedule, deposit, and other information stored in the smart contract, they can accept it by putting a digital signature with their own private keys. Since then, the status of the smart contract will be set as *SCHEDULE*, meaning that the doctor and the patient have made a promise that they will have surgery, and the public can know whether they followed the schedule well. At the same time, the deposit will be charged from the patient's account.

2.3.3 Further Consulting

After the first consultation, doctors and patients should have a plan for the following surgery. In this stage, they need to decide the medicine used for the surgery as well as link the smart contract to the insurance claims with clear terms.

All valid medicines used in the surgery, such as anesthetics, should be recorded on the blockchain as soon as they're manufactured. Therefore, the doctor and the patient should be able to check the information of the medicine and make sure that it's valid for the surgery. The insurance claims will be written by the insurance company in a smart contract as well. Compared to those written on paper, coded insurance is clearer and more automated, because there is no ambiguity and can be triggered automatically. After this stage, both surgeon and patient need to put a digital signature again and the status of the smart contract will be converted to *CONFIRMED*.

2.3.4 Surgery

In this stage, the surgeon is responsible for recording any important information related to the surgery on the smart contract. After the surgery is done, the surgeon can set the smart contract as *EXECUTED*.

2.3.5 After Surgery

After the surgery, the doctor will encrypt the surgery summary and further prescription with the user's public key and upload the data on the blockchain. Then the patient can check the surgery summary and write a review. If the patient is not satisfied with the summary or the result, then he can write the concern into the review so that it can be referred to by other patients. When the review is completed, the patient can set the smart contract as *CLOSED* by putting in another digital signature, then the deposit will be transferred to the doctor automatically and the summary will be linked to the user's account. If the surgery's outcome matches any term in the insurance, it'll automatically trigger the claims.

2.3.6 Accidents

Neither the surgeon nor the doctor can be sure that everything will go well during the surgery; however, we must have a clear and convincing protocol to deal with it. If an accident happens, the surgeon and doctor should either deal with it with the insurance term or reach an agreement and record the history on the smart contract. Then they need to update the smart contract's state together with their digital signatures. If the accident is far worse, since the doctor's and smart contract's information is public, there should be enough evidence for investigation.

Since updating the smart contract requires both the patient's and surgeon's signatures, the surgery can't be closed without everyone being unsatisfied. In this way, previous surgeries can be valid evidence to show how the surgeon is. The better the surgeon is, the more *CLOSED* smart contract with the good review he should have. If there is a large fraction of unclosed smart contracts, then the public can trace the recorded information and see why it's unclosed. With immutability and authenticity accuracy, the system provides a fair evaluation system for the patients, and it encourages the surgeons to behave well.

3 How Blockchain Avoids Accidents

3.1 Secure Data

As the healthcare industry becomes increasingly digitized, a vast amount of data, including medical history records, surgeon certificates, and medicine manufacturing information, are stored in databases. These data are crucial for successful surgeries. However, traditional databases manage data in a centralized approach, making them vulnerable to compromise. If attackers breach one node, they can erase or modify all data records, resulting in significant losses from healthcare data breaches (Seh et al. 2020).

Blockchain systems offer a more effective solution to ensure data security. Firstly, blockchain is resistant to single-point failure, as malicious individuals would need to compromise more than 1/3 of the nodes to invade the system, making it significantly harder. Moreover, the data management team can enhance security by adding more nodes to the blockchain system. Additionally, the use of cryptography and hash functions in blockchain systems ensures that nobody can control another user's account without knowledge of the private key. Thus, even though medical data is sensitive, users need not worry about data breaches affecting their surgeries.

3.2 Authenticated Surgeon

A significant number of plastic surgery disasters are caused by unqualified surgeons performing surgeries without licenses. However, patients may be unaware of this and can be lured by advertisements. Traditional systems face challenges in authenticating surgeons since they cannot verify the identity of the publisher of the advertisement or the surgeon.

In contrast, identifying activities on a blockchain system is straightforward. As every user can only interact with the system using a public key, qualified surgeons can be required to register their public keys along with their licenses. With this setup, the system can automatically verify the surgeon's qualifications and prevent them from conducting unauthorized surgeries.

3.3 Traceable Records

When choosing a surgeon, patients often heavily rely on recommendations provided by the system. These recommendations may be based on algorithms or peer reviews. A popular traditional system called Zocdoc offers higher accessibility to doctors based on reviews, patient locations, and other criteria (Siddiqui and Rashid 2013). Zocdoc's popularity underscores the demand for trust and convenience in the health-care industry. However, a common issue with Zocdoc is the credibility of reviews and recommendations, as doctors may post fake reviews to boost their profiles and attract more customers, eroding trust in reviews. Relying on low-quality reviews may misguide patients and lead to wrong surgeon choices and even accidents.

In a blockchain system, fake reviews are eliminated as each review is linked to an actual surgery that took place. Users can trace all details of the surgery associated with each review, including the surgeon's qualifications, transaction history, medicine records, surgery information, and more, enabling reliable surgeon recommendations for patients.

Beyond reviews, a blockchain system can record comprehensive details of medicines, including manufacturing procedures, license IDs, and instructions. This information allows patients and surgeons to verify the qualifications of medicines and ensure their proper usage. Such transparency is crucial in reducing the number of disasters caused by incorrect medicine use. Furthermore, in the unfortunate event of a disaster, the blockchain system facilitates easier identification of the responsible party.

4 Limitations and Future Work

4.1 Performance

For now, the prototype is developed on the Ethereum Virtual Machine, which is suffering from high transaction costs as well as latency. The blockchain's performance will be critical for scalability as well as user experience, because patients may not be willing to spend a large amount of extra time or fee for a decentralized system. Luckily, there are a lot of projects aiming at improving the blockchain's performance by using different consensus algorithms such as Algorand (Gilad et al. 2017).

4.2 Acceptance

Blockchain and cryptocurrency are still a controversial topic all over the world, and terms like private key and smart contract make blockchain look unapproachable for normal people. To encourage more people to use the blockchain system, more development in the ecosystem is required. On the other hand, a native token is usually issued on the blockchain to smoothen the transactions. For users, the volatility of the token's price can be a concern, and more infrastructures like exchanges may be required.

4.3 Compatibility

Since our solution has much more function, compatibility can be a concern. For example, some patients' medical history records may be stored in the traditional database right now, and it may require a large effort to export them to the blockchain. In addition, traditional insurance terms are too complex to be translated into clear smart contract codes. When it comes to disputes, accidents, or claims, people may still prefer human service. However, the system can still work without a fully automated insurance smart contract; we just need to record what happened during the surgery and show it to an insurance company.

5 Conclusion

In this chapter, we have provided an overview of the current status of plastic surgery and provided examples of surgery disasters caused by human error. To reduce the number of such disasters, a blockchain system can be used to enhance the trust between different stakeholders in a surgery. We've shown the practicability by developing a prototype on an Ethereum Debug Virtual Machine with a systematic process and working interfaces.

The blockchain can lower the human error risk in surgeries in three ways. Firstly, it reinforces data security, so that sensitive data such as a patient's medical history cannot be altered. Secondly, the system can avoid surgery performed by unqualified surgeons by authenticating their identity. Thirdly, blockchain can make customer reviews and medicine manufacturing records as transparent as possible, enabling the patients and families to further verify whether they're making the right choice of surgeons and medicines.

Although blockchain can play an effective role in reducing surgery accidents, it still has some limitations. We've discussed drawbacks in performance, acceptable, and compatibility, and more future work is required to address this concern. Besides the plastic surgery industry, blockchain systems can provide trust and transparency in any social business such as the entire healthcare sector (Mukkamala et al. 2018). If the system is designed properly, it can make the information access to stakeholders as accurate and transparent as possible.

Acknowledgements The authors are grateful to Joe Ortigara, Noah Chatman, Yue Hu, and Ziwei Chen for their contribution to this idea at the University of Michigan. Their efforts for market analysis and prototype development were helpful in developing this chapter.

References

BeauCare Clinics First Cosmetic Hospital China Data Research Center, Chinese Association of Plastic Surgery Medical Risk Control Center. Black needle in Chinese underground cosmetic industry. 2018

Christidis K, Devetsikiotis M (2016) Blockchains and smart contracts for the internet of things. IEEE Access 4:2292–2303

Gilad Y, Hemo R, Micali S, Vlachos G, Zeldovich N (2017) Algorand: scaling byzantine agreements for cryptocurrencies. pp 51–68

International Society of Aesthetic Plastic Surgeries. Isaps international survey on aesthetic/cosmetic procedures. 2020

Mukkamala RR, Vatrapu R, Ray PK, Sengupta G, Halder S (2018) Blockchain for social business: principles and applications. IEEE Eng Manag Rev 46(4):94–99

Nakamoto S (2008). Bitcoin: a peer-to-peer electronic cash system

Patra KP, De Jesus O (2023) Sentinel Event. [Updated 2023 Mar 29]. In: StatPearls [Internet]. Treasure Island (FL): StatPearls Publishing. https://www.ncbi.nlm.nih.gov/books/NBK564388/

Savrin R. Wrong site surgery statistics. The Sullivan Group. https://blog.thesullivangroup.com/wrong-site-surgery-statistics

Seh AH, Zarour M, Alenezi M, Sarkar AK, Agrawal A, Kumar R, Khan RA (2020) Healthcare data breaches: insights and implications. Healthcare 8(2)

Siddiqui Z, Rashid R (2013) Cancellations and patient access to physicians: Zocdoc and the evolution of e-medicine. Dermatol Online J 19(4)

Simmons GJ (1979) Symmetric and asymmetric encryption. ACM Comput Surv 11(4):305–330

Statista (2019) Market size of the aesthetic medical and surgical industry in China from 2012 to 2019, with a forecast until 2023

Healthcare 4.0 (and 5.0): Digitising Health and Care

Colin Pye, Abhijit Das, and Anirban Dutta

1 Introduction

This chapter looks at the emergence (and origin) of Healthcare 4.0, how, within whole systems of Health and Care, the elements of 'Flow' (patient and person-centric) are relevant, how digitisation challenges may be addressed (maximising a partnership way of working, in the widest sense) and the recognition and management of risks, from both a patient and clinical perspective and a technological perspective.

Over the past years Healthcare 4.0, the fourth healthcare revolution, has gained considerable interest and has started to gather momentum and has some exciting future directions (Gupta and Singh 2023). Healthcare 4.0 (taken from Industry 4.0) incorporates work on Artificial Intelligence (AI), Data Analytics and Wearable Technology, for example. Artificial Intelligence (AI) has the potential to support more detailed and consistent analysis of scans for example whilst wearable technology, based on solid human-centred design principles (Bause et al. 2020), has the potential to support more individuals within a community setting.

Within the context of Healthcare 4.0, an important factor relates to lessons from 'Design Engineering in the Age of Industry 4.0' (Jiao et al. 2021). Furthermore, there are also lessons from industry on 'Flow' across whole systems (Flanigan 2023). The Flow of patients and individuals across 'Whole Health Systems', the coordination of

C. Pye (✉)
Preston, UK
e-mail: colin.pye1@btinternet.com

A. Das
Royal Preston Hospital, Sharoe Green Lane, Preston PR29HT, UK
e-mail: abhijit.das@lthtr.nhs.uk

A. Dutta
University of Bimingham, Birmingham, UK
e-mail: a.dutta.1@bham.ac.uk

P. K. Ray et al. (eds.), *Technology Innovation for Sustainable Development of Healthcare and Disaster Management*, Disaster Risk Reduction,
https://doi.org/10.1007/978-981-97-2049-1_15

clinical 'assets' and the use of data analytics become more important as the management of specific clinical conditions becomes more seamless across hospital and community settings through the implementation of AI and wearable technology.

There are also lessons on digitising Health and Care with respect to 'disaster planning'. 'Disasters' from a patient and individual's perspective and 'disasters' from a technological perspective and the impact it has on reverting to manual systems. This will be looked at further later in the chapter.

2 Healthcare 4.0 (and 5.0)—The Position for Health and Care

Industry 4.0 is in fact the Fourth Industrial Revolution, much of which has been written (Popov et al. 2140). However, it's important to understand Industry 4.0 in order to understand Healthcare 4.0.

The Industry message has been documented in a number of stages and recognised over the last 250 or so years: Industry 1.0 in the 1780s (with industrial production based on machines driven by water and steam); Industry 2.0 in the 1870s (with electric driving mass-production using assembly lines); Industry 3.0 in the 1970s (with electronics and computers being used to drive automation); Industry 3.5 in the 1980s (with Globalisation driving off-shore production to low-cost economies); Industry 4.0 today (with connected devices, artificial intelligence and data analytics—technologies automating processes further); Industry 5.0 future (with the focus on man–machine cooperation and collaborative robotics leading to the upskilling of workers and mass customisation and personalisation for customers).

Throughout this timeline, the focus has been on ensuring that products and services are delivered more efficiently and effectively. Manufacturing assembly line 'Flow' has driven the focus on automating processes that could be automated but always within the context of automating manufacturing 'cells' (individual steps on the assembly line) yet with an objective of not automating an individual process to the extent of creating 'bottlenecks' elsewhere. This 'holistic', 'whole system' approach is an important lesson for Health and Care.

For Health and Care, it could be argued that Healthcare (Industry) 5.0 is already recognised—that of 'personalisation'. Health and Care professionals do start by treating the patient as an individual. However, this tends to be within the 'silo' of treatment from the professionals involved. This is not a criticism but a fact of the way the system is set up. Each speciality is treated in the best interest of the patient, yet patients often have co-morbidities and complex conditions. The 'Flow' of the patient for an individual treatment is excellent yet the holistic view is often missed. This is how Health and Care are different from the linear, sequential, Flow of products and services, yet 'Flow' is relevant. For Health and Care, the 'Flow' impact is at the 'micro' level (for example a department within a hospital), the 'meso' level (for example between hospital departments) and the 'macro' level (across all organisations that are involved in the whole health system within a specific geography).

3 The Current Position for Health and Care

The Healthcare sector, globally, faces ongoing pressures in the management of patient Flows for all aspects of care, increasingly moving towards a 'lot-size of one' that addresses specific care for the individual. In addition, the dedicated and hard-working workforce deserves all of the support that they can be given. They need the tools, services and environment to enable them to deliver the care to match their skills and dedication. Globally (and in the UK since the NHS inception in 1948), the 'system' model of care has largely remained the same. It is inevitable that the sector must evolve and change in order to deliver the system-wide transformation required. It must change so that

- The healthcare delivery model can be redesigned in order to deliver services that are fit for future needs and are sustainable and affordable.
- The workforce can become more agile, work in different ways and potentially become multi-disciplinary, recognising that the 'Flow' and 'pathways' of patients through the healthcare system will be different.
- The experience of the health sector, the academic sector and industry can be maximised and can be fused together in a partnership that will allow accelerated digitisation, innovation and research and development.

In the UK, the recent Health and Social Care White Paper (https://www.gov.uk/government/publications/health-and-social-care-integration-joining-up-care-for-people-places-and-populations), along with plans for other Health and Care reforms, has outlined specific areas. The purpose of this chapter is to take account of recent initiatives, the current situation and local needs in order to propose an example 'Flow' method and partnership way of working that encapsulates health sector needs, academic research and rigour and industry experience in order to deliver an accelerated and sustainable change in the way that patient-centric health care is delivered, taking account of patient (and person-centric) Flow. One example is that of FND, as will be looked at later.

Change is certain and digitisation is happening. It is clear that health care is changing in the way it is delivered and the way outcomes are measured. The COVID-19 pandemic has been, and still is, a catalyst for change having been a major disruptive event in its own right. Tensions between planned and unplanned care have become more obvious. Precious resources have been moved from planned care in order to support unplanned care. Digital solutions, such as remote consultation technology, have been accelerated into mainstream implementation. Yet the healthcare sector still lags behind other industry sectors in many areas of the take-up of technology as applied to the digitisation of the system itself. This is not necessarily due to the fact that there is reticence on behalf of the patients (*the 'Consumers'*) as their behaviours and expectations are increasingly driven by their own personal experiences of the digitisation of other services (ranging from retail through banking to local government). Rather it is because of the way digitisation of health care is approached. It needs to be applied in a holistic, and sustainable, way, across the whole system whilst acknowledging the 'Flow' and productivity, efficiency and effectiveness drivers and benefits realisation and return on investment.

There are many clear success stories in the way that technology has been applied to health care. These include the technology in medical screening programs and deployments in elective care procedures, operations and diagnostics to ongoing patient monitoring through solutions in telecare and telemedicine that, in certain implementations, make extensive use of the breakthroughs gained in sensor technology (sometimes implanted) and wearable devices. There are also many commercial organisations that already offer solutions in many fields and deliver valuable products and services that address specific needs. This then raises a fundamental point; if Healthcare 4.0—Digitisation of the Healthcare System is the answer, what are the fundamental questions it needs to address for it to truly transform the way the system functions?

The aforementioned solutions present answers to specific, symptomatic, areas yet do not cover the wider, holistic points that need to be addressed. These points include collaboration, coherence and convergence yet also need to include the productivity, efficiency and effectiveness of the healthcare system itself. Merely adding new technology to old ways of working usually leads to an increase in cost. A full understanding of what is happening within the system is essential in order to address blockages and bottlenecks. In general, healthcare systems are never short of data, an acknowledged key currency of any organisation, yet the data only becomes valuable when it can be turned into useful information that can be applied to inform knowledge and then relevant action can be taken. The information itself needs to be accurate, relevant and timely. An understanding of clinical pathways and patient pathways is critical in order to deal with 'Flow' and resource needs and hence eradicate bottlenecks. The pathways may be referred to as 'swimlanes' incorporating both patient and clinical aspects. Identification of 'swimlanes' is essential so that they can be prioritised if trying to 'boil the ocean' of healthcare digitisation is to be avoided. As collaboration becomes the necessary way forward and Integrated Care Systems and place-based care seek to resolve the 'right patient','right care' and 'right place' challenges, a full understanding of 'micro', 'meso' and 'macro' patient Flow is needed. 'Micro' patient Flow is the processes and interactions within an organisation, 'meso' is between departments within a hospital and 'macro' patient Flow is the processes and interactions between organisations and the system stakeholders. This is particularly pertinent as the screening-intervention-treatment process maps to planned-unplanned-planned care.

3.1 *Recognising Gaps and the Need to Map Processes*

There are a number of critical factors that require addressing and need to be understood in order to successfully deploy digitisation in health and get the most benefit. The delivery of health services is, inevitably, complex. This is due to many factors ranging from patient co-morbidities to complexities in diagnostics, operations and treatments. Healthcare systems comprise, broadly, of balancing 4 elements:

- The needs of the Patient—to ensure a successful outcome
- The needs of the Clinical Staff—to deliver the right medical resources
- The needs of the Operational Staff—to deliver the right operational and asset needs
- The needs of Finance—to ensure a viable and sustainable service and ongoing investment.

The move towards Healthcare 5.0 has the requirement to

- create new knowledge, particularly with respect to interactive healthcare systems,
- create the next generation model of interactive healthcare systems based on the Integrated Care Systems. This must address a specific 'toolkit' that includes, and takes account of, 'micro' and 'macro' (and 'meso') patient Flow, information needs and geographic demographics, current deployments and benefits realisation and Return on Investment (RoI),
- address the deployment needs of a digitised healthcare service within a holistic framework
- be inkeeping with Industry 5.0 design principles and, in particular, allow for commercial 'exploitation' of solutions.

A key area must be a model that can be used across different geographic healthcare settings (and can be applied to 'swimlanes') to

- address priorities for digitising health care within a specific demography (including signposting technology gaps within a commercial model),
- incorporate modules that include process mapping for 'Flow', benefits realisation and RoI, logging of population health for specific demographics.

3.2 *Moving to a Digitised Healthcare System*

A Digitised Healthcare System (Digitisation of the Healthcare System) requires many variables to be addressed including addressing the current status of deployment within a setting that, although common in ultimate objectives, exists at different stages of development. The key fundamental steps include

- Where are we now?
- Where do we want to be?
- How do we get there?

Where are we now?

Many (targeted or even fragmented) initiatives are already in existence ranging from Global Digital Exemplars and Healthcare Information and Management Systems Society (HIMSS) and Electronic Medical Record Adoption Model (EMRAM) stages. These measures help to score hospitals on their digital maturity with a view to aiding the move to digital transformation. Some 'model hospital' studies (https://www.nric.org.uk/node/54145) have taken place in the past and commercial deployments exist but an overall deployment and implementation model that takes fully into account 'Flow' across the whole health system does not widely exist.

Where do we want to be?

Successful digitisation of the healthcare system is dependent upon a holistic understanding of the key drivers. It is this that will deliver the proactive, and manageable, transformation of healthcare services rather than symptomatically 'tweaking' elements. Being 'outcomes driven' across the whole health system is an important factor.

How do we get there?

By fully analysing the healthcare drivers, it is possible to build an implementation model that takes account of specific demographics, based on regional and local needs, and delivers a sustainable phased, focused and prioritised digital healthcare system that is supported by benefits realisation and return on investment and driven by intelligent information and data analytics.

In the UK, for example, the move from Sustainability and Transformation Partnerships (STPs) through Integrated Care Systems (ICSs) to an Integrated Care Partnership (ICP) within an Integrated Care System:

Integrated care systems (ICSs) have evolved from STPs and take the lead in planning and commissioning care for their populations and providing system leadership. They bring together NHS providers and commissioners and local authorities to work in partnership to improve Health and Care in their area.

Integrated care partnerships (ICPs) are alliances of NHS providers that work together to deliver care by agreeing to collaborate rather than compete. These providers include hospitals, community services, mental health services and General Practitioners (GPs). Social care (Local Government) and independent and third-sector providers (charitable organisations) may also be involved.

NOTE: Accountable Care Organisations (ACOs) are also relevant and are established when commissioners award a long-term contract to a single organisation to provide a range of Health and Care services to a defined population following a competitive procurement. This organisation may subcontract with other providers to deliver the contract (https://www.kingsfund.org.uk/publications/accountable-care-explained).

It is possible to think of Integrated Care Systems as having micro, meso and macro levels. For example, the micro level may be a department within a hospital, the meso level being between departments and the macro level between entities (such as an acute hospital, a community hospital and a nursing or care organisation) within a whole health system. Why should this be necessary? Because in general (and in particular for managing complex, long-term conditions), Health and Care do not end with discrete, elective, episodes of care.

'Micro' Mapping within Stakeholders (example below for Outpatient Flow) exists (https://www.kingsfund.org.uk/publications/accountable-care-explained) and, in itself is very important, but there is a growing need to map Flow across whole health systems.

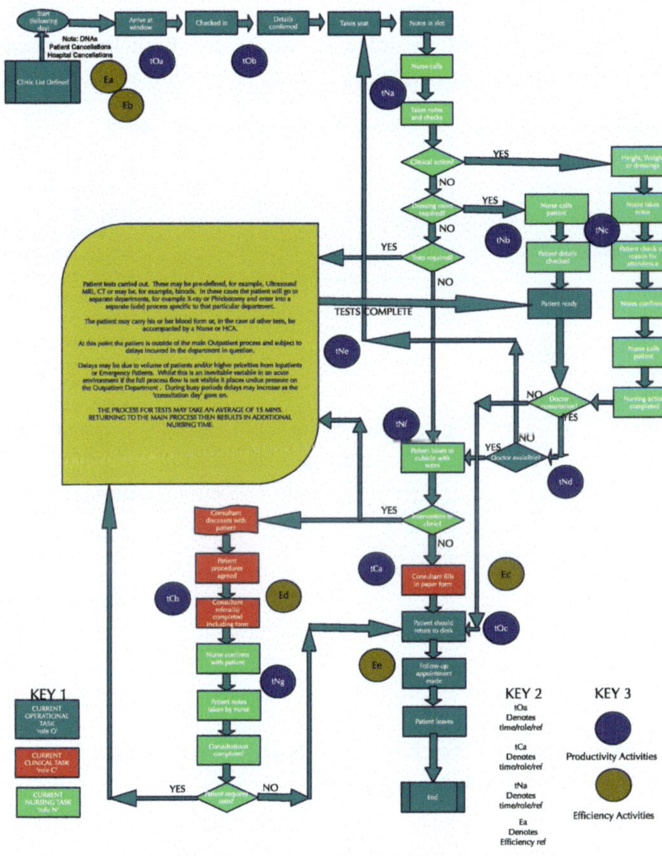

Patient Flow from arrival at hospital for Outpatient appointment, taking into account any pre-defined or on-the-day tests required through to post consultation Outcome and discharge, follow-up with further appointment or elective care requiring pre-operative assessment.

At each point in the Flow process, above, technology and people may be reviewed so as to improve productivity and efficiency.

4 Planning for 'Step-Up' and 'Step-Down' Care

In order to deliver a redesigned healthcare delivery model that is sustainable in every sense integrated care inevitably means integrated systems. In turn, this means common platforms that will track the 'Flow' of a patient's or persons' needs as follows a 'step-up' process (as more care is required through complexities or exacerbated health issues) or 'step-down' as the individual recovers. This model can allow for community support at every level from 'signposting' through to clinical support (and 'virtual wards') in a community setting.

To achieve this, the technology must be integrated with the process and the (healthcare professional) people (https://www.intouchwithhealth.co.uk/).

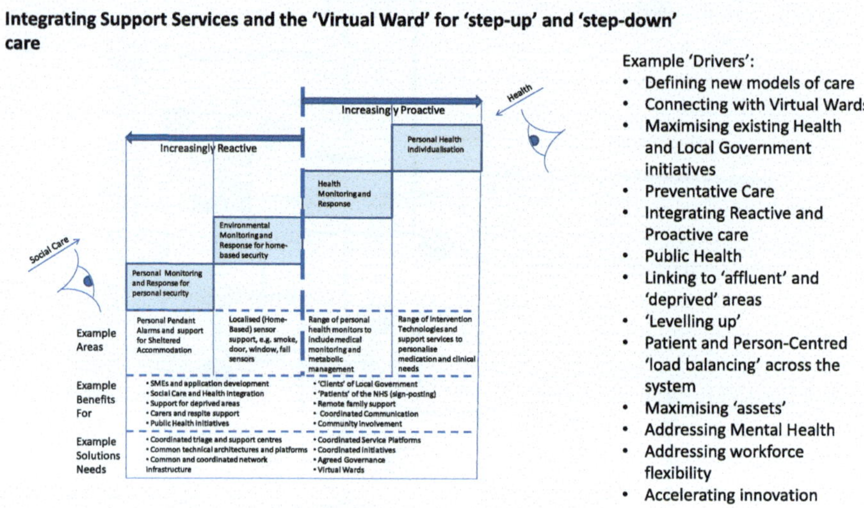

5 The Relationship of Technology Solutions to the Needs of Health and Care Systems

Merely adding new technology to old ways of working usually leads to an increase in cost. A full understanding of what is happening within the system is essential in order to address blockages and bottlenecks. The relationship between 'the needs', 'the potential (technological) 'building blocks' and the 'enablers' required to make it happen is essential. The figure, below, illustrates this point and, in particular, the point that the 'Flow' of a patient through the system helps to define the.

- Needs
- Gaps
- Priorities

- Benefits
- Return on Investment.

However, the 'needs' are required to be matched to the technology building blocks and supported by specific 'enablers' to ensure that the redesigned system is sustainable in all aspects.

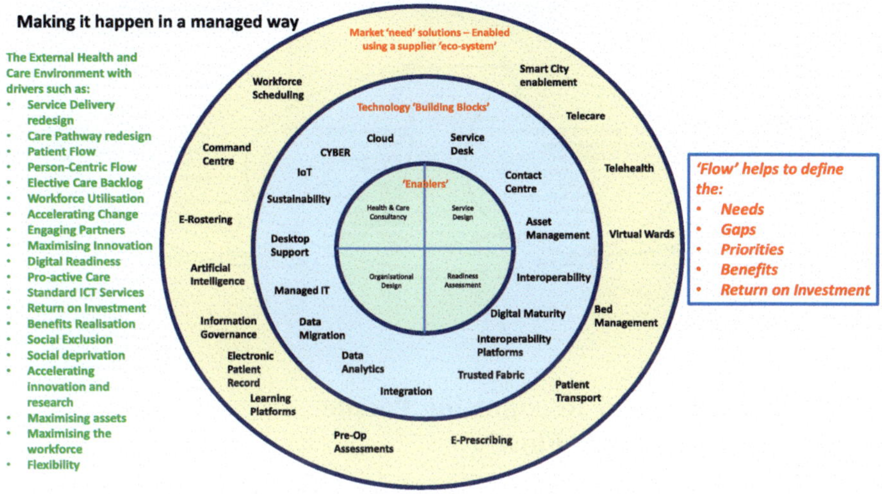

5.1 *Academic, Health and Care and Private Sector Partnerships*

Digitisation of the Healthcare system will not just happen! 'Silo' solutions can (and have) been applied to various elements from advancements in Picture Archiving and Scanning (PACS), networks and computer systems to utilising technology in patient communications. However, if the analogy is drawn with the manufacturing sector, the Health and Care 'factory' has had various 'cells' modernised that have resulted in advances in efficiency and throughput for those immediate areas but caused bottlenecks elsewhere. Examples at the front-end are that of patients being unable to be transferred from ambulances into the hospital because at the back-end patients are unable to be discharged from the hospital into the community. In order to achieve true Health and Care transformation, a partnership way of working should be employed that is not merely that of the hospital and community but includes the academic and private (technology) sectors. The figure, below, illustrates what each party can bring.

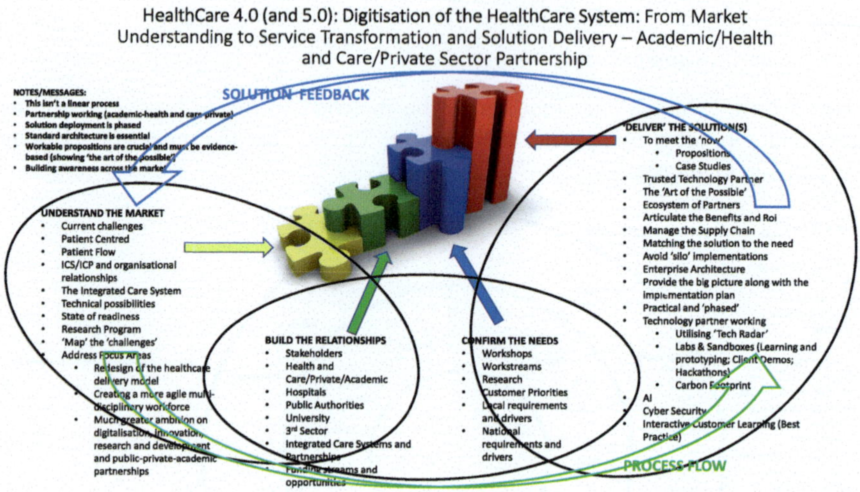

6 The Importance of Recognising People and Person-Centric 'Flow'

'Flow' is the critical factor by which the path to Health and Care digitisation can be mapped and, importantly, implemented in a way that is clinically acceptable, improves patient care, is operationally viable and financially sustainable.

The movement of individuals through a System (their 'Flow'), and the ease by which this happens, is greatly affected by the following:

- Their specific needs (including acute, chronic and public health—to address reactive and proactive care)
- Access to their information (that takes account of security and access rights)
- Access to the required 'assets' at each stage (be they clinical, carers, equipment or information)
- Recognising defined Outcomes (for ongoing care, discharge or population health and wellbeing)
- Defining the required Services (recognising the 'what', the 'where' and the 'how').

In the context of Integrated Care Systems (ICSs), the 'Flow' can be thought of as having three core elements:

1. **Care Pathways**: *the way an individual follows and Flows through a specific element of their care; for example a specific neurological, cancer pathway or pathway that supports chronic disease states (Clinical Pathways)*
2. **Patient Flow**: *the way an individual Flows through a specific organisation within an ICS; for example an Acute ((hospital) or Community organisation as an 'outpatient' or 'inpatient'*

3. **Person-Centric Flow**: *the way an individual follows a route and Flows across a system; for example multiple organisations of all types within an ICS.*

Identifying aspects of 'Flow' will drive out the 'needs' of the System. It also highlights the 'pinch-points'; however, whilst secure and ubiquitous technology is a critical enabler, people and processes are also key.

6.1 People, Processes and Technology

In order to realise the vision of a redesigned healthcare delivery model, it is essential that all assets, skills and experience, from all parties and stakeholders, are taken advantage of and brought to bear. No one organisation has all of the answers; however, collectively new ways of working are within reach. In order to move forward, the needs of organisations must be recognised and accepted in a balanced way. The figure below illustrates the suggested structure, needs and skills and abilities of each area. The Health and Care community brings the patient 'needs', clinical and health professionals' experience and operational environment, the academic community brings the research rigour, critical evaluation and health economists, and industry brings experience and 'the art of the possible' from health and other sectors, innovation, and sustainable scalability within a workable ecosystem of suppliers.

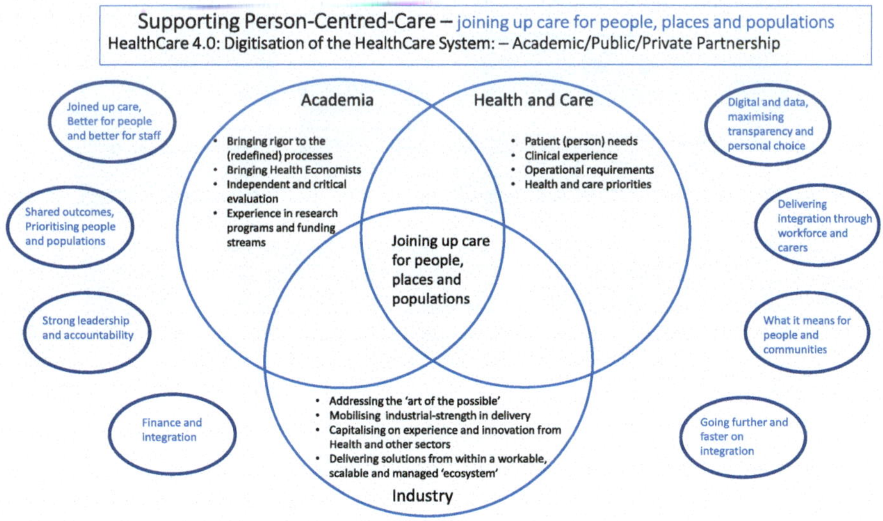

6.2 An Example of 'Flow' Applied to Functional Neurological Disorder

Shown below is the Pathway Map from the National Neurosciences Advisory Group (NNAG)—February 2023 (Optimal Clinical Pathways 2023). The Pathway shows five scenarios for patient 'Flows' through the system (Optimal clinical pathway for adults with Functional Neurological Disorder FND 2023). These are.

I. Onset of Neurological Symptoms
II. Self-Referral (without established local service)
III. Self-Referral (patients already known to a service)
IV. Independent Service
V. Transition.

The NNAG has stated that *'The symptoms, disability, and co-morbidities of people with FND are highly heterogeneous: they vary widely from person to person. A successful pathway for people with FND requires a range of services available which people can be referred depending on need and complexity.'*

This relates to the 'lot-size of one' that addresses specific care for the individual.

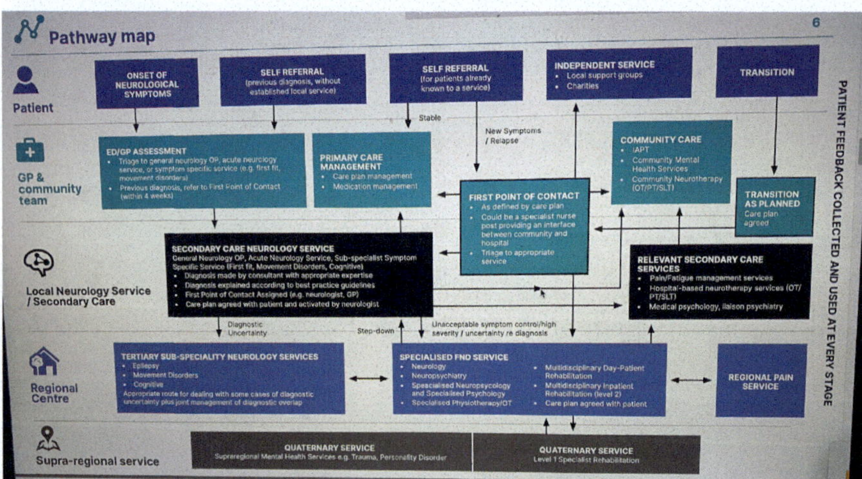

For the implementation of Healthcare technology, and hence the move towards Healthcare 5.0, to be successful, patient 'Flow' needs to take account of those co-morbidities and patients' complex conditions. The diagram, below, illustrates an example of a specific patient scenario. The patient scenario ('Cameo') illustrates the situation for the patient and the diagram illustrates how, at each example step, technology 'building blocks' may be employed in the journey.

Cameo (example): *'Rose is 50. She was diagnosed with epilepsy when she was 18. Whilst this is under control her medication adversely affects her calcium and vitamin D levels. She takes vitamin D supplements. She has recently been diagnosed with*

Breast Cancer and, after surgery, lymph node removal and radiotherapy has been put on a drug that also adversely affects her calcium. She has undertaken Dexa Scans with her Oncology Consultant but has recently had new neurological symptoms and has contacted her GP.'

The example taken is that of *Self-Referral (patients already known to a service)—NNAG patient example III.*

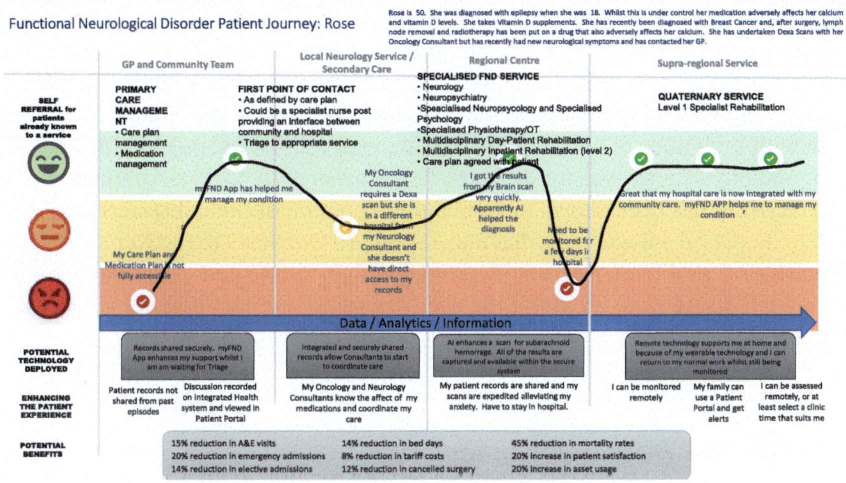

7 Planning for Disasters

7.1 From the Patient and Clinical Standpoint

Disaster is defined as 'an event and its consequences that result in a serious disruption of the functioning of a community and cause widespread human, material, economic or environmental losses that exceed the capacity of the affected area to respond to without external assistance to save lives, preserve property and maintain the stability and integrity of the affected area' (Medina 2016). Since Charles Myers (1873–1946) coined the term 'shell shock' to define the prolonged suffering of soldiers returning from the First World War, the psychological and physical result of distressing experiences during disasters, known as trauma, has been of academic interest. During that period, around 80 000 soldiers had the diagnosis, which encompassed a wide variety of physical and psychological presentations, including nightmares, anxiety, tremors, gait disorders and panic attacks (Stone 2016). Moreover, the psychological and physical effects of the trauma can affect more than just the primary experiencers of such trauma and can span over several generations known as transgenerational transmission of trauma (TTT) (Yehuda and Lehrner 2018). Hence, mitigation of neuropsychiatric conditions should be an important part of disaster preparedness.

Since the phenomenon of 'shell shock' became well known, the neuropsychiatric impacts of disasters are well described and they include post-traumatic stress disorders (PTSD), depressive or anxiety disorders, somatic complaints and neurological conditions also known as functional neurological disorders (FND) (Mel et al. 2008). In the short term, the trauma of exposure to disasters can trigger psychological distress, somatic complaints, sleep problems and psychosocial and behavioural problems including FND. Most of these effects are mild to moderate, but about 30% of symptomatic cases require intervention (Gil-Rivas and Kilmer 2016). The impact is more severe with women and people with a low socioeconomic status, and elderly people with high rates of chronic illnesses are especially susceptible to the psychological and physical stresses of disasters (Al-Rousan et al. 2015). Although mental health recovery has been shown to be mainly a function of time, disasters have been linked to increased suicide rates that may take up to two years to recover and the risk of behavioural problems and suicide remains high in affected communities many years after a disaster, highlighting the need for sustained post-disaster surveillance (Orui et al. 2014).

7.2 From the Technology Standpoint

The old 'adage' applies in that if a problem can occur it will occur. As digitised processes become more embedded in the Whole System, there is more reliance on them as the 'normal way of working' but also more vulnerability when faults occur. Over the years, technology has become more reliable and, furthermore, system design allows for designed-in redundancy; however, there should be preparation for 'fall-back' to more manual ways of working. Experience has shown, from those organisations that are ahead of the Health and Care sector in their digitisation journey, that implementing technology solutions alongside process change and people involvement can result in improved efficiencies and more effective performance and increase capacity and throughput. However, should the digitised system (or key parts of the system) fail, the result is a drop back to a reduced throughput. In the case of Health and Care, this would mean fewer elective operations, less outpatient appointments or less diagnostic services for example.

A relevant illustration, from another sector, would be that of the experience of the (partial) failure of the UK Air Traffic Control System. The system fell back to well-planned and rehearsed manual processes and, whilst remaining safe, was only able to handle a reduced 'throughput' of aircraft (the analogy for Health and Care would be a reduced throughput of patients). Interestingly, whilst this is a preliminary report there were immediate actions and areas for further investigation (https://publicapps.caa.co.uk/docs/33/NERL%20Major%20Incident%20Investigation%20Preliminary%20Report.pdf).

8 Conclusions

Digitisation of Health and Care is already happening and has been for a number of years—with considerable success. This ranges from the implementation of new scanning technology through to patient records and, importantly, the advance of 'wearable devices'. However, despite this, digitisation of Health Care is taking place in 'pockets'. Using a manufacturing analogy, digitisation is taking place in 'Cells'. In the immediate environment, it's good but in the 'Whole System' it can create bottlenecks unless looked at holistically. Furthermore, technology alone is NOT the answer. It needs to go hand-in-hand with Process and People. Clearly, lessons can be learned from other sectors so as to address the 'Whole System' which includes the needs of

a. The Patient
b. The Clinician
c. Operational Management
d. Finance

For transformation to be sustainable.

Within the context of the digitisation of Health Care (comprising Health and Care), 'Flow' (patient and people-centric) is the key to unlocking the what, where, how, when and why technology can be applied—and to whom. The six honest men principle applies! In this respect partnership working applies; however, partnership working does not just mean those organisations in Health and Care (including charitable organisations and the 3rd Sector) but, at a higher level, should include Academia—to address research, health economics and analysis and Industry—to show the 'art of the possible' and the 'commercials'.

Within any Health and Care environment, globally, areas of deprivation and affluence are present to a greater or lesser extent; therefore, demographics should be foremost in planning and should draw on data analytics. As part of this planning, social exclusion should be inherent in any plan to digitise, acknowledging that not everyone has access to what the latest technology can offer. A strategic 'route-map' should be documented but implemented within an achievable plan that is not over-ambitious to 'boil the ocean'. It should include CYBER and Disaster planning scenarios and, more importantly, be applied from the patient and clinical perspective taking account of the 'human' factors. Digitisation is inclusive. It should not be done 'to' people but 'with' people. Clinicians have the knowledge, whilst the Patients have the need.

References

Accountable care explained, https://www.kingsfund.org.uk/publications/accountable-care-explained, last accessed 2023/12/07

Al-Rousan TM, Rubenstein LM, Wallace RB (2015) Preparedness for natural disasters among older US adults: a nationwide survey. Am J Public Health. 105 Suppl 4, S621–626, S614–620. https://doi.org/10.2105/AJPH.2013.301559r

Bause MF, Forbes H, Nickpour F, Schaefer D (2020) Towards a health 4.0 framework for the design of wearables: leveraging human-centered and robust design. Procedia CIRP 91:639–645. https://doi.org/10.1016/j.procir.2020.02.222.

de Mel S, McKenzie D, Woodruff C (2008) Mental health recovery and economic recovery after the tsunami: high-frequency longitudinal evidence from Sri Lankan small business owners. Soc Sci Med 66:582–595. https://doi.org/10.1016/j.socscimed.2007.10.006

Flanigan R (2023) What is a manufacturing flow management process?, https://www.method.me/blog/manufacturing-flow-management-process/, last accessed 2023/12/06

Gil-Rivas V, Kilmer RP (2016) Building community capacity and fostering disaster resilience. J Clin Psychol 72:1318–1332. https://doi.org/10.1002/jclp.22281

Gupta A, Singh A (2023) Healthcare 4.0: recent advancements and futuristic research directions. Wirel Pers Commun 129:933–952. https://doi.org/10.1007/s11277-022-10164-8

Intouch With Health | Experts in Patient Flow Management Systems, https://www.intouchwithhealth.co.uk/, last accessed 2023/12/07

Health and social care integration: joining up care for people, places and populations, https://www.gov.uk/government/publications/health-and-social-care-integration-joining-up-care-for-people-places-and-populations, last accessed 2023/12/06

Jiao R, Commuri S, Panchal J, Milisavljevic-Syed J, Allen JK, Mistree F, Schaefer D (2021) Design engineering in the age of industry 4.0. J Mech Des. 143. https://doi.org/10.1115/1.4051041

Medina A (2016) Promoting a culture of disaster preparedness. J Bus Contin Emer Plan 9:281–290

NATS Major Incident Preliminary Report. 2023. https://publicapps.caa.co.uk/docs/33/NERL%20Major%20Incident%20Investigation%20Preliminary%20Report.pdf

Operational productivity and performance in English NHS acute hospitals: Unwarranted variations | National Resource for Infection Control (NRIC), https://www.nric.org.uk/node/54145, last accessed 2023/12/06

Optimal clinical pathway for adults with Functional Neurological Disorder FND, https://www.nnag.org.uk/optimal-clinical-pathway-adults-fnd-functional-neurological-disorder, last accessed 2023/12/07

Optimal Clinical Pathways, https://www.nnag.org.uk/optimum-clinical-pathways, last accessed 2023/12/07

Orui M, Harada S, Hayashi M (2014) Changes in suicide rates in disaster-stricken areas following the Great East Japan Earthquake and their effect on economic factors: an ecological study. Environ Health Prev Med 19:459–466. https://doi.org/10.1007/s12199-014-0418-2

Popov VV, Kudryavtseva EV, Kumar Katiyar N, Shishkin A, Stepanov SI, Goel S (2022) Industry 4.0 and digitalisation in healthcare. Materials (Basel) 15:2140. https://doi.org/10.3390/ma15062140

Stone J (2016) Neurologic approaches to hysteria, psychogenic and functional disorders from the late 19th century onwards. Handb Clin Neurol 139:25–36. https://doi.org/10.1016/B978-0-12-801772-2.00003-5

Yehuda R, Lehrner A (2018) Intergenerational transmission of trauma effects: putative role of epigenetic mechanisms. World Psychiatry 17:243–257. https://doi.org/10.1002/wps.20568

Portable Neuroimaging-Based Digital Twin Model for Individualized Interventions in Type 2 Diabetes

Fei Zhao, Machiko Tomita, and Anirban Dutta

1 Overview of Type 2 Diabetes and Dementia

Approximately 500 million individuals worldwide live with diabetes, with nearly 80% residing in low- and middle-income nations. Type 2 diabetes, afflicting nine out of ten people with the condition, exhibits its fastest surge in low- and middle-income countries (https://www.cdc.gov/globalhealth/healthprotection/ncd/diabetes.html). Persistent high blood sugar levels inflict damage on various bodily systems, particularly nerves and blood vessels. The repercussions of diabetes encompass heart disease, stroke, kidney failure, blindness, and lower limb amputation. Recent studies have unveiled links between diabetes and conditions such as dementia, hearing loss, and certain forms of cancer. The substantial global burden of diabetes exerts adverse economic effects on individuals, healthcare systems, and entire nations. Dementia is more prevalent among individuals with Type 2 diabetes (Lai et al. 2023) where elevated levels of blood sugar, blood pressure, and cholesterol in individuals harm blood vessels and contribute to severe cardiovascular issues, such as heart attacks and strokes. Therefore, the presence of diabetes escalates the likelihood of premature mortality, and associated complications including Alzheimer's disease and related dementias (AD/ADRD) can significantly diminish the quality of life.

World Health Organization reports that dementia is a global concern (https://www.who.int/news-room/fact-sheets/detail/dementia). Dementia is characterized by

F. Zhao · M. Tomita
School of Public Health and Health Professions, University at Buffalo, Buffalo, NY 14214, USA
e-mail: fzhao6@buffalo.edu

M. Tomita
e-mail: machikot@buffalo.edu

A. Dutta (✉)
University of Birmingham, Birmingham, UK
e-mail: a.dutta.1@bham.ac.uk

P. K. Ray et al. (eds.), *Technology Innovation for Sustainable Development of Healthcare and Disaster Management*, Disaster Risk Reduction,
https://doi.org/10.1007/978-981-97-2049-1_16

a decline in cognitive functioning beyond what is expected due to normal aging. It impacts various aspects of a person's mental abilities, including memory, comprehension, calculation, language, and judgment. This cognitive decline may also be accompanied by emotional and behavioral changes. There are several different types of dementia, as defined by the Royal College of Psychiatrists. These subtypes include Alzheimer's disease (AD), vascular dementia, dementia with Lewy bodies, Parkinson's disease (PD) dementia, and frontotemporal dementia. AD is the most prevalent form, accounting for 60% of all dementia cases. It typically involves a gradual onset of memory impairments, with tau deposits in the brain affecting neurotransmission. Vascular dementia has two common subtypes: stroke-related vascular dementia and subcortical dementia, which results from disrupted blood supply to the lower regions of the brain due to blockages in smaller blood vessels. Chronic conditions like hypertension and hyperlipidemia heighten the risk of vascular dementia, and it often presents with memory loss, language difficulties, and other physical symptoms. Dementia with Lewy bodies is characterized by the buildup of Lewy body deposits in the brain, leading to memory problems, confusion, sleep disturbances, visual hallucinations, and Parkinsonism features. Frontotemporal dementia typically results from damage to the frontal areas of the brain and is associated with personality changes, behavioral alterations, and language difficulties. Addressing AD/ADRD is a paramount and pressing task in the twenty-first century. While the quest for a cure or effective treatment is crucial, it's not the sole solution to combat this epidemic.

To align the longevity of brain function with overall lifespan, we need to expand our endeavors by integrating prevention, behavioral health, and risk-reduction strategies into individuals' lives. This involves incorporating a focus on brain health into both the cultural fabric and the healthcare system. Here, the advancements in big data neuroimaging, cloud computing, virtual reality, and the Internet of Things have established technological infrastructure for the utilization of digital twin technology (Sun et al. 2023) in home-based monitoring of vascular and cognitive health. A digital twin is a virtual portrayal of a physical entity, individual, or procedure, situated within a digital replica of its surroundings. Then, organoids represent in vitro models of tissue, constructed from cells, and mirror numerous facets of the intricate structure and functionality found in their in vivo tissue counterparts (Zhao et al. 2021, 2022). These structures can be disassembled and examined to conduct essential mechanistic "phase zero" studies (Dutta et al. 2020) related to disease development, regeneration, and repair in human tissues. Additionally, organoids find applications in diagnostics, disease modeling, drug discovery, and personalized medicine. Therefore, in vitro organoid models of human organs can be combined with digital twin technology, called BioDigital Twin model, that can equip clinicians and caregivers with an enhanced framework to explore the onset and progression of Type 2 diabetes-related multi-organ vascular dysfunction in finer detail, enabling more accurate diagnoses and timely treatment approaches.

In this chapter, we first present an innovative BioDigital Twin modeling approach for the neurovascular system in Type 2 diabetes. Here, portable neuroimaging with operational modal analysis is presented as a novel approach to modeling the neurovascular system for personalized interventions in Type 2 diabetes. Then, we present preliminary results on portable neuroimaging and operational modal analysis of the neurovascular system affected by 2-month mild to moderate exercise intervention at home in sedentary Type 2 diabetes patients (Zhao et al. 2023).

2 BioDigital Twin Modeling for Personalized Interventions

Diabetes is presently categorized into Type 1 and Type 2 diabetes. Ahlqvist et al. (2018) utilized six variables (glutamate decarboxylase antibodies, age at diagnosis, body mass index, HbA1c, and homoeostatic model assessment 2 estimates of β-cell function and insulin resistance) in relation to subsequent data from patient records regarding the onset of complications and the prescription of medication. Ahlqvist et al. (2018) pinpointed five consistent clusters among patients with diabetes, each exhibiting distinct patient characteristics and varying risks of diabetic complications. Furthermore, to substantiate the clustering, genetic associations within the clusters deviated from those observed in conventional Type 2 diabetes. Then, Xue et al. (2023) showed that in contrast to cluster 1, distinguished by an early onset of Type 2 diabetes, patients in cluster 5, marked by impaired renal function, exhibited the highest susceptibility to cardiovascular disease (CVD) events. Following closely were individuals in cluster 4, characterized by poor glycemic control, and those in cluster 3, characterized by severe obesity. Cluster 2, identified by a late onset of Type 2 diabetes, demonstrated higher, though not statistically significant, risks of CVD compared to cluster 1, except for stroke. Here, arterial stiffness and diabetes represented significant health concerns, and their reported association underscores the need for a comprehensive understanding to facilitate effective disease prevention and management (Zheng et al. 2020). Zheng et al. (2020) conducted a large cohort study marking the first exploration of the bidirectional relationship between these two issues. The temporal analysis conducted in this study (Zheng et al. 2020) indicated that changes in arterial stiffness precede alterations in blood glucose levels, rather than the reverse. This discovery contributes to the appreciation of the role of healthy vascular aging in preventing chronic diseases. Subsequent studies are necessary to validate these findings in diverse populations and delve into the underlying biological and pathological pathways for biomarkers possibly starting with metabolic syndrome (Stehouwer et al. 2008).

It has been proposed that these "cardiometabolic factors", well-recognized for their impact on vascular health, might also exert an influence on brain health (Zhao et al. 2023). Consequently, technological innovations such as functional near-infrared spectroscopy (fNIRS), electroencephalogram (EEG), modal analysis to extract resonant frequencies, and damping, from the vascular signal, could potentially provide a biomarker of early vascular dysfunction including arterial stiffening in metabolic syndrome (Stehouwer et al. 2008). Here, our current understanding is of the connection between insulin resistance and heightened vascular stiffness (Hill et al. 2021).

Moreover, some individuals have coined terms like "Type 3 diabetes (T3D)" to describe AD where insulin deficiency is central to cognitive decline. With recent studies offering continued evidence linking T3D to AD, technological innovations are crucial to elucidate their relationship where brain energy failure may be the key in dementia syndromes (Yassine et al. 2022). Then, scientific validation is based on the shared characteristics that both involve the mishandling of amyloid-β (Aβ) precursor protein toxicity and the impaired clearance of Aβ, often attributed to compromised insulin signaling. Here, insulin resistance plays a pivotal role in the disruption of bioenergetics and the progression toward AD; however, it remains to be determined whether the restoration of brain energy pathways can effectively reverse synaptic loss and mitigate cognitive decline (Yassine et al. 2022).

While animal models are valuable, human-induced pluripotent stem cell (iPSC)-derived neurovascular organoid models offer a subject-specific understanding in Type 2 diabetes for personalized medicine. Previous work has confirmed that organoids derived from human stem cells accurately mimic the structure and function of human blood vessels in diabetic vasculopathy (Wimmer et al. 2019). By incorporating an affordable 3D printed bioreactor (Khan et al. 2021) that includes imaging wells suitable for the long-term culture of organoids, as well as microfluidic channels for medium flow and preheating, continuous monitoring alongside simultaneous electrophysiology and visible/near-infrared spectroscopy can be performed for "phase zero" studies (Dutta et al. 2020). These technological innovations lead to individualized BioDigital Twin modeling as shown in Fig. 1. Here, the aim of "phase zero" studies is to optimize the interventions in vitro where real-time monitoring of glucose and oxygen exchange can be achieved using a dual polymer sensor embedded in the extracellular matrix (Zhang et al. 2013). For example, supplementing basal insulin to an oral glucose-lowering regimen can be tested in vitro using neurovascular organoids (Salmon et al. 2022) to model the effects on stiffness and electrophysiology for computational model-based investigations (see Dutta et al. 2022 for a different disease application). Figure 1 shows a BioDigital Twin approach that involves collecting and analyzing vast amounts of biological data, such as genomic information, physiological parameters including (*Continuous Glucose Monitoring—NIDDK*, n.d.), and behavioral activity from wearables. By merging this biological data with advanced digital modeling techniques, a BioDigital Twin provides a dynamic and comprehensive representation of an individual's health and well-being. This technology holds great promise in health care, enabling precision medicine, early disease detection, and personalized treatment strategies. BioDigital Twins have the potential to revolutionize how we understand and manage health by offering a real-time, holistic view of an individual's biological processes, fostering proactive and tailored interventions for improved health outcomes.

Therapeutic approaches including caloric restriction (Johnson et al. 2016) and routine physical activity (Bird and Hawley 2017) may hold the potential (Zhao et al. 2019) in Type 2 diabetes with Mild Cognitive Impairment (MCI) that represents a less severe condition, potentially slowing down its progressive nature or even preventing future complications. Moreover, the relationship between vascular and mitochondrial

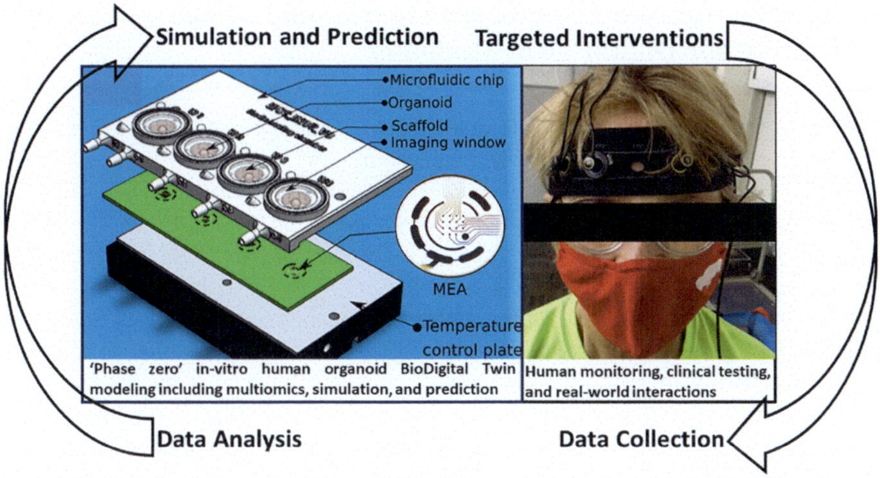

Fig. 1 BioDigital Twin is a cutting-edge concept that integrates biological information with digital technology to create a personalized, virtual replica of an individual's physiological system (organoid bioreactor shown from ISMO Bio-Photonics Pvt. Ltd., India)

dysfunction and the central pathological aspect of AD, which is neuronal dysfunction, is intimately intertwined (Ashleigh et al. 2023). Here, mitochondrial dysfunction could expedite the advancement of insulin resistance and subsequent organ dysfunction by elevating the generation of reactive oxygen species (Szendroedi et al. 2012). Therefore, exploring mitochondria and related proteins might offer potential avenues for innovative approaches in crafting therapies that could alter the course of the disease. Mitochondrial Cascade Hypothesis posits that the origins of AD are rooted in mitochondrial dysfunction. Here, technological innovations including imaging mitochondrial cytochrome-c-oxidase activity (M.d and Mann 2018) are necessary for developing biomarkers for point-of-care screening and monitoring since individuals with MCI are typically functional and do not experience the same level of memory impairment seen in dementia. The diagnostic criteria for MCI, initially proposed by Peterson et al. (1999), include domains such as memory complaints, normal daily activities, regular cognitive functioning, abnormal memory based on the patient's age, and the absence of clinical signs of dementia. Therefore, digital twin modeling is crucial with no single causative factor for MCI, and individuals with Type 2 diabetes and MCI should be monitored, as it could be an early indicator of AD. Future "phase zero" studies can apply photobiomodulation targeting Mitochondrial Oxidative Phosphorylation System (https://neuromodec.org/nyc-neuromodulation-online-2020/P18.html) in conjunction with non-invasive electrical stimulation to modulate the excitatory/inhibitory (E/I) ratio and neuroenergetics in neurovascular organoid (Salmon et al. 2022).

3 Portable Neuroimaging-based Operational Modal Analysis for Digital Twin Modeling

Neuroimaging can play a crucial role in the differentiation of subtypes of dementia. Studies have demonstrated the value of neuroimaging techniques such as structural MRI and F-18 fluorodeoxyglucose-positron emission tomography (FDG-PET) in early detection and prediction of disease progression (Smailagic et al. 2018). Then, MRI (Inglese et al. 2022) can reveal a biomarker called "Alzheimer's Predictive Vector" including gray matter loss in the entorhinal and hippocampal areas in early Alzheimer's disease, coupled with hypometabolism and hypoperfusion in certain brain regions (Ferrando and Damian 2021). The presence of hypometabolism and hypoperfusion in specific areas, along with entorhinal atrophy, may predict the conversion from MCI to Alzheimer's disease (Matsuda 2007). Computer-based approaches, such as automatic brain segmentation methods combined with evolutionary algorithms, have been employed to characterize and differentiate individuals with early-stage MCI from those who are cognitively normal; however, these investigations come with significant costs, making it crucial to establish and showcase its precision while standardizing the procedure (Smailagic et al. 2015).

Affordable Functional Near-Infrared Spectroscopy (fNIRS) has gained attention as an alternative to functional MRI for neuroimaging. It offers advantages such as non-invasiveness, cost-effectiveness, and portability (Sakai 2022). fNIRS measures hemodynamic responses to neural activity based on the differential absorption properties of biological chromophores like hemoglobin. Unlike MRI, fNIRS does not require the individual to remain completely still during scans and can be conducted in naturalistic environments. It is particularly useful for examining brain changes during various neurocognitive tasks and physical activities. While fNIRS has limitations, such as primarily assessing the frontal areas of the brain, it is valuable for patients with metallic implants. So, fNIRS has found application in psychiatry research (Li et al. 2023), including the study of psychiatric disorders like schizophrenia, depression, bipolar disorder, panic disorder, obsessive-compulsive disorder, autism spectrum disorder, and attention deficit hyperactivity disorder.

A wearable portable neuroimaging approach (Zhao et al. 2023)—see Fig. 2—can measure "task-related activations", and lends well to feedback for self-management through lifestyle modifications as well as interventions using neuromodulation technologies, including photobiomodulation (Bhattacharya and Dutta 2019), and artificial intelligence (Rahul et al. 2021). Specifically, the significance of the dynamical modeling of the neuroimaging signals to advance the development of digital twin models of the neurovascular system needs further investigation. Here, the stiffness of arteries and its correlation with the clustering of risk factors for metabolic syndrome is known (Lopes-Vicente et al. 2017). Then, various physiologically relevant frequency bands have already been identified in the literature, such as those linked to cardiac and respiratory functions, and smooth muscle autonomic activity in peripheral blood circulation (Stefanovska et al. 1999). In patients with Type 2 diabetes, the effects of chronic hyperglycemia can lead to small fiber neuropathy (Gibbons and Freeman

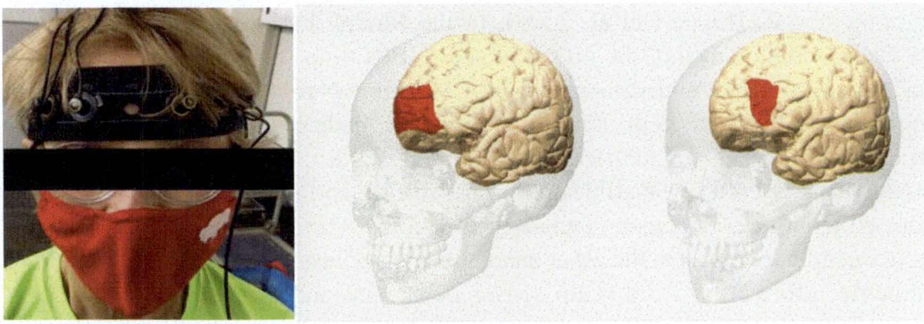

Fig. 2 Prefrontal functional near-infrared spectroscopy using a headband sensor targeting the middle frontal gyrus (orbital part) and the superior frontal gyrus (dorsolateral part) bilaterally

2015) besides interacting with arterial stiffening and hypertension (Tian et al. 2022) affecting cerebral blood flow oscillations that can be detected with fNIRS (Zhao et al. 2023). The low oscillatory frequency at capillaries, associated with the Fahræus–Lindqvist effect, may play a vital role in microvascular networks, given its nonlinear dependence on apparent blood viscosity in relation to hematocrit and vessel diameter (Geddes et al. 2010). Here, the mechanistic understanding of microvascular and macrovascular flow in metabolic syndrome (Carter et al. 2023), which is COX-mediated (Niwa et al. 2001), is necessary. For example, physical stimulation of blood vessels within the brain tissue during microvascular flow can lead to very low frequency (0.01–0.3 Hz) astrocyte-regulated stretch-mediated TRPV4-COX-1 feedback vessel oscillations (Haidey et al. 2021). Importantly, young adults with metabolic syndrome may show a decline in COX-mediated vasodilation specifically in the anterior circulation, while the posterior circulation remains unaffected (Carter et al. 2023). Moreover, a large cohort study has linked diabetes onset at a younger age to an increased risk of subsequent dementia (Barbiellini Amidei et al. 2021) which needs mechanistic analysis.

Operational Modal Analysis is a technique used in mechanical engineering to determine the dynamic properties of a system while it is subjected to external loads or naturally occurring excitations (Neu et al. 2017). Operational Modal Analysis aims to identify the system's natural frequencies, mode shapes, and damping ratios without the need for controlled laboratory tests. This method typically involves measuring the system's response to real-world environmental conditions and the collected data is then used to extract modal parameters, providing insights into the system's dynamic behavior, which can be essential for structural health monitoring, damage detection, and design validation. In our novel application of Operational Modal Analysis in human vascular health monitoring, we investigated neurovascular system response to cognitive task-related perturbations (Zhao et al. 2023). The fNIRS raw data was preprocessed in HOMER3 for transforming data from raw intensity to optical density, assessing the data quality, converting good quality channel data from optical density to hemoglobin levels, and eliminating the influence of skin blood flow artifacts from the brain signal (Huppert et al. 2009). In the Modal Toolkit (https://code.vt.edu/vibes-lab/modal-analysis), the preprocessed fNIRS time series data was organized in a 2D matrix format with dimensions N × M, where N represents the count of time samples, and M represents the number of channels. Scalp sensor/channel coordinates can be defined as N × 3 matrices in the Modal Toolkit. In the Modal Toolkit, the *timedat* class offered a versatile solution for storing and managing time series data collected from a set of sensors including functional near-infrared spectroscopy and electroencephalogram. On the other hand, the *geoviz* class serves as a tool for storing geometric information from scalp sensor placement and visualizing multi-modal sensor data in a three-dimensional context. Covariance-Based Stochastic Subspace Identification (SSI COV) stands as a widely favored technique in Operational Modal Analysis and serves as the cornerstone of the Modal Toolkit. The third custom class within the Modal Toolkit, *modeObj*, was designed for efficiently storing modal data in a compact format. SSI COV offered an uncertainty quantification feature, albeit with the trade-off of longer computation time.

Interpreting SSI stabilization diagrams can be challenging, so automated methods are crucial. Therefore, a multi-stage clustering was applied to identifying physiological clusters using the Modal Toolkit (Zhao et al. 2023). This is achieved by grouping similar modes into clusters, retaining only the most significant clusters, and ultimately selecting a representative mode from within those clusters. Figure 3 presents an illustrative example where the top panel shows the SSI stabilization diagram with the selected modes shown with black circles in the panel below (titled Selected Modes). The multiple stages of the clustering process are shown where Stage 1 effectively identifies and eliminates spurious or systemic modes using 2-means clustering based on user given criteria. For example, after 2-means clustering, the cluster with the most short separation channel modes is removed (short separation fNIRS channels are reference channels for skin blood flow artifacts). Then, hierarchical clustering is employed to categorize modes into "mode clusters" by considering the proximity of their poles (using non-dimensional distance abs(a1 − a2)/max(abs(a1), abs(a2)). Modal distance estimated by MAC, that is the modal assurance criterion, represents the measure of the correlation between two modal vectors. The total number of clusters following hierarchical clustering based on Pole distance and MAC distance is determined based on the criterion that their centroids are separated by a distance greater than three times the standard deviation. Figure 2 panels show these stages of the hierarchical clustering, and the selection of the final mode clusters relies on the assumption that the sets with the highest number of modes correspond to the physiological modes (small clusters are removed). The final mode clusters of prefrontal fNIRS measured task-evoked oscillations exhibited improvement post-intervention in the sedentary Type 2 diabetes Intervention group.

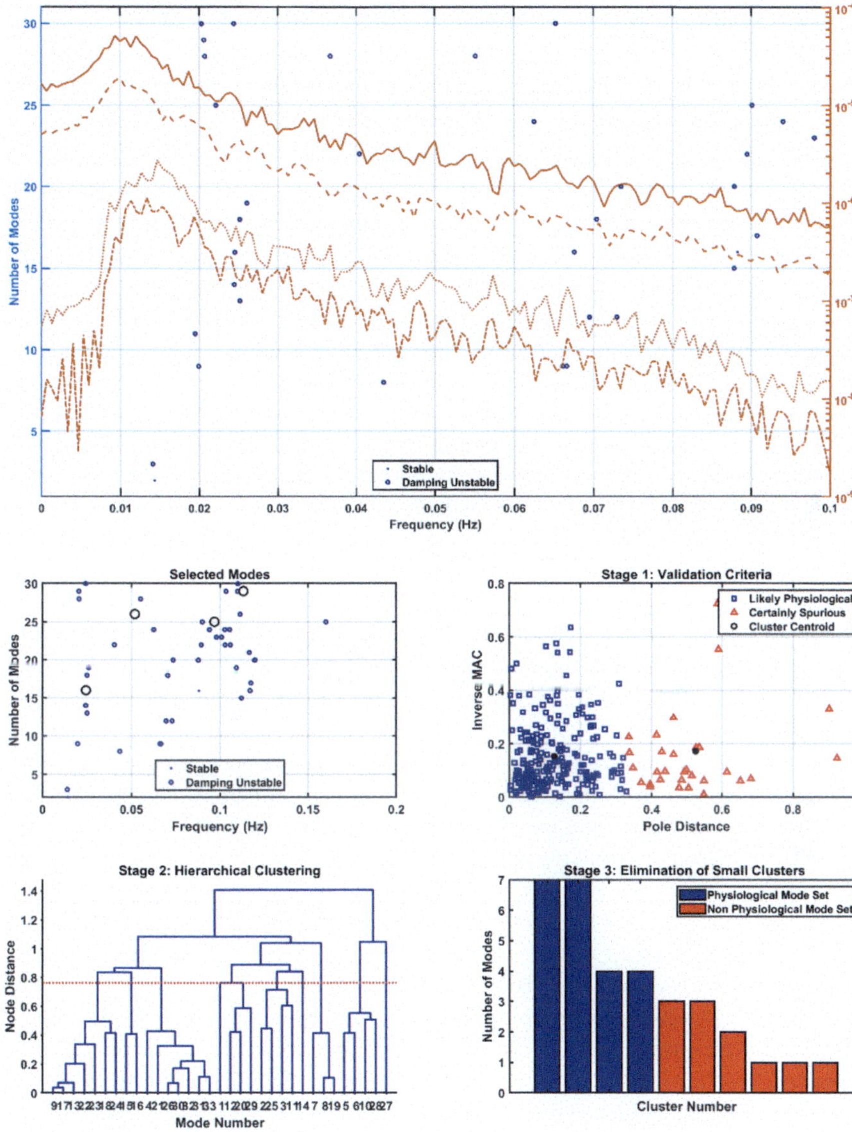

Fig. 3 Illustrative operational modal analysis of fNIRS time series data with maximum 30 modes for the power spectrum <0.1Hz shown in the stabilization diagram shown. A multi-stage clustering was applied for identifying physiological clusters using the Modal Toolkit (https://code.vt.edu/vibes-lab/modal-analysis)

3.1 Operational Modal Analysis of Home-Based Exercise Effects

To address dementia risk linked to vascular and lifestyle-related risk factors that can be modified, an extensive, extended-duration, randomized controlled trial indicated that a multidomain intervention of diet, exercise, cognitive training, and vascular risk monitoring intervention has the potential to enhance or preserve cognitive performance in older individuals at risk from the general population (Ngandu et al. 2015). In principal accordance, to address cognitive impairment in individuals with Type 2 diabetes, a comprehensive aerobic and resistance exercise regimen was designed in accordance with the recommendations of the American College of Sports Medicine (Zhao et al. 2023). Notably, motor learning performance and cortical excitability changes demonstrated significant improvement during moderate-intensity aerobic exercise when compared to the other conditions (Kuo et al. 2023). Our clinical trial on 2-month moderate-intensity exercise intervention was prospectively registered at ClinicalTrials.gov (NCT04626453 and NCT04812288), involving two groups: an Intervention group comprising older sedentary adults with Type 2 diabetes and a Control group consisting of healthy older adults, encompassing both active and sedentary individuals. Importantly, moderate-intensity exercise intervention was individualized using muscle functional near-infrared spectroscopy during baseline evaluation (Zhao et al. 2019) which is crucial to avoid metaboreflex activation that can affect cerebral blood flow distribution (Lam et al. 2019; Pinna et al. 2021; Ogoh et al. 2019).

Remarkably, the completion rate for the 2-month exercise program was high, with participants averaging 89.14% of the exercise sessions completed. This highlighted the practicality, feasibility, and excellent tolerance of the individualized moderate-intensity home-based program during the COVID-19 pandemic. Furthermore, it was a safe program, requiring minimal equipment and no supervision so sustainable in a low-resource setting. Our study introduced measures (Zhao et al. 2019) that demonstrated significant ($p < 0.05$) decreases in muscle oxygen saturation (SmO_2) during the bilateral heel rise task (BHR) and the 6-minute walk task (6MWT) in the Type 2 diabetes Intervention group at the post-intervention follow-up compared to the pre-intervention baseline. Additionally, post-intervention changes from the pre-intervention baseline showed statistically significant ($p < 0.05$, $q < 0.05$) effects on prefrontal activation, including both oxyhemoglobin and deoxyhemoglobin levels at the right superior frontal gyrus, dorsolateral, during the Mini-Cog task (https://mini-cog.com/scoring-the-mini-cog/). Operational modal analysis of prefrontal fNIRS data from the headband sensors (see Fig. 2) (Dutta et al. 2021) provided further insights into the effects of the 2-month exercise intervention on very-low-frequency oscillations (<0.05 Hz) during the Mini-Cog task.

We applied an automated multi-stage clustering approach, utilizing the Modal Toolkit available at https://code.vt.edu/vibes-lab/modal-analysis (instead of the peak-picking method in 'modalfit' from Matlab), to identify physiologically rele-vant clusters from fNIRS sensor data during cognitive tasks (Zhao et al. 2023). A robust stochastic subspace identification (SSI) algorithm, specifically reference-based SSI-Cov algorithm (Peeters and De Roeck 1999), was used to get neurovascular system and output matrices (Van Overschee and De Moor 1996) from noisy hemodynamics data. Here, it is crucial to select reference channels as projection channels that capture a wide range of dynamic data e.g. the longest separation source-detector channels in a multi-distance fNIRS measurement from the vascular tree – see top left panel of Fig. 4. Note that if certain modes are not detectable by the reference channels, the algorithm will not be able to estimate them in the case of reference-based SSI-Cov algorithm (Peeters and De Roeck 1999). Top right panel of Fig. 4 shows the Stable and Damping Unstable modes <0.2 Hz found from the stabilization diagram of multi-distance frequency domain (FD) NIRS (Imagent, ISS, USA) oxyhaemoglobin concentration measurements in a young and healthy subject during the resting state which were found comparable to the natural frequencies below 0.2 Hz found from physiological modelling (very low frequency: 0.04 Hz, low frequency: 0.06 Hz and 0.16 Hz) of Pathway 1 (Arora et al. 2021). Specifically, three resting state FDNIRS oxyhaemoglobin concentration oscillatory clusters were delineated in the bottom left panel of Fig. 4 corresponding to very low frequency, low frequency, and the respiration related cluster ~0.2 Hz in young and healthy subject for bench testing the modal analysis before applying this novel method in the T2DM patient studies (ClinicalTrials.gov NCT04626453 and NCT04812288). The mean of the complex modes for frequencies from 0.019 Hz, 0.022 Hz, 0.028 Hz, 0.053 Hz, 0.069 Hz, 0.089 Hz, 0.13 Hz, 0.19 Hz, 0.20 Hz are $0.1381 + 0.2543i$, $0.4181 + 0.0950i$, $0.6087 - 0.0313i$, $0.4682 + 0.1152i$, $0.4661 - 0.2487i$, $0.4480 + 0.0912i$, $0.0145 - 0.1356i$, $0.0600 + 0.0122i$, $0.1434 + 0.2187i$, respectively. Interestingly, when we plotted the complex modes using vector compass plot in the bottom right panel of Fig. 4, (respiratory) mode 0.20Hz (':k') aligned with the very low frequency capillary mode 0.019Hz ('r') while low frequency smooth muscle arteriole modes (from penetrating arterioles?) 0.053Hz ('b') and 0.089Hz (':b') aligned with the very low frequency capillary modes 0.022Hz ('–r') and 0.028Hz (':r'). Then, the (other respiratory) mode 0.19Hz ('−k') aligned with the low frequency smooth muscle arteriole (Mayer wave in pial vessels?) mode 0.13 Hz ('k'). Then, the low frequency mode, 0.069 Hz ('−b') was found separate. Investigation of physiological relevance of these vector alignment of complex modes will need grey-box modelling (Arora et al. 2021) and patient studies on various cerebrovascular dysfunction (e.g., in T2DM, small vessel disease, etc.); however, bulky FDNIRS system (Imagent, ISS, USA) was found inconvenient to use in the T2DM patient studies (NCT04626453 and NCT04812288) so we used Octamon+ (Artinis, Netherlands) where standard SSI was used with all measurement channels used as projection channels (Zhao et al. 2023).

In the moderate-intensity exercise intervention studies in T2DM (NCT04626453 and NCT04812288), two separate cluster of hemodynamic oscillations <0.05 Hz in the Sedentary Healthy Control Group and multiple (>2) cluster of oscillations <0.05 Hz in the Sedentary type 2 diabetes Intervention Group were found at baseline. The clustering of hemodynamic oscillations improved with more fused oscillations post-intervention that was comparable to the Active healthy Control Group (that has one cluster of oscillations <0.1 Hz) (Zhao et al. 2023). A positive impact

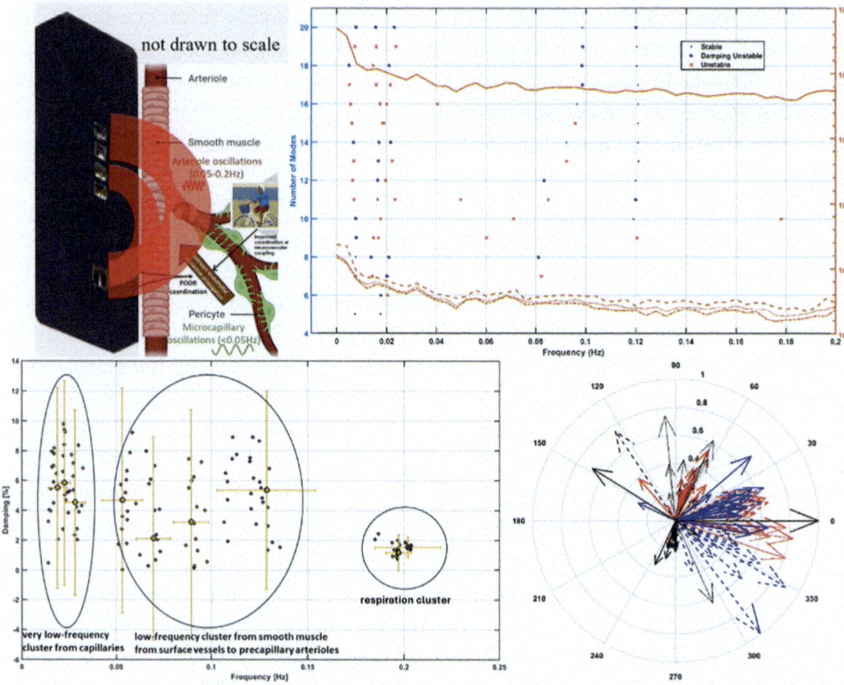

Fig. 4 Top Left panel: multi-distance FDNIRS oxyhemoglobin concentration measurement from the vascular tree with longest separation source-detector channels used as reference channel for reference-based SSI-Cov algorithm to determine physiologically relevant mode clusters. Top Right panel: Stable and Damping Unstable modes (e.g., Stable 0.0073 Hz and 0.017 Hz < 0.05 Hz and 0.092 Hz and 0.121 Hz > 0.05 Hz) that are <0.2 Hz were found from stabilization diagram for the number of modes from 5 to 20. The stabilization diagram was followed by a multi-stage clustering to identify physiological clusters (for frequency clusters from low to high: 0.019 Hz, 0.022 Hz, 0.028 Hz, 0.053 Hz, 0.069 Hz, 0.089 Hz, 0.13 Hz, 0.19 Hz, 0.20 Hz) in a young and healthy subject at resting state shown in the Bottom Left panel. The complex modes are shown using a vector compass plot in the Bottom Right panel for frequencies: 0.019Hz ('r'), 0.022Hz ('−r'), 0.028Hz ('r'), 0.053Hz ('b'), 0.069Hz ('−b'), 0.089Hz (':b'), 0.13Hz ('k'), 0.19Hz ('−k'), 0.20Hz (':k'))

of the moderate-intensity exercise intervention is postulated on the coordination of microcapillary dilatations (<0.05 Hz) with arteriole dilatations (0.05–0.2 Hz) which may be attributed to potential upstream endothelial vasoactive signaling to vascular smooth muscle cells with ~0.1 Hz indicating synchroni-zation of the intermittent release of calcium within vascular smooth muscle cells (Aalkjær et al. 2011). Improved coordination of microcapillary dilatations (<0.05 Hz) with arteriole dilatations (0.05–0.2 Hz) is postulated to improve micro-vascular perfusion (aerobic capacity) that possible effected significantly ($p < 0.01$) improved 6MWT distance in type 2 diabetes Intervention group at the post-intervention follow-up compared to the pre-intervention baseline. In summary, modal analysis of fNIRS signals offered valuable insights into the coordination of microcapillary dilatations at frequencies below 0.05 Hz with arteriole dilations ranging from 0.05 to 0.2 Hz that might be associated with endothelial vasoactive signaling to vascular smooth muscle cells. This improved synchronization in the Sedentary type 2 diabetes Intervention Group could be indicative of coordinated intermittent release of calcium within vascular smooth muscle cells (~0.1 Hz). Therefore, low-cost portable near-infrared spectroscopy-based muscle and brain measures (Zhao et al. 2019, 2023) hold practical implications for therapists, enabling them to personalize exercise regimens and monitor the effects based on individual responses.

In fact, modal analysis and eigenvalue placement serve as robust techniques for designing controllers and observers for non-invasive brain stimulation in the realm of linear systems (Dutta 2015; Arora and Dutta 2022). Human-in-the-loop optimization (Arora and Dutta 2022) can be used to attain specific performance and stability objectives while leveraging the interplay between controllability and observability that can be challenging as discussed in prior work (Dutta 2015). Specifically, we draw upon the developments facilitating a modal decomposition (Proctor et al. 2016) that separates the system into distinct, decoupled controllable, and uncontrollable (observable and unobservable) subspaces, as expounded in Whalen et al. (2015). Here, inherent vascular vibrational modes, as defined by their eigenvalues and eigenvectors from modal analysis, encapsulate a distinct system response to an input from non-invasive brain stimulation. By scrutinizing these modes, one can gain insights into the neurovascular system's behavior and its potential for not only transcranial electrical stimulation (tES) as an adjuvant treatment, e.g. to target oscillatory modes of perivascular transport and paravascular clearance (van Veluw et al. 2020) but also for monitoring of intervention effects including physical and cognitive exercise interventions in home settings.

An affordable wearable device is necessary for sustainable cognitive health care in home settings leveraging a non-invasive brain stimulation approach with fNIRS-EEG joint imaging for closed-loop control of neuroenergetics (von Lühmann et al. 2017). A Hybrid Mobile, Modular, Multimodal Biosignal Acquisition (M3BA) architecture, as described in reference (von Lühmann et al. 2017), offers a compact and customizable wireless platform for simultaneous, high-precision acquisition of electrophysiological signals (EEG, ECG) and optical signals (fNIRS). It provides exceptional performance, including low EEG input-referred noise (1.39 μV at 500 samples per second), low NIRS noise equivalent power (NEP750 nm = 5.92 pWpp, NEP850 nm = 4.77 pWpp), full input linearity, lock-in amplification, driven right leg circuitry, modularity, Bluetooth connectivity, and low power consumption. Each module within this architecture can accommodate up to 6 electrophysiological and 6 optical measurement channels, along with a 3D accelerometer. It allows for flexible biopotential reference configurations, such as simultaneous EEG and ECG measurements with different references. Moreover, it can be seamlessly integrated into a Wireless Body Area/Sensor Network (WBAN/WBSN). Extensive performance testing and in vivo experiments including animal studies have confirmed the capabilities of the M3BA architecture that has been used in a human neonatal feasibility study in low-resource settings (Hagan et al. 2023). M3BA architecture is well-suited for custom applications involving the measurement of neurovascular coupling, vasomotor reactivity, and cerebral autoregulation under Non-Invasive Brain Stimulation (NIBS) including tES and photobiomodulation.

Our particular focus for NIBS has been on facilitating low-frequency hemodynamic oscillations ranging from 0.01 to 1 Hz (Arora and Dutta 2022). These low-frequency oscillations play a crucial role in small vessel function in the aging brain (Schroeter et al. 2004) and may exhibit dysfunction in individuals with Type 2 diabetes (Huang et al. 2021). During our study involving 19 elderly individuals aged 60 years and older with Type 2 diabetes and 38 age-matched control subjects (Zhao et al. 2022), we found a significant reduction in relative power within the 0.021–0.052 Hz frequency band during the Mini-Cog three-item recall test in elderly participants with Type 2 diabetes. In fact, low-frequency arteriolar oscillations also play a significant role in the removal of solutes from brain activity, so focusing on facilitating these inherent vasomotion patterns in individuals with Alzheimer's disease could present a promising and early therapeutic approach to prevent the buildup of Aβ in the brain (van Veluw et al. 2020). Spontaneous vasomotion typically exhibits low-frequency fluctuations, approximately around 0.1 Hz, that are impaired in Alzheimer's disease (Di Marco et al. 2015). Here, sensory stimulation-related increase in blood flow through vasodilation (Dutta et al. 2021) is understood to occur along penetrating arterioles to pial arterioles which have a similar pattern to spontaneous vasomotion (Munting et al. 2023). Therefore, the goal is to facilitate vasomotion in Type 2 diabetes using closed-loop non-invasive brain stimulation (von Lühmann et al. 2017; Arora and Dutta 2022). Future research needs to delve into the intricate relationship between brain glucose hypometabolism, a prominent feature of AD, and the role of mitochondrial dysfunction and oxidative stress (Dewanjee et al. 2022) since photobiomodulation (Bhattacharya and Dutta 2019) and electric field can synergistically effect beneficial neuroenergetic function. Then, improving vasomotion with sensory stimulation (Munting et al. 2023) may address low transport across the blood-brain barrier, including low glucose and insulin, and the shift from glycolysis to fatty acid oxidation, especially in astrocytes (Demarest et al. 2020). Here, the potential of tES in reducing blood glucose through an insulin-independent mechanism is tied to brain energy metabolism (Kistenmacher et al. 2017), and the impact of cognitive load (and Locus Coeruleus Norepinephrine response) on blood glucose sensitivity (Scholey et al. 2001) should be investigated in the future.

4 Conclusion

In conclusion, our chapter underscores the significant global impact of diabetes, contributing substantially to the rise in AD/ADRD disability-adjusted life years. The proposed digital twin model, leveraging advancements in big data neuroimaging, cloud computing, and the Internet of Things, emerges as an innovative approach for monitoring vascular brain health. Then, integration of human-induced pluripotent stem cell organoid models represents a significant advancement, providing a subject-specific understanding of Type 2 diabetes and its impact on blood vessels and neurovascular units. The incorporation of 3D printed bioreactors facilitates continuous monitoring, electrophysiology, and spectroscopy, enabling "phase zero" studies. Then, an innovative bioreactor approach, combining photobiomodulation and non-invasive electrical stimulation in neurovascular organoids, aims to optimize the therapeutic interventions including multi-modal stimulation to modulate neuroenergetics as an adjuvant treatment to multi-domain interventions for dementia prevention in Type 2 diabetes. Therefore, the BioDigital twin approach emerges as a powerful tool by integrating genomic information, physiological parameters, and behavioral patterns. This technology holds great promise in health care, offering a dynamic and comprehensive representation of an individual's health and well-being.

References

Aalkjær C, Boedtkjer D, Matchkov V (2011) Vasomotion—what is currently thought? Acta Physiol 202(3):253–269. https://doi.org/10.1111/j.1748-1716.2011.02320.x

Ahlqvist E, Storm P, Käräjämäki A, Martinell M, Dorkhan M, Carlsson A, Vikman P, Prasad RB, Aly DM, Almgren P, Wessman Y, Shaat N, Spégel P, Mulder H, Lindholm E, Melander O, Hansson O, Malmqvist U, Lernmark Å et al (2018) Novel subgroups of adult-onset diabetes and their association with outcomes: a data-driven cluster analysis of six variables. Lancet Diabet Endocrinol 6(5):361–369. https://doi.org/10.1016/S2213-8587(18)30051-2

Amidei CB, Fayosse A, Dumurgier J, Machado-Fragua MD, Tabak AG, Van Sloten T, Kivimäki M, Dugravot A, Sabia S, Singh-Manoux A (2021) Association between age at diabetes onset and subsequent risk of dementia. JAMA. Apr 27;325(16):1640-1649. PMID: 33904867; PMCID: PMC8080220. https://doi.org/10.1001/jama.2021.4001

Arora Y, Dutta A (2022) Human-in-the-loop optimization of transcranial electrical stimulation at the point of care: a computational perspective. Brain Sci 12(10), Article 10. https://doi.org/10.3390/brainsci12101294

Arora Y, Walia P, Hayashibe M, Muthalib M, Chowdhury SR, Perrey S, Dutta A (2021) Grey-box modeling and hypothesis testing of functional near-infrared spectroscopy-based cerebrovascular reactivity to anodal high-definition tDCS in healthy humans. PLOS Comput Biol 17(10):e1009386. https://doi.org/10.1371/journal.pcbi.1009386

Ashleigh T, Swerdlow RH, Beal MF (2023) The role of mitochondrial dysfunction in Alzheimer's disease pathogenesis. Alzheimer's Dementia 19(1):333–342. https://doi.org/10.1002/alz.12683

Bhattacharya M, Dutta A (2019) Computational modeling of the photon transport, tissue heating, and Cytochrome C oxidase absorption during transcranial near-infrared stimulation. Brain Sci 9(8):179. https://doi.org/10.3390/brainsci9080179

Bird SR, Hawley JA (2017) Update on the effects of physical activity on insulin sensitivity in humans. BMJ Open Sport Exer Med 2(1):e000143. https://doi.org/10.1136/bmjsem-2016-000143

Carter KJ, Ward AT, Kellawan JM, Harrell JW, Peltonen GL, Roberts GS, Al-Subu A, Hagen SA, Serlin RC, Eldridge MW, Wieben O, Schrage WG (2023) Reduced basal macrovascular and microvascular cerebral blood flow in young adults with metabolic syndrome: potential mechanisms. J Appl Physiol (Bethesda, Md.: 1985) 135(1):94–108. https://doi.org/10.1152/japplphysiol.00688.2022

Continuous Glucose Monitoring—NIDDK (n.d.) National Institute of Diabetes and Digestive and Kidney Diseases. https://www.niddk.nih.gov/health-information/diabetes/overview/managing-diabetes/continuous-glucose-monitoring. Accessed 19 Nov 2023

Demarest TG, Varma VR, Estrada D, Babbar M, Basu S, Mahajan UV, Moaddel R, Croteau DL, Thambisetty M, Mattson MP, Bohr VA (2020) Biological sex and DNA repair deficiency drive Alzheimer's disease via systemic metabolic remodeling and brain mitochondrial dysfunction. Acta Neuropathol 140(1):25–47. https://doi.org/10.1007/s00401-020-02152-8

Dewanjee S, Chakraborty P, Bhattacharya H, Chacko L, Singh B, Chaudhary A, Javvaji K, Pradhan SR, Vallamkondu J, Dey A, Kalra RS, Jha NK, Jha SK, Reddy PH, Kandimalla R (2022) Altered glucose metabolism in Alzheimer's disease: role of mitochondrial dysfunction and oxidative stress. Free Radical Biol Med 193(Pt 1):134–157. https://doi.org/10.1016/j.freeradbiomed.2022.09.032

Di Marco LY, Farkas E, Martin C, Venneri A, Frangi AF (2015) Is vasomotion in cerebral arteries impaired in Alzheimer's disease? J Alzheimer's Dis 46(1):35–53. https://doi.org/10.3233/JAD-142976

Dutta A (2015) Bidirectional interactions between neuronal and hemodynamic responses to transcranial direct current stimulation (tDCS): challenges for brain-state dependent tDCS. Front Syst Neurosci 9:107. https://doi.org/10.3389/fnsys.2015.00107

Dutta A, Karanth SS, Bhattacharya M, Liput M, Augustyniak J, Cheung M, Stachowiak EK, Stachowiak MK (2020) A proof of concept 'phase zero' study of neurodevelopment using brain organoid models with Vis/near-infrared spectroscopy and electrophysiology. Sci Rep 10(1), Article 1. https://doi.org/10.1038/s41598-020-77929-8

Dutta A, Zhao F, Cheung M, Das A, Tomita M, Chatterjee K (2021) Cerebral and muscle near-infrared spectroscopy during lower-limb muscle activity—volitional and neuromuscular electrical stimulation. In: 2021 43rd annual international conference of the IEEE Engineering in Medicine & Biology Society (EMBC), pp 6577–6580. https://doi.org/10.1109/EMBC46164.2021.9629721

Dutta A, Biber J, Bae Y, Augustyniak J, Liput M, Stachowiak E, Stachowiak MK (2022) Model-based investigation of elasticity and spectral exponent from atomic force microscopy and electrophysiology in normal versus Schizophrenia human cerebral organoids. In: Annual international conference of the IEEE Engineering in Medicine and Biology Society, 2022, pp 1585–1589.https://doi.org/10.1109/EMBC48229.2022.9871376

Ferrando R, Damian A (2021) Brain SPECT as a biomarker of neurodegeneration in dementia in the era of molecular imaging: still a valid option? Front Neurol 12. https://www.frontiersin.org/articles/10.3389/fneur.2021.629442

Geddes JB, Carr RT, Wu F, Lao Y, Maher M (2010) Blood flow in microvascular networks: a study in nonlinear biology. Chaos 20(4):045123. https://doi.org/10.1063/1.3530122

Gibbons CH, Freeman R (2015) Treatment-induced neuropathy of diabetes: an acute, iatrogenic complication of diabetes. Brain 138(1):43–52. https://doi.org/10.1093/brain/awu307

Hagan B, Mujumdar R, Sahoo JP, Das A, Dutta A (2023) Technical feasibility of multimodal imaging in neonatal hypoxic-ischemic encephalopathy from an ovine model to a human case series. Front Pediatr 11. https://www.frontiersin.org/articles/10.3389/fped.2023.1072663

Haidey JN, Peringod G, Institoris A, Gorzo KA, Nicola W, Vandal M, Ito K, Liu S, Fielding C, Visser F, Nguyen MD, Gordon GR (2021) Astrocytes regulate ultra-slow arteriole oscillations via stretch-mediated TRPV4-COX-1 feedback. Cell Rep 36(5):109405. https://doi.org/10.1016/j.celrep.2021.109405

Hill MA, Yang Y, Zhang L, Sun Z, Jia G, Parrish AR, Sowers JR (2021) Insulin resistance, cardiovascular stiffening and cardiovascular disease. Metabol Clin Exp 119:154766. https://doi.org/10.1016/j.metabol.2021.154766

Huang X, Wen Z, Qi C-X, Tong Y, Shen Y (2021) Dynamic changes of amplitude of low-frequency fluctuations in patients with diabetic retinopathy. Front Neurol 12:611702. https://doi.org/10.3389/fneur.2021.611702

Huppert TJ, Diamond SG, Franceschini MA, Boas DA (2009) HomER: a review of time-series analysis methods for near-infrared spectroscopy of the brain. Appl Opt 48(10):D280–D298

Inglese M, Patel N, Linton-Reid K, Loreto F, Win Z, Perry RJ, Carswell C, Grech-Sollars M, Crum WR, Lu H, Malhotra PA, Aboagye EO (2022) A predictive model using the mesoscopic architecture of the living brain to detect Alzheimer's disease. Commun Med 2(1), Article 1. https://doi.org/10.1038/s43856-022-00133-4

Johnson ML, Distelmaier K, Lanza IR, Irving BA, Robinson MM, Konopka AR, Shulman GI, Nair KS (2016) Mechanism by which caloric restriction improves insulin sensitivity in sedentary obese adults. Diabetes 65(1):74–84. https://doi.org/10.2337/db15-0675

Khan I, Prabhakar A, Delepine C, Tsang H, Pham V, Sur M (2021) A low-cost 3D printed microfluidic bioreactor and imaging chamber for live-organoid imaging. Biomicrofluidics 15(2):024105. https://doi.org/10.1063/5.0041027

Kistenmacher A, Manneck S, Wardzinski EK, Martens JC, Gohla G, Melchert UH, Jauch-Chara K, Oltmanns KM (2017) Persistent blood glucose reduction upon repeated transcranial electric stimulation in men. Brain Stimul 10(4):780–786. https://doi.org/10.1016/j.brs.2017.03.011

Kuo H-I, Hsieh M-H, Lin Y-T, Nitsche MA (2023) Acute aerobic exercise at different intensities modulates motor learning performance and cortical excitability in sedentary individuals. eNeuro 10(11). https://doi.org/10.1523/ENEURO.0182-23.2023

Lai HTM, Chang K, Sharabiani MTA, Valabhji J, Gregg EW, Middleton L, Majeed A, Pearson-Stuttard J, Millett C, Bottle A, Vamos EP (2023) Twenty-year trajectories of cardio-metabolic factors among people with type 2 diabetes by dementia status in England: a retrospective cohort study. Eur J Epidemiol 38(7):733–744. https://doi.org/10.1007/s10654-023-00977-7

Lam E, Greenhough E, Nazari P, White MJ, Bruce RM (2019) Muscle metaboreflex activation increases ventilation and heart rate during dynamic exercise in humans. Exp Physiol 104(10):1472–1481. https://doi.org/10.1113/EP087726

Li R, Hosseini H, Saggar M, Balters SC, Reiss AL (2023) Current opinions on the present and future use of functional near-infrared spectroscopy in psychiatry. Neurophotonics 10(1):013505. https://doi.org/10.1117/1.NPh.10.1.013505

Lopes-Vicente WRP, Rodrigues S, Cepeda FX, Jordão CP, Costa-Hong V, Dutra-Marques ACB, Carvalho JC, Alves MJNN, Bortolotto LA, Trombetta IC (2017) Arterial stiffness and its association with clustering of metabolic syndrome risk factors. Diabetol Metab Syndr 9:87. https://doi.org/10.1186/s13098-017-0286-1

von Lühmann A, Addesa J, Chandra S, Das A, Hayashibe M, Dutta A (2017) Neural interfacing non-invasive brain stimulation with NIRS-EEG joint imaging for closed-loop control of neuroenergetics in ischemic stroke. In: 2017 8th International IEEE/EMBS conference on neural engineering (NER), pp 349–353. https://doi.org/10.1109/NER.2017.8008362

Matsuda H (2007) The role of neuroimaging in mild cognitive impairment. Neuropathol Off J Jpn Soc Neuropathol 27(6):570–577. https://doi.org/10.1111/j.1440-1789.2007.00794.x

Md LH, Mann JJ (2018) Test–retest reliability of brain mitochon-drial cytochrome-c-oxidase assessed by functional near-infrared spectroscopy. J Biomed Optics 23(5):056006. https://doi.org/10.1117/1.JBO.23.5.056006

Munting LP, Bonnar O, Kozberg MG, Auger CA, Hirschler L, Hou SS, Greenberg SM, Bacskai BJ, van Veluw SJ (2023) Spontaneous vasomotion propagates along pial arterioles in the awake mouse brain like stimulus-evoked vascular reactivity. J Cereb Blood Flow Metab 43(10):1752–1763. https://doi.org/10.1177/0271678X231152550

Neu E, Janser F, Khatibi AA, Orifici AC (2017) Fully automated operational modal analysis using multi-stage clustering. Mech Syst Signal Process 84:308–323. https://doi.org/10.1016/j.ymssp.2016.07.031

Ngandu T, Lehtisalo J, Solomon A, Levälahti E, Ahtiluoto S, Antikainen R, Bäckman L, Hänninen T, Jula A, Laatikainen T, Lindström J, Mangialasche F, Paajanen T, Pajala S, Peltonen M, Rauramaa R, Stigsdotter-Neely A, Strandberg T, Tuomilehto J et al (2015) A 2 year multidomain intervention of diet, exercise, cognitive training, and vascular risk monitoring versus control to prevent cognitive decline in at-risk elderly people (FINGER): a randomised controlled trial. Lancet (london, England) 385(9984):2255–2263. https://doi.org/10.1016/S0140-6736(15)60461-5

Niwa K, Haensel C, Ross ME, Iadecola C (2001) Cyclooxygenase-1 participates in selected vasodilator responses of the cerebral circulation. Circ Res 88(6):600–608. https://doi.org/10.1161/01.RES.88.6.600

Ogoh S, Sato K, Hirasawa A, Sadamoto T (2019) The effect of muscle metaboreflex on the distribution of blood flow in cerebral arteries during isometric exercise. J Physiol Sci 69(2), Article 2. https://doi.org/10.1007/s12576-018-0653-1

Peeters B, De Roeck G (1999) Reference-based stochastic subspace identification for output-only modal analysis. Mech Syst Signal Process 13(6):855–878. https://doi.org/10.1006/mssp.1999.1249

Petersen RC, Smith GE, Waring SC, Ivnik RJ, Tangalos EG, Kokmen E (1999) Mild cognitive impairment: clinical characterization and outcome. Arch Neurol 56(3):303–308. https://doi.org/10.1001/archneur.56.3.303

Pinna V, Doneddu A, Roberto S, Magnani S, Ghiani G, Mulliri G, Sanna I, Serra S, Hosseini Kakhak SA, Milia R, Fadda D, Lecis R, Guicciardi M, Crisafulli A (2021) Combined mental task and metaboreflex impair cerebral oxygenation in patients with type 2 diabetes mellitus. Am J Physiol Regul Integr Compar Physiol 320(4):R488–R499. https://doi.org/10.1152/ajpregu.00288.2020

Proctor JL, Brunton SL, Kutz JN (2016) Dynamic mode decomposition with control. SIAM J Appl Dyn Syst 15(1):142–161. https://doi.org/10.1137/15M1013857

Rahul FNU, Dutta A, Subedi A, Makled B, Norfleet J, Intes X, De S (2021) A deep learning model for a priori estimation of spatiotemporal regions for neuroimaging guided non-invasive brain stimulation. Brain Stimul Basic Transl Clin Res Neuromods 14(6):1689. https://doi.org/10.1016/j.brs.2021.10.323

Sakai J (2022) Functional near-infrared spectroscopy reveals brain activity on the move. Proc Natl Acad Sci 119(25):e2208729119. https://doi.org/10.1073/pnas.2208729119

Salmon I, Grebenyuk S, Fattah ARA, Rustandi G, Pilkington T, Verfaillie C, Ranga A (2022) Engineering neurovascular organoids with 3D printed microfluidic chips. Lab Chip 22(8):1615–1629. https://doi.org/10.1039/D1LC00535A

Scholey AB, Harper S, Kennedy DO (2001) Cognitive demand and blood glucose. Physiol Behav 73(4):585–592. https://doi.org/10.1016/s0031-9384(01)00476-0

Schroeter ML, Schmiedel O, von Cramon DY (2004) Spontaneous low-frequency oscillations decline in the aging brain. J Cereb Blood Flow Metab 24(10):1183–1191. https://doi.org/10.1097/01.WCB.0000135231.90164.40

Smailagic N, Lafortune L, Kelly S, Hyde C, Brayne C (2018) 18F-FDG PET for prediction of conversion to Alzheimer's disease dementia in people with mild cognitive impairment: an updated systematic review of test accuracy. J Alzheimer's Dis JAD 64(4):1175–1194. https://doi.org/10.3233/JAD-171125

Smailagic N, Vacante M, Hyde C, Martin S, Ukoumunne O, Sachpekidis C (2015) 18F-FDG PET for the early diagnosis of Alzheimer's disease dementia and other dementias in people with mild cognitive impairment (MCI). Cochrane Database Syst Rev 2015(1):CD010632. https://doi.org/10.1002/14651858.CD010632.pub2

Stefanovska A, Bracic M, Kvernmo HD (1999) Wavelet analysis of oscillations in the peripheral blood circulation measured by laser Doppler technique. IEEE Trans Biomed Eng 46(10):1230–1239. https://doi.org/10.1109/10.790500

Stehouwer CDA, Henry RMA, Ferreira I (2008) Arterial stiffness in diabetes and the metabolic syndrome: a pathway to cardiovascular disease. Diabetologia 51(4):527–539. https://doi.org/10.1007/s00125-007-0918-3

Sun T, He X, Li Z (2023) Digital twin in healthcare: recent updates and challenges. Digit Health 9:20552076221149652. https://doi.org/10.1177/20552076221149651

Szendroedi J, Phielix E, Roden M (2012) The role of mitochondria in insulin resistance and type 2 diabetes mellitus. Nat Rev Endocrinol 8(2), Article 2. https://doi.org/10.1038/nrendo.2011.138

Tian X, Zuo Y, Chen S, Zhang Y, Zhang X, Xu Q, Wu S, Wang A (2022) Hypertension, arterial stiffness, and diabetes: a prospective cohort study. Hypertension 79(7):1487–1496. https://doi.org/10.1161/HYPERTENSIONAHA.122.19256

Van Overschee P, De Moor B (1996) Subspace Identification for Linear Systems. Springer US. https://doi.org/10.1007/978-1-4613-0465-4

van Veluw SJ, Hou SS, Calvo-Rodriguez M, Arbel-Ornath M, Snyder AC, Frosch MP, Greenberg SM, Bacskai BJ (2020) Vasomotion as a driving force for paravascular clearance in the awake mouse brain. Neuron 105(3):549-561.e5. https://doi.org/10.1016/j.neuron.2019.10.033

von Luhmann A, Wabnitz H, Sander T, Muller K-R (2017) M3BA: a mobile, modular, multimodal biosignal acquisition architecture for miniaturized EEG-NIRS-based hybrid BCI and monitoring. IEEE Trans Biomed Eng 64(6):1199–1210. https://doi.org/10.1109/TBME.2016.2594127

Whalen AJ, Brennan SN, Sauer TD, Schiff SJ (2015) Observability and controllability of nonlinear networks: the role of symmetry. Phys Rev X 5(1):011005. https://doi.org/10.1103/PhysRevX.5.011005

Wimmer RA, Leopoldi A, Aichinger M, Wick N, Hantusch B, Novatchkova M, Taubenschmid J, Hämmerle M, Esk C, Bagley JA, Lindenhofer D, Chen G, Boehm M, Agu CA, Yang F, Fu B, Zuber J, Knoblich JA, Kerjaschki D, Penninger JM (2019) Human blood vessel organoids as a model of diabetic vasculopathy. Nature 565(7740):505–510. https://doi.org/10.1038/s41586-018-0858-8

Xue Q, Li X, Wang X, Ma H, Heianza Y, Qi L (2023) Subtypes of Type 2 diabetes and incident cardiovascular disease risk: UK biobank and all of us cohorts. Mayo Clin Proc 98(8):1192–1204. https://doi.org/10.1016/j.mayocp.2023.01.024

Yassine HN, Solomon V, Thakral A, Sheikh-Bahaei N, Chui HC, Braskie MN, Schneider LS, Talbot K (2022) Brain energy failure in dementia syndromes: opportunities and challenges for glucagon-like peptide-1 receptor agonists. Alzheimer's Dementia 18(3):478–497. https://doi.org/10.1002/alz.12474

Zhang L, Su F, Buizer S, Lu H, Gao W, Tian Y, Meldrum D (2013) A dual sensor for real-time monitoring of glucose and oxygen. Biomaterials 34(38). https://doi.org/10.1016/j.biomaterials.2013.09.031

Zhao F, Cheung M, Dutta A, Fisher N, Tomita M (2019) Exercises to determine older adults' muscle oxygenation change rate by various physical performance levels. Arch Phys Med Rehabil 100(12):e178–e179. https://doi.org/10.1016/j.apmr.2019.10.049

Zhao F, Tomita M, Dutta A (2023) Operational modal analysis of near-infrared spectroscopy measure of 2-month exercise intervention effects in sedentary older adults with diabetes and cognitive impairment. Brain Sci 13(7):1099. https://doi.org/10.3390/brainsci13071099

Zhao X, Xu Z, Xiao L, Shi T, Xiao H, Wang Y, Li Y, Xue F, Zeng W (2021) Review on the vascularization of organoids and organoids-on-a-chip. Front Bioeng Biotechnol 9. https://www.frontiersin.org/articles/10.3389/fbioe.2021.637048

Zhao Z, Chen X, Dowbaj AM, Sljukic A, Bratlie K, Lin L, Fong ELS, Balachander GM, Chen Z, Soragni A, Huch M, Zeng YA, Wang Q, Yu H (2022) Organoids. Nat Rev Methods Primers 2(1), Article 1. https://doi.org/10.1038/s43586-022-00174-y

Zheng M, Zhang X, Chen S, Song Y, Zhao Q, Gao X, Wu S (2020) Arterial stiffness preceding diabetes. Circ Res 127(12):1491–1498. https://doi.org/10.1161/CIRCRESAHA.120.317950

SPRINGER NATURE

GPSR Compliance

The European Union's (EU) General Product Safety Regulation (GPSR) is a set of rules that requires consumer products to be safe and our obligations to ensure this.

If you have any concerns about our products, you can contact us on ProductSafety@springernature.com

In case Publisher is established outside the EU, the EU authorized representative is:

Springer Nature Customer Service Center GmbH
Europaplatz 3
69115 Heidelberg, Germany

The manufacturer's authorised representative in the EU is Springer Nature Customer Service Centre GmbH, Europaplatz 3, 69115 Heidelberg, Germany. If you have any concerns regarding our products, please contact ProductSafety@springernature.com

Printed and bound by CPI Group (UK) Ltd, Croydon, CR0 4YY
30/09/2026
02225240-0006